SCHAUM'S OUTLINE OF

THEORY AND PROBLEMS

of

DISCRETE MATHEMATICS

•

by

SEYMOUR LIPSCHUTZ, Ph.D.

Professor of Mathematics
Temple University

SCHAUM'S OUTLINE SERIES

McGRAW-HILL BOOK COMPANY

New York   St. Louis   San Francisco   Auckland   Bogotá   Düsseldorf   Johannesburg
London   Madrid   Mexico   Montreal   New Delhi   Panama   Paris
São Paulo   Singapore   Sydney   Tokyo   Toronto

07-037981-5

8 9 10 11 12 13 14 15 SH SH 8 7 6 5

Library of Congress Cataloging in Publication Data

Lipschutz, Seymour.

   Schaum's outline of theory and problems of discrete mathematics.

   (Schaum's outline series)
   Includes index.

   1. Algebra, Abstract—Outlines, syllabi, etc.
2. Combinatorial analysis—Outlines, syllabi, etc.
3. Logic, Symbolic and mathematical—Outlines, syllabi, etc. I. Title.

QA162.L56   511'.02'02   76-50619
ISBN 0-07-037981-5

Preface

Discrete mathematics, the study of finite systems, has become increasingly important as the computer age has advanced. The digital computer is basically a finite structure, and many of its properties can be understood and interpreted within the framework of finite mathematical systems. This book, in presenting the more essential material, may be used as a textbook for a formal course in discrete mathematics or as a supplement to all current standard texts.

The first three chapters cover the standard material on sets, relations and functions. Next comes a chapter on vectors and matrices. We then have three chapters on graphs, directed graphs, and trees. Finally, there are individual chapters on combinatorial analysis, algebraic systems, posets and lattices, propositional calculus, and Boolean algebra. The chapters on graph theory include discussions on planarity, traversability, the four color theorem, minimal paths, and finite automata. The chapter on algebraic systems also treats formal languages and grammars. Furthermore, the chapter on Boolean algebra includes discussions of circuits and the use of Karnaugh maps for finding minimal disjunctive normal forms. We emphasize that the chapters have been written so that their order can be changed without difficulty and without loss of continuity.

Each chapter begins with a clear statement of pertinent definitions, principles and theorems together with illustrative and other descriptive material. This is followed by sets of solved and supplementary problems. The solved problems serve to illustrate and amplify the material, and also include proofs of theorems. The supplementary problems furnish a complete review of the material in the chapter. Finally, each chapter contains a set of computer programming problems directly related to the content of the chapter. More material has been included than can be covered in most first courses. This has been done to make the book more flexible, to provide a more useful book of reference and to stimulate further interest in the topics.

I wish to thank many of my friends and colleagues, especially Arthur Poe, for invaluable suggestions and critical review of the manuscript. I also wish to express my gratitude to the staff of the McGraw-Hill Schaum's Outline Series division, particularly to David Beckwith, for their unfailing cooperation.

SEYMOUR LIPSCHUTZ

Temple University
September 1976

CONTENTS

CONTENTS

Chapter 1

Set Theory

1.1 SETS AND ELEMENTS

The concept of a *set* appears in all branches of mathematics. Intuitively, a set is any well-defined list or collection of objects, and will be denoted by capital letters $A, B, X, Y, \ldots$. The objects comprising the set are called its *elements* or *members* and will be denoted by lower case letters $a, b, x, y, \ldots$. The statement "p is an element of A" or, equivalently, "p belongs to A" is written

$$p \in A$$

The negation of $p \in A$ is written $p \notin A$.

The fact that a set is completely determined when its members are specified is formally stated as the principle of extension.

Principle of Extension: Two sets A and B are equal if and only if they have the same members.

As usual, we write $A = B$ if the sets A and B are equal, and we write $A \neq B$ if the sets are not equal.

There are essentially two ways to specify a particular set. One way, if it is possible, is to list its members. For example,

$$A = \{a, e, i, o, u\}$$

denotes the set A whose elements are the letters a, e, i, o, u. Note that the elements are separated by commas and enclosed in braces { }. The second way is to state those properties which characterize the elements in the set. For example,

$$B = \{x : x \text{ is an integer}, x > 0\}$$

which reads "B is the set of x such that x is an integer and x is greater than 0," denotes the set B whose elements are the positive integers. A letter, usually x, is used to denote a typical member of the set; the colon is read as "such that" and the comma as "and."

EXAMPLE 1.1.

(a) The set A above can also be written as

$$A = \{x : x \text{ is a letter in the English alphabet}, x \text{ is a vowel}\}$$

Observe that $b \notin A$, $e \in A$ and $p \notin A$.

(b) We could not list all the elements of the above set B although frequently we specify the set by writing

$$B = \{1, 2, 3, \ldots\}$$

where we assume that everyone knows what we mean. Observe that $8 \in B$ but $-6 \notin B$.

(c) Let $E = \{x : x^2 - 3x + 2 = 0\}$. In other words, E consists of those numbers which are solutions of the equation $x^2 - 3x + 2 = 0$, sometimes called the solution set of the given equation. Since the solutions of the equation are 1 and 2, we could also write $E = \{1, 2\}$.

(d) Let $E = \{x : x^2 - 3x + 2 = 0\}$, $F = \{2, 1\}$ and $G = \{1, 2, 2, 1, 6/3\}$. Then $E = F = G$. Observe that a set does not depend on the way in which its elements are displayed. A set remains the same if its elements are repeated or rearranged.

Some sets will occur very often in the text and so we use special symbols for them. Unless otherwise specified, we will let:

$\mathbf{N}$ = the set of positive integers: $1, 2, 3, \ldots$

$\mathbf{Z}$ = the set of integers: $\ldots, -2, -1, 0, 1, 2, \ldots$

$\mathbf{Q}$ = the set of rational numbers

$\mathbf{R}$ = the set of real numbers

$\mathbf{C}$ = the set of complex numbers

Even if we can list the elements of a set, it may not be practical to do so. For example, we would not list the members of the set of people born in the world during the year 1976 although theoretically it is possible to compile such a list. That is, we describe a set by listing its elements only if the set contains a few elements; otherwise we describe a set by the property which characterizes its elements.

The fact that we can describe a set in terms of a property is formally stated as the *principle of abstraction*.

Principle of Abstraction: Given any set U and any property P, there is a set A such that the elements of A are exactly those members of U which have the property P.

1.2   UNIVERSAL SET, EMPTY SET

In any application of the theory of sets, the members of all sets under investigation usually belong to some fixed large set called the *universal set* or *universe of discourse*. For example, in plane geometry, the universal set consists of all the points in the plane; and in human population studies the universal set consists of all the people in the world. We will let the symbol

$$U$$

denote the universal set unless otherwise stated or implied.

For a given set U and a property P, there may not be any elements of U which have property P. For example, the set

$$S = \{x : \ x \text{ is a positive integer}, \ x^2 = 3\}$$

has no elements since no positive integer has the required property.

The set with no elements is called the *empty set* or *null set* and is denoted by

$$\varnothing$$

From the principle of extension, it follows that there is only one empty set. In other words, if S and T are both empty, then $S = T$ since they have exactly the same elements, namely, none.

1.3   SUBSETS

If every element in a set A is also an element of a set B, then A is called a subset of B. We also say that A is *contained in* B or that B *contains* A. This relationship is written

$$A \subset B \qquad \text{or} \qquad B \supset A$$

If A is not a subset of B, i.e. if at least one element of A does not belong to B, we write $A \not\subset B$ or $B \not\supset A$.

EXAMPLE 1.2.

(a) Consider the sets

$$A = \{1, 3, 4, 5, 8, 9\} \qquad B = \{1, 2, 3, 5, 7\} \qquad C = \{1, 5\}$$

Then $C \subset A$ and $C \subset B$ since 1 and 5, the elements of C, are also members of A and B. But $B \not\subset A$ since some of its elements, e.g. 2 and 7, do not belong to A. Furthermore, since the elements of A, B and C must also belong to the universal set U, we have that U must at least contain the set $\{1, 2, 3, 4, 5, 7, 8, 9\}$.

(b) Let **N**, **Z**, **Q** and **R** be defined as in Section 1.1.   Then

$$\mathbf{N} \subset \mathbf{Z} \subset \mathbf{Q} \subset \mathbf{R}$$

(c) The set $E = \{2, 4, 6\}$ is a subset of the set $F = \{6, 2, 4\}$, since each number 2, 4 and 6 belonging to E also belongs to F. In fact, $E = F$. In a similar manner it can be shown that every set is a subset of itself.

Every set A is a subset of the universal set U since, by definition, all the members of A belong to U. Also the empty set $\emptyset$ is a subset of A.

As noted above, every set A is a subset of itself since, trivially, the elements of A belong to A.

If every element of a set A belongs to a set B, and every element of B belongs to a set C, then clearly every element of A belongs to C. In other words, if $A \subset B$ and $B \subset C$, then $A \subset C$.

If $A \subset B$ and $B \subset A$ then A and B have the same elements, i.e. $A = B$. Conversely, if $A = B$ then $A \subset B$ and $B \subset A$ since every set is a subset of itself.

We state these results formally as

Theorem 1.1:   (i)   For any set A, we have $\emptyset \subset A \subset U$.

(ii)   For any set A, we have $A \subset A$.

(iii)   If $A \subset B$ and $B \subset C$, than $A \subset C$.

(iv) $A = B$ if and only if $A \subset B$ and $B \subset A$.

If $A \subset B$, then it is still possible that $A = B$. Where $A \subset B$ but $A \neq B$, we say A is a *proper subset* of B. (Some authors write $A \subseteq B$ to say that A is a subset of B, and use $A \subset B$ to mean that A is a proper subset of B.)   For example, if

$$A = \{1, 3\} \qquad B = \{1, 2, 3\} \qquad C = \{1, 3, 2\}$$

then A and B are both subsets of C; but A is a proper subset of C whereas B is not a proper subset of C since $B = C$.

1.4 VENN DIAGRAMS

A Venn diagram is a pictorial representation of sets by sets of points in the plane. The universal set U is represented by the interior of a rectangle, and the other sets are represented by disks lying within the rectangle. If $A \subset B$, then the disk representing A will be entirely within the disk representing B as in Fig. 1-1(a). If A and B are disjoint, i.e. have no elements in common, then the disk representing A will be separated from the disk representing B as in Fig. 1-1(b).

However, if A and B are two arbitrary sets, it is possible that some objects are in A but not B, some are in B but not A, some are in both A and B, and some are in neither A nor B; hence in general we represent A and B as in Fig. 1-1(c).

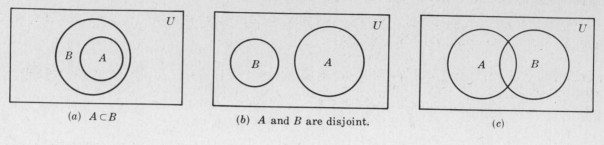

(a) $A \subset B$ (b) A and B are disjoint.   (c)

Fig. 1-1

1.5 SET OPERATIONS

The *union* of two sets A and B, denoted by $A \cup B$, is the set of all elements which belong to A or to B:

$$A \cup B = \{x : x \in A \text{ or } x \in B\}$$

Here "or" is used in the sense of and/or.

The *intersection* of two sets A and B, denoted by $A \cap B$, is the set of elements which belong to both A and B:

$$A \cap B = \{x : x \in A \text{ and } x \in B\}$$

If $A \cap B = \emptyset$, that is, if A and B do not have any elements in common, then A and B are said to be *disjoint* or *nonintersecting*.

The *relative complement* of a set B with respect to a set A or, simply, the *difference* of A and B, denoted by $A \setminus B$, is the set of elements which belong to A but which do not belong to B:

$$A \setminus B = \{x : x \in A, x \notin B\}$$

The set $A \setminus B$ is read "A minus B". Many texts denote $A \setminus B$ by $A - B$.

As noted before, all sets under consideration at a particular time are subsets of a fixed universal set U. The *absolute complement* or, simply, *complement* of a set A, denoted by A^c, is the set of elements which belong to U but which do not belong to A:

$$A^c = \{x : x \in U, x \notin A\}$$

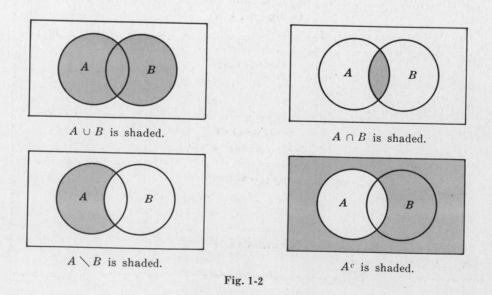

$A \cup B$ is shaded. $A \cap B$ is shaded.

$A \setminus B$ is shaded. A^c is shaded.

Fig. 1-2

That is, A^c is the difference of the universal set U and A. Some texts denote this set by A'.

We illustrate the above operations by shading, respectively, the sets $A \cup B$, $A \cap B$, $A \setminus B$ and A^c in the Venn diagrams in Fig. 1-2.

EXAMPLE 1.3. Let $A = \{1, 2, 3, 4\}$ and $B = \{3, 4, 5, 6\}$, where $U = \{1, 2, 3, \ldots\}$, the set of positive integers. Then:

$$A \cup B = \{1, 2, 3, 4, 5, 6\} \qquad A \cap B = \{3, 4\}$$
$$A \setminus B = \{1, 2\} \qquad\qquad A^c = \{5, 6, 7, \ldots\}$$

Our next theorem shows the relationship between set inclusion and the operations of union and intersection.

Theorem 1.2:   The following are equivalent: $A \subset B$, $A \cap B = A$, $A \cup B = B$.

(*Note*: This theorem is proved in Problem 1.21. Other conditions equivalent to $A \subset B$ are given in Problem 1.31.)

1.6 ALGEBRA OF SETS, DUALITY

Sets under the above operations satisfy various laws or identities which are listed in Table 1.1. In fact we formally state:

Theorem 1.3:   Sets satisfy the laws in Table 1.1.

Table 1.1 Laws of the Algebra of Sets

Idempotent Laws	
1a. $A \cup A = A$	1b. $A \cap A = A$
Associative Laws	
2a. $(A \cup B) \cup C = A \cup (B \cup C)$	2b. $(A \cap B) \cap C = A \cap (B \cap C)$
Commutative Laws	
3a. $A \cup B = B \cup A$	3b. $A \cap B = B \cap A$
Distributive Laws	
4a. $A \cup (B \cap C) = (A \cup B) \cap (A \cup C)$	4b. $A \cap (B \cup C) = (A \cap B) \cup (A \cap C)$
Identity Laws	
5a. $A \cup \emptyset = A$	5b. $A \cap U = A$
6a. $A \cup U = U$	6b. $A \cap \emptyset = \emptyset$
Involution Law	
7. $(A^c)^c = A$	
Complement Laws	
8a. $A \cup A^c = U$	8b. $A \cap A^c = \emptyset$
9a. $U^c = \emptyset$	9b. $\emptyset^c = U$
DeMorgan's Laws	
10a. $(A \cup B)^c = A^c \cap B^c$	10b. $(A \cap B)^c = A^c \cup B^c$

We discuss two methods of proving equations involving set operations. The first is to break down what it means for an object x to be an element of each side, and the second is to use Venn diagrams. For example, consider the first of DeMorgan's laws,

$$(A \cup B)^c = A^c \cap B^c$$

Method 1. We first show that $(A \cup B)^c \subset A^c \cap B^c$. If $x \in (A \cup B)^c$, then $x \notin A \cup B$. Thus $x \notin A$ and $x \notin B$, and so $x \in A^c$ and $x \in B^c$. Hence $x \in A^c \cap B^c$.

Next we show that $A^c \cap B^c \subset (A \cup B)^c$. Let $x \in A^c \cap B^c$. Then $x \in A^c$ and $x \in B^c$, so $x \notin A$ and $x \notin B$. Hence $x \notin A \cup B$, so $x \in (A \cup B)^c$.

We have proven that every element of $(A \cup B)^c$ belongs to $A^c \cap B^c$ and that every element of $A^c \cap B^c$ belongs to $(A \cup B)^c$. Together, these inclusions prove that the sets have the same elements, i.e. that $(A \cup B)^c = A^c \cap B^c$.

(*Remark*: Implicitly, we have used an analogous logical law which is discussed in Chapter 11.)

Method 2. From the Venn diagram for $A \cup B$ in Fig. 1-2, we see that $(A \cup B)^c$ is represented by the shaded area in Fig. 1-3(a). To find $A^c \cap B^c$, the area in both A^c and B^c, we shade A^c with strokes in one direction and B^c with strokes in another direction as in Fig. 1-3(b). Then $A^c \cap B^c$ is represented by the cross-hatched area which is shaded in Fig. 1-3(c). Since $(A \cup B)^c$ and $A^c \cap B^c$ are represented by the same area, they are equal.

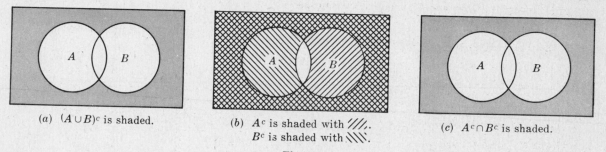

(a) $(A \cup B)^c$ is shaded.   (b) A^c is shaded with ///.   (c) $A^c \cap B^c$ is shaded.
                                              B^c is shaded with \\\.

Fig. 1-3

The reader may have wondered why the identities in Table 1.1 are arranged in pairs, as, for example, 2a and 2b. We now consider the principle behind this arrangement. Suppose E is an equation of set algebra. The *dual* E^* of E is the equation obtained by replacing each occurrence of $\cup$, $\cap$, U and $\emptyset$ in E by $\cap$, $\cup$, $\emptyset$ and U respectively. For example, the dual of

$$(U \cap A) \cup (B \cap A) = A \qquad \text{is} \qquad (\emptyset \cup A) \cap (B \cup A) = A$$

Observe that the pairs of laws in Table 1.1 are duals of each other. It is a fact of set algebra, called the *principle of duality*, that if any equation E is an identity then its dual E^* is also an identity.

1.7 FINITE SETS, COUNTING PRINCIPLE

A set is said to be finite if it contains exactly m distinct elements where m denotes some nonnegative integer. Otherwise, a set is said to be infinite. For example, the empty set $\emptyset$ and the set of letters of the English alphabet are finite sets, whereas the set of even positive integers, $\{2, 4, 6, \ldots\}$, is infinite.

If a set A is finite, we let $n(A)$ denote the number of elements of A. Some texts use $\#(A)$ instead of $n(A)$.

Lemma 1.4:   If A and B are disjoint finite sets, then $A \cup B$ is finite and

$$n(A \cup B) = n(A) + n(B)$$

Proof. In counting the elements of $A \cup B$, first count those that are in A. There are $n(A)$ of these. The only other elements of $A \cup B$ are those that are in B but not in A. But since A and B are disjoint, no element of B is in A, so there are $n(B)$ elements that are in B but not in A. Therefore, $n(A \cup B) = n(A) + n(B)$.

We also have a formula for $n(A \cup B)$ even when they are not disjoint. This is proved in Problem 1.22.

Theorem 1.5:   If A and B are finite sets, then $A \cup B$ and $A \cap B$ are finite and

$$n(A \cup B) = n(A) + n(B) - n(A \cap B)$$

We can apply this result to obtain a similar formula for three sets.

Corollary 1.6:   If A, B and C are finite sets, then so is $A \cup B \cup C$, and

$$n(A \cup B \cup C) = n(A) + n(B) + n(C) - n(A \cap B) - n(A \cap C) - n(B \cap C) + n(A \cap B \cap C)$$

Mathematical induction (Section 1.10) may be used to further generalize this result to any finite number of sets.

EXAMPLE 1.4.   Suppose that 100 of the 120 mathematics students at a college take at least one of the languages French, German, and Russian. Also suppose:

  65 study French,

  45 study German,

  42 study Russian,

  20 study French and German,

  25 study French and Russian,

  15 study German and Russian.

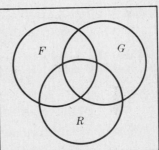

Let F, G and R denote the sets of students studying French, German and Russian respectively. We wish to find the number of students who study all three languages, and to fill in the correct number of students in each of the eight regions of the Venn diagram shown in Fig. 1-4.

Fig. 1-4

By Corollary 1.6,

$$n(F \cup G \cup R) = n(F) + n(G) + n(R) - n(F \cap G) - n(F \cap R) - n(G \cap R) + n(F \cap G \cap R)$$

Now, $n(F \cup G \cup R) = 100$ because 100 of the students study at least one of the languages. Substituting,

$$100 = 65 + 45 + 42 - 20 - 25 - 15 + n(F \cap G \cap R)$$

and so, $n(F \cap G \cap R) = 8$, i.e. 8 students study all three languages.

We now use this result to fill in the Venn diagram. We have:

  8 study all three languages,

  $20 - 8 = 12$ study French and German but not Russian,

  $25 - 8 = 17$ study French and Russian but not German,

  $15 - 8 = 7$ study German and Russian but not French,

  $65 - 12 - 8 - 17 = 28$ study only French,

  $45 - 12 - 8 - 7 = 18$ study only German,

  $42 - 17 - 8 - 7 = 10$ study only Russian,

  $120 - 100 = 20$ do not study any of the languages.

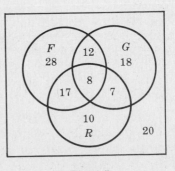

Fig. 1-5

Accordingly, the completed diagram is Fig. 1-5. Observe that $28 + 18 + 10 = 56$ students study only one of the languages.

1.8 CLASSES OF SETS, POWER SETS

Given a set A, we might wish to talk about some of its subsets. Thus we would be considering a set of sets. Whenever such a situation occurs, to avoid confusion we will speak of a *class* of sets or *collection* of sets rather than a set of sets. If we wish to consider some of the sets in a given class of sets, then we speak of a *subclass* or *subcollection*.

EXAMPLE 1.5. Suppose $A = \{1, 2, 3, 4\}$. Let $\mathcal{A}$ be the class of subsets of A which contain exactly three elements of A. Then

$$\mathcal{A} = [\{1, 2, 3\}, \{1, 2, 4\}, \{1, 3, 4\}, \{2, 3, 4\}]$$

The elements of $\mathcal{A}$ are the sets $\{1, 2, 3\}$, $\{1, 2, 4\}$, $\{1, 3, 4\}$, and $\{2, 3, 4\}$.

Let $\mathcal{B}$ be the class of subsets of A which contain 2 and two other elements of A. Then

$$\mathcal{B} = [\{1, 2, 3\}, \{1, 2, 4\}, \{2, 3, 4\}]$$

The elements of $\mathcal{B}$ are the sets $\{1, 2, 3\}$, $\{1, 2, 4\}$, and $\{2, 3, 4\}$. Thus $\mathcal{B}$ is a subclass of $\mathcal{A}$, since every element of $\mathcal{B}$ is also an element of $\mathcal{A}$. (To avoid confusion, we will sometimes enclose the sets of a class in brackets instead of braces.)

For a given set A, we may speak of the class of all subsets of A. This class is called the *power set* of A, and is denoted by $\mathcal{P}(A)$. If A is finite then so is $\mathcal{P}(A)$. Moreover, the number of elements in $\mathcal{P}(A)$ is 2 raised to the power $n(A)$:

$$n(\mathcal{P}(A)) = 2^{n(A)}$$

For this reason, the power set of A is sometimes denoted by 2^A.

EXAMPLE 1.6. Suppose $A = \{1, 2, 3\}$. Then

$$\mathcal{P}(A) = [\emptyset, \{1\}, \{2\}, \{3\}, \{1, 2\}, \{1, 3\}, \{2, 3\}, A]$$

Note that the empty set $\emptyset$ belongs to $\mathcal{P}(A)$ since $\emptyset$ is a subset of A. Similarly, A belongs to $\mathcal{P}(A)$. Note also that $\mathcal{P}(A)$ has $2^3 = 8$ elements.

1.9 ARGUMENTS AND VENN DIAGRAMS

Many verbal statements can be translated into equivalent statements about sets which can be described by Venn diagrams. Hence Venn diagrams can very often be used to determine the validity of an argument.

EXAMPLE 1.7. Examine the following argument adapted from a book on logic by Lewis Carroll, the author of *Alice in Wonderland*:

S_1: My saucepans are the only things I have that are made of tin.

S_2: I find all your presents very useful.

S_3: None of my saucepans is of the slightest use.

S: Your presents to me are not made of tin.

Here the statements S_1, S_2, and S_3 above the horizontal line denote the assumptions, and the statement S below the line denotes the conclusion. We will determine whether or not the conclusion S follows logically from the assumptions, that is, whether or not S is a valid conclusion.

Now by S_1 the tin objects are contained in the set of saucepans; hence draw the Venn diagram of Fig. 1-6.

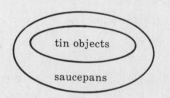

Fig. 1-6

By S_3 the set of saucepans and the set of useful things are disjoint; hence draw Fig. 1-7.

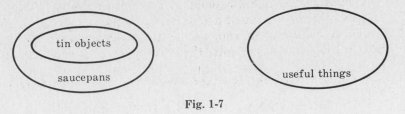

Fig. 1-7

By S_2 the set of "your presents" is a subset of the set of useful things; hence draw Fig. 1-8.

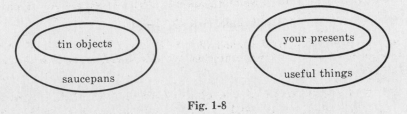

Fig. 1-8

The conclusion is clearly valid by the above Venn diagram because the set of "your presents" is disjoint from the set of tin objects.

1.10 MATHEMATICAL INDUCTION

An essential property of the set

$$\mathbf{N} = \{1, 2, 3, \ldots\}$$

which is used in many proofs, follows:

Principle of Mathematical Induction I: Let P be a proposition defined on the positive integers **N**, i.e. $P(n)$ is either true or false for each n in **N**. Suppose P has the following two properties.

   (i) $P(1)$ is true.

   (ii) $P(n+1)$ is true whenever $P(n)$ is true.

Then P is true for every positive integer.

We shall not prove this principle. In fact, this principle is usually given as one of the axioms when **N** is developed axiomatically.

EXAMPLE 1.8.   Let P be the proposition that the sum of the first n odd numbers is n^2; that is,

$$P(n):   1 + 3 + 5 + \cdots + (2n-1) = n^2$$

(The nth odd number is $2n - 1$, and the next odd number is $2n + 1$.) Observe that $P(n)$ is true for $n = 1$,

$$P(1):   1 = 1^2$$

Assuming $P(n)$ is true, we add $2n + 1$ to both sides of $P(n)$, obtaining:

$$1 + 3 + 5 + \cdots + (2n-1) + (2n+1) = n^2 + (2n+1) = (n+1)^2$$

which is $P(n+1)$. That is, $P(n+1)$ is true whenever $P(n)$ is true. By the principle of mathematical induction, P is true for all n.

There is a form of the principle of mathematical induction which is sometimes more convenient to use. Although it appears different, it is really equivalent to the principle of induction.

Principle of Mathematical Induction II: Let P be a proposition defined on the positive integers $\mathbf{N}$ such that:

(i) $P(1)$ is true.

(ii) $P(n)$ is true whenever $P(k)$ is true for all $1 \leqq k < n$.

Then P is true for every positive integer.

Solved Problems

SETS, SUBSETS

1.1. Which of these sets are equal: $\{r, t, s\}$, $\{s, t, r, s\}$, $\{t, s, t, r\}$, $\{s, r, s, t\}$?

They are all equal. Order and repetition do not change a set.

1.2. List the elements of the following sets; here $\mathbf{N} = \{1, 2, 3, \ldots\}$.

(a) $A = \{x : x \in \mathbf{N}, 3 < x < 12\}$

(b) $B = \{x : x \in \mathbf{N}, x \text{ is even}, x < 15\}$

(c) $C = \{x : x \in \mathbf{N}, 4 + x = 3\}$.

(a) A consists of the positive integers between 3 and 12; hence
$$A = \{4, 5, 6, 7, 8, 9, 10, 11\}$$

(b) B consists of the even positive integers less than 15; hence
$$B = \{2, 4, 6, 8, 10, 12, 14\}$$

(c) There are no positive integers which satisfy the condition $4 + x = 3$; hence C contains no elements. In other words, $C = \emptyset$, the empty set.

1.3. Prove that $A = \{2, 3, 4, 5\}$ is not a subset of $B = \{x : x \in \mathbf{N}, x \text{ is even}\}$.

It is necessary to show that at least one element in A does not belong to B. Now $3 \in A$ and, since B consists of even numbers, $3 \notin B$; hence A is not a subset of B.

1.4. Consider the following sets:

$$\emptyset, \quad A = \{1\}, \quad B = \{1, 3\}, \quad C = \{1, 5, 9\}, \quad D = \{1, 2, 3, 4, 5\},$$
$$E = \{1, 3, 5, 7, 9\}, \quad U = \{1, 2, \ldots, 8, 9\}$$

Insert the correct symbol $\subset$ or $\not\subset$ between each pair of sets:

(a) $\emptyset, A$ (c) B, C (e) C, D (g) D, E

(b) A, B (d) B, E (f) C, E (h) D, U

(a) $\emptyset \subset A$ because $\emptyset$ is a subset of every set.

(b) $A \subset B$ because 1 is the only element of A and it belongs to B.

(c) $B \not\subset C$ because $3 \in B$ but $3 \notin C$.

(d) $B \subset E$ because the elements of B also belong to E.

(e) $C \not\subset D$ because $9 \in C$ but $9 \notin D$.

(f) $C \subset E$ because the elements of C also belong to E.

(g) $D \not\subset E$ because $2 \in D$ but $2 \notin E$.

(h) $D \subset U$ because the elements of D also belong to U.

SET OPERATIONS

In Problems 1.5–1.7, assume $U = \{1, 2, \ldots, 8, 9\}$ and

$$A = \{1, 2, 3, 4, 5\} \qquad C = \{5, 6, 7, 8, 9\} \qquad E = \{2, 4, 6, 8\}$$
$$B = \{4, 5, 6, 7\} \qquad D = \{1, 3, 5, 7, 9\} \qquad F = \{1, 5, 9\}$$

1.5. Find: (a) $A \cup B$ and $A \cap B$ (d) $D \cup E$ and $D \cap E$

               (b) $B \cup D$ and $B \cap D$ (e) $E \cup E$ and $E \cap E$

               (c) $A \cup C$ and $A \cap C$ (f) $D \cup F$ and $D \cap F$

Recall that the union $X \cup Y$ consists of those elements in either X or Y (or both), and that the intersection $X \cap Y$ consists of those elements in both X and Y.

(a) $A \cup B = \{1, 2, 3, 4, 5, 6, 7\}$ $A \cap B = \{4, 5\}$

(b) $B \cup D = \{1, 3, 4, 5, 6, 7, 9\}$ $B \cap D = \{5, 7\}$

(c) $A \cup C = \{1, 2, 3, 4, 5, 6, 7, 8\ 9\} = U$ $A \cap C = \{5\}$

(d) $D \cup E = \{1, 2, 3, 4, 5, 6, 7, 8, 9\} = U$ $D \cap E = \emptyset$

(e) $E \cup E = \{2, 4, 6, 8\} = E$ $E \cap E = \{2, 4, 6, 8\} = E$

(f) $D \cup F = \{1, 3, 5, 7, 9\} = D$ $D \cap F = \{1, 5, 9\} = F$

Observe that $F \subset D$; so by Theorem 1.2 we must have $D \cup F = D$ and $D \cap F = F$.

1.6. Find: (a) A^c and B^c (c) $A \setminus B$ and $B \setminus A$

               (b) D^c and E^c (d) $D \setminus E$ and $F \setminus D$

Recall that the complement X^c consists of those elements in the universal set U which do not belong to X, and the difference $X \setminus Y$ consists of those elements in X which do not belong to Y.

(a) $A^c = \{6, 7, 8, 9\}$ (c) $A \setminus B = \{1, 2, 3\}$

     $B^c = \{1, 2, 3, 8, 9\}$ $B \setminus A = \{6, 7\}$

(b) $D^c = \{2, 4, 6, 8\} = E$ (d) $D \setminus E = \{1, 3, 5, 7, 9\} = D$

     $E^c = \{1, 3, 5, 7, 9\} = D$ $F \setminus D = \emptyset$

1.7. Find: (a) $A \cap (B \cup E)$ (c) $(A \cap D) \setminus B$

               (b) $(A \setminus E)^c$ (d) $(B \cap F) \cup (C \cap E)$

(a) First compute $B \cup E = \{2, 4, 5, 6, 7, 8\}$. Then $A \cap (B \cup E) = \{2, 4, 5\}$.

(b) $A \setminus E = \{1, 3, 5\}$. Then $(A \setminus E)^c = \{2, 4, 6, 7, 8, 9\}$.

(c) $A \cap D = \{1, 3, 5\}$. Now $(A \cap D) \setminus B = \{1, 3\}$.

(d) $B \cap F = \{5\}$ and $C \cap E = \{6, 8\}$. So $(B \cap F) \cup (C \cap E) = \{5, 6, 8\}$.

1.8. Given: $A = \{\text{Ann, Alice, Audrey}\}$

$\qquad\qquad B = \{\text{Ann, Audrey, Ellen}\}$

$\qquad\qquad C = \{\text{Alice, Audrey, Betty, Ellen}\}$

$\qquad\qquad D = \{\text{Ann, Alice, Betty, Ellen}\}$

Find: (a) $A \cup B$ and $C \cup D$ (c) $B \setminus C$ and $D \setminus A$

         (b) $A \cap C$ and $B \cap D$ (d) $(A \cap D) \cup (C \setminus B)$

(a) $A \cup B = \{\text{Ann, Alice, Audrey, Ellen}\}$

     $C \cup D = \{\text{Ann, Alice, Audrey, Betty, Ellen}\}$

(b) $A \cap C = \{\text{Alice, Audrey}\}$

     $B \cap D = \{\text{Ann, Ellen}\}$

(c) $B \setminus C = \{\text{Ann}\}$

     $D \setminus A = \{\text{Betty, Ellen}\}$

(d)   First compute $A \cap D = \{\text{Ann, Alice}\}$ and $C \setminus B = \{\text{Alice, Betty}\}$. Then $(A \cap D) \cup (C \setminus B) = \{\text{Ann, Alice, Betty}\}$.

1.9. Show that we can have $A \cap B = A \cap C$ without $B = C$.

Let $A = \{1, 2\}$, $B = \{2, 3\}$ and $C = \{2, 4\}$. Then $A \cap B = \{2\}$ and $A \cap C = \{2\}$. Thus $A \cap B = A \cap C$ but $B \neq C$.

VENN DIAGRAMS

1.10. Shade the sets (a) $A \cap B^c$ and (b) $(B \setminus A)^c$ in the Venn diagram of Fig. 1-9.

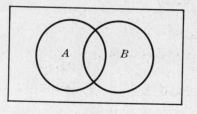

Fig. 1-9

(a) As in Fig. 1-10(a), shade A with strokes in one direction (/////), and shade B^c, the area outside B, with strokes in another direction (\\\\\); the cross-hatched area is the intersection $A \cap B^c$ shown in Fig. 1-10(b). Observe that $A \cap B^c = A \setminus B$.

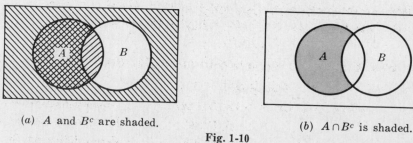

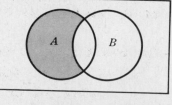

(a) A and B^c are shaded.   (b) $A \cap B^c$ is shaded.

Fig. 1-10

(b) As in Fig. 1-11(a) shade $B \setminus A$, the area of B which does not lie in A; then $(B \setminus A)^c$ is the area outside of $B \setminus A$, as shown in Fig. 1-11(b).

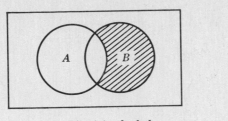

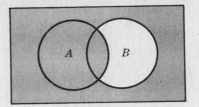

    (a) $B \setminus A$ is shaded.   (b) $(B \setminus A)^c$ is shaded.

Fig. 1-11

1.11. Illustrate the distributive law $A \cap (B \cup C) = (A \cap B) \cup (A \cap C)$ with Venn diagrams.

    Draw three intersecting circles labeled A, B and C, as in Fig. 1-12(a). Now, as in Fig. 1-12(b), shade A with strokes in one direction and shade $B \cup C$ with strokes in another direction; the cross-hatched area is $A \cap (B \cup C)$, as in Fig. 1-12(c). Next shade $A \cap B$ and then $A \cap C$, as in Fig. 1-12(d); the total area shaded is $(A \cap B) \cup (A \cap C)$, as in Fig. 1-12(e).

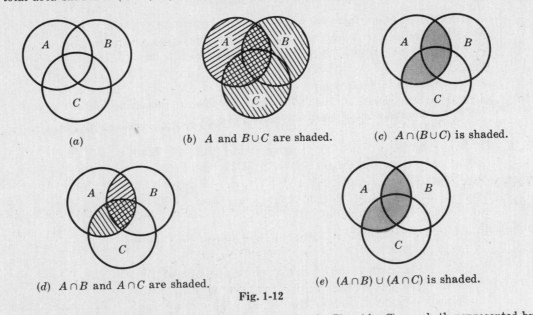

      (a)   (b) A and $B \cup C$ are shaded.   (c) $A \cap (B \cup C)$ is shaded.

    (d) $A \cap B$ and $A \cap C$ are shaded.   (e) $(A \cap B) \cup (A \cap C)$ is shaded.

Fig. 1-12

    As expected by the distributive law, $A \cap (B \cup C)$ and $(A \cap B) \cup (A \cap C)$ are both represented by the same set of points.

ALGEBRA OF SETS, DUALITY

1.12. Write the dual of each set equation:

    (a) $(A \cup B) \cap (A \cup B^c) = A \cup \emptyset$,   (b) $(A \cap U) \cup (B \cap A) = A$

    Replace each occurrence of $\cup$, $\cap$, U and $\emptyset$ by $\cap$, $\cup$, $\emptyset$ and U respectively:

    (a) $(A \cap B) \cup (A \cap B^c) = A \cap U$,   (b) $(A \cup \emptyset) \cap (B \cup A) = A$

1.13. Use the laws in Table 1.1 on page 5 to prove the identity $(U \cap A) \cup (B \cap A) = A$.

Statement	Reason
$(U \cap A) \cup (B \cap A) = (A \cap U) \cup (A \cap B)$	Commutative law 3a.
$\qquad\qquad\qquad\quad = A \cap (U \cup B)$	Distributive law 4b.
$\qquad\qquad\qquad\quad = A \cap (B \cup U)$	Commutative law 3a.
$\qquad\qquad\qquad\quad = A \cap U$	Identity law 6a.
$\qquad\qquad\qquad\quad = A$	Identity law 5b.

1.14. Prove the identity $(\emptyset \cup A) \cap (B \cup A) = A$.

This is the dual of the identity proved in Problem 1.13 and hence is true by the principle of duality. In other words, replacing each step in the proof in Problem 1.13 by dual statements gives a proof of this identity.

FINITE SETS, COUNTING PRINCIPLE

1.15. Determine which of the following sets are finite.

    (a) $A = \{$seasons in the year$\}$

    (b) $B = \{$states in the Union$\}$

    (c) $C = \{$positive integers less than 1$\}$

    (d) $D = \{$odd integers$\}$

    (e) $E = \{$positive integral divisors of 12$\}$

    (f) $F = \{$cats living in the United States$\}$

(a) A is finite because there are four seasons in the year, i.e. $n(A) = 4$.

(b) B is finite because there are 50 states in the Union, i.e. $n(B) = 50$.

(c) There are no positive integers less than 1; hence C is empty. Thus C is finite and $n(C) = 0$.

(d) D is infinite.

(e) The positive integral divisors of 12 are 1, 2, 3, 4, 6 and 12. Hence E is finite and $n(E) = 6$.

(f) Although it may be difficult to count the number of cats living in the United States, there is still a finite number of them. Hence F is finite.

1.16. In a survey of 60 people, it was found that 25 read *Newsweek* magazine, 26 read *Time* and 26 read *Fortune*. Also 9 read both *Newsweek* and *Fortune*, 11 read both *Newsweek* and *Time*, 8 read both *Time* and *Fortune*, and 8 read no magazine at all.

(a) Find the number of people who read all three magazines.

(b) Fill in the correct number of people in each of the eight regions of the Venn diagram of Fig. 1-13. Here N, T and F denote the set of people who read *Newsweek*, *Time* and *Fortune* respectively.

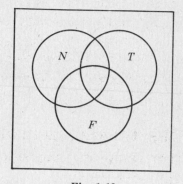

Fig. 1-13

(c) Determine the number of people who read exactly one magazine.

(a) Let $x = n(N \cap T \cap F)$, the number of people who read all three magazines. Note $n(N \cup T \cup F) = 52$ because 8 people read none of the magazines. We have

$$n(N \cup T \cup F) = n(N) + n(T) + n(F) - n(N \cap T) - n(N \cap F) - n(T \cap F) + n(N \cap T \cap F)$$

Hence, $52 = 25 + 26 + 26 - 11 - 9 - 8 + x$ or $x = 3$.

(b) The required Venn diagram, Fig. 1-14, is obtained as follows:

3 read all three magazines,

$11 - 3 = 8$ read *Newsweek* and *Time* but not all three magazines,

$9 - 3 = 6$ read *Newsweek* and *Fortune* but not all three magazines,

$8 - 3 = 5$ read *Time* and *Fortune* but not all three magazines,

$25 - 8 - 6 - 3 = 8$ read only *Newsweek*,

$26 - 8 - 5 - 3 = 10$ read only *Time*,

$26 - 6 - 5 - 3 = 12$ read only *Fortune*.

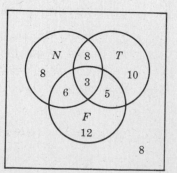

Fig. 1-14

(c) $8 + 10 + 12 = 30$ read only one magazine.

CLASSES OF SETS

1.17. Determine the power set $\mathcal{P}(A)$ of $A = \{a, b, c, d\}$.

The elements of $\mathcal{P}(A)$ are the subsets of A. Hence:

$$\mathcal{P}(A) = [A, \{a, b, c\}, \{a, b, d\}, \{a, c, d\}, \{b, c, d\}, \{a, b\}, \{a, c\}, \{a, d\},$$
$$\{b, c\}, \{b, d\}, \{c, d\}, \{a\}, \{b\}, \{c\}, \{d\}, \emptyset]$$

We note that $\mathcal{P}(A)$ has $2^4 = 16$ elements.

1.18. Consider the set $A = [\{1, 2, 3\}, \{4, 5\}, \{6, 7, 8\}]$.

(a) What are the elements of A?

(b) Determine whether each of the following is true or false:

    (i) $1 \in A$ (iv) $\{\{4, 5\}\} \subset A$

    (ii) $\{1, 2, 3\} \subset A$ (v) $\emptyset \notin A$

    (iii) $\{6, 7, 8\} \in A$ (vi) $\emptyset \subset A$

(a) A is a class of sets; its elements are the sets $\{1, 2, 3\}$, $\{4, 5\}$ and $\{6, 7, 8\}$.

(b) (i)   False. 1 is not one of the elements of A.

    (ii)   False. $\{1, 2, 3\}$ is not a subset of A; it is one of the elements of A.

    (iii) True. $\{6, 7, 8\}$ is one of the elements of A.

    (iv) True. $\{\{4, 5\}\}$, the set consisting of the element $\{4, 5\}$, is a subset of A.

    (v)   False. The empty set $\emptyset$ is not an element of A, i.e. it is not one of the three sets listed in the problem statement.

    (vi) True. The empty set is a subset of every set; even a class of sets.

MISCELLANEOUS PROBLEMS

1.19. Prove the proposition P that the sum of the first n positive integers is $\frac{1}{2}n(n+1)$; that is,

$$P(n): \quad 1 + 2 + 3 + \cdots + n = \tfrac{1}{2}n(n+1)$$

The proposition holds for $n = 1$ since

$$P(1): \quad 1 = \tfrac{1}{2}(1)(1+1)$$

Assuming $P(n)$ is true, we add $n+1$ to both sides of $P(n)$, obtaining

$$1 + 2 + 3 + \cdots + n + (n+1) = \tfrac{1}{2}n(n+1) + (n+1)$$
$$= \tfrac{1}{2}[n(n+1) + 2(n+1)]$$
$$= \tfrac{1}{2}[(n+1)(n+2)]$$

which is $P(n+1)$. That is, $P(n+1)$ is true whenever $P(n)$ is true. By the principle of induction, P is true for all n.

1.20. Consider the following assumptions:

              S_1: Poets are happy people.

              S_2: Every doctor is wealthy.

              S_3: No one who is happy is also wealthy.

Determine the validity of each of the following conclusions:

(a) No poet is wealthy.

(b) Doctors are happy people.

(c) No one can be both a poet and a doctor.

By S_1 the set of poets is contained in the set of happy people, and by S_3 the set of happy people is disjoint from the set of wealthy people.   Hence draw the Venn diagram of Fig. 1-15.

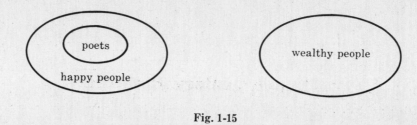

Fig. 1-15

By S_2 the set of doctors is contained in the set of wealthy people.   So draw the Venn diagram of Fig. 1-16.   From this diagram it is obvious that (a) and (c) are valid conclusions whereas (b) is not valid.

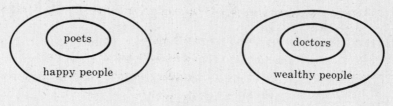

Fig. 1-16

1.21. (a) Prove $(A \cap B) \subset A \subset (A \cup B)$ and $(A \cap B) \subset B \subset (A \cup B)$.

(b) Prove Theorem 1.2: The following are equivalent: $A \subset B$, $A \cap B = A$, $A \cup B = B$.

(a) Since every element in $A \cap B$ is in both A and B, it is certainly true that if $x \in (A \cap B)$ then $x \in A$; hence $(A \cap B) \subset A$.   Furthermore, if $x \in A$ then $x \in (A \cup B)$ (by the definition of $A \cup B$), so $A \subset (A \cup B)$.   Putting these together gives $(A \cap B) \subset A \subset (A \cup B)$.   Similarly, $(A \cap B) \subset B \subset (A \cup B)$.

(b) Suppose $A \subset B$ and let $x \in A$.   Then $x \in B$, hence $x \in A \cap B$ and $A \subset A \cap B$.   By part (i), $(A \cap B) \subset A$.   Therefore $A \cap B = A$.   On the other hand, suppose $A \cap B = A$ and let $x \in A$. Then $x \in (A \cap B)$, hence $x \in A$ and $x \in B$.   Therefore, $A \subset B$.   Both results show that $A \subset B$. is equivalent to $A \cap B = A$.

Suppose again that $A \subset B$.   Let $x \in (A \cup B)$.   Then $x \in A$ or $x \in B$.   If $x \in A$, then $x \in B$ because $A \subset B$.   In either case, $x \in B$.   Therefore $A \cup B \subset B$.   By part (i), $B \subset A \cup B$. Therefore $A \cup B = B$.   Now suppose $A \cup B = B$ and let $x \in A$.   Then $x \in A \cup B$ by definition of union of sets.   Hence $x \in B = A \cup B$.   Therefore $A \subset B$.   Both results show that $A \subset B$ is equivalent to $A \cup B = B$.

Thus $A \subset B$, $A \cap B = A$ and $A \cup B = B$ are equivalent.

1.22. Prove Theorem 1.5: If A and B are finite sets, then $A \cup B$ and $A \cap B$ are finite and

$$n(A \cup B) \, = \, n(A) + n(B) - n(A \cap B)$$

Clearly $A \cup B$ and $A \cap B$ are finite if A and B are finite.

By Problem 1.30, $A \cup B = B \cup (A \setminus B)$; and B and $A \setminus B$ are disjoint. Applying Lemma 1.4,

$$n(A \cup B) = n(B) + n(A \setminus B) \qquad (1)$$

Also by Problem 1.30, $A = (A \setminus B) \cup (A \cap B)$; and $A \setminus B$ and $A \cap B$ are disjoint. Then by Lemma 1.4,

$$n(A) = n(A \setminus B) + n(A \cap B) \quad \text{or} \quad n(A \setminus B) = n(A) - n(A \cap B)$$

Substituting for $n(A \setminus B)$ in (1) gives the required result.

Supplementary Problems

SETS, SUBSETS

1.23. Which of the following sets are equal?

$$\{1,2\}, \{1,3\}, \{2,1\}, \{3,1,3\}, \{1,2,1\}$$

$A = \{x : x^2 - 4x + 3 = 0\} \qquad C = \{x : x \in \mathbf{N}, x < 3\}$

$B = \{x : x^2 - 3x + 2 = 0\} \qquad D = \{x : x \in \mathbf{N}, x \text{ is odd}, x < 5\}$

1.24. List the elements of the following sets if the universal set is $U = \{a, b, c, \ldots, y, z\}$. Which of the sets, if any, are equal?

$A = \{x : x \text{ is a vowel}\}$

$B = \{x : x \text{ is a letter in the word "little"}\}$

$C = \{x : x \text{ precedes } f \text{ in the alphabet}\}$

$D = \{x : x \text{ is a letter in the word "title"}\}$

1.25. Let $A = \{1, 2, \ldots, 8, 9\}$, $B = \{2, 4, 6, 8\}$, $C = \{1, 3, 5, 7, 9\}$, $D = \{3, 4, 5\}$ and $E = \{3, 5\}$. Which sets can equal X if we are given the following information?

(a) X and B are disjoint.   (b) $X \subset D$ but $X \not\subset B$.   (c) $X \subset A$ but $X \not\subset C$.   (d) $X \subset C$ but $X \not\subset A$.

1.26. Consider the following sets:

$$\varnothing, \quad A = \{a\}, \quad B = \{c, d\}, \quad C = \{a, b, c, d\}, \quad D = \{a, b\}, \quad E = \{a, b, c, d, e\}$$

Insert the correct symbol, $\subset$ or $\not\subset$, between each pair of sets:

(a) $\varnothing, A$ (c) A, B (e) B, C (g) C, D

(b) D, E (d) D, A (f) D, C (h) B, D

SET OPERATIONS

1.27. Given: $A = \{a, b, c, d, e\}$ $C = \{b, c, e, g, h\}$

$B = \{a, b, d, f, g\}$ $D = \{d, e, f, g, h\}$

Find:

(a) $A \cup B$ (d) $A \cap (B \cup D)$ (g) $(A \cup D) \setminus C$

(b) $B \cap C$ (e) $B \setminus (C \cup D)$ (h) $B \cap C \cap D$

(c) $C \setminus D$ (f) $(A \cap D) \cup B$ (i) $(C \setminus A) \setminus D$

1.28. The Venn diagram of Fig. 1-17 shows sets A, B, and C. Shade the following sets:

(a) $A \cap B \cap C$ (d) $A \setminus (B \cup C)$ (g) $A^c \cap (B \cup C)$

(b) $A^c \cup B \cup C$ (e) $(A \cup B) \setminus C$ (h) $(A^c \cap B) \setminus C$

(c) $A \cup (B \cap C)$ (f) $A^c \cap B^c \cap C$

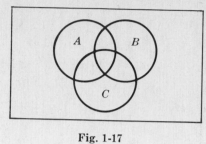

Fig. 1-17

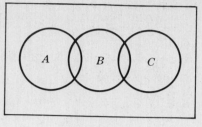

Fig. 1-18

1.29.   Repeat the preceding problem with the Venn diagram of Fig. 1-18.

1.30.   Let A and B be any sets. Prove:

(a) $A \cup B = B \cup (A \setminus B)$ and $B \cap (A \setminus B) = \emptyset$

(b) $A = (A \setminus B) \cup (A \cap B)$ and $(A \setminus B) \cap (A \cap B) = \emptyset$

1.31.   Prove: (a) $A \subset B$ if and only if $A \cap B^c = \emptyset$.

(b) $A \subset B$ if and only if $A^c \cup B = U$.

(c) $A \subset B$ if and only if $B^c \subset A^c$.

(d) $A \subset B$ if and only if $A \setminus B = \emptyset$.

(Compare with Theorem 1.2.)

1.32.   Prove the absorption laws:

(a) $A \cup (A \cap B) = A$,   (b) $A \cap (A \cup B) = A$.

1.33.   The formula $A \setminus B = A \cap B^c$ defines the difference operation in terms of the operations of intersection and complement. Find a formula that defines the union of two sets, $A \cup B$, in terms of the operations of intersection and complement.

1.34.   (a)   Prove: $A \cap (B \setminus C) = (A \cap B) \setminus (A \cap C)$.

(b)   Give an example to show that $A \cup (B \setminus C) \neq (A \cup B) \setminus (A \cup C)$.

ALGEBRA OF SETS, DUALITY

1.35.   Write the dual of each set equation:

(a) $A \cup (A \cap B) = A$

(b) $(A \cap B) \cup (A^c \cap B) \cup (A \cap B^c) \cup (A^c \cap B^c) = U$

1.36.   Use the laws in Table 1.1 to prove each set identity:

(a) $(A \cap B) \cup (A \cap B^c) = A$

(b) $A \cup (A \cap B) = A$

(c) $A \cup B = (A \cap B^c) \cup (A^c \cap B) \cup (A \cap B)$

FINITE SETS, COUNTING PRINCIPLE

1.37.   Determine which of the following sets are finite:

(a)   The set of lines parallel to the x axis.

(b)   The set of letters in the English alphabet.

(c)   The set of numbers which are multiples of 5.

(d)   The set of animals living on the earth.

(e)   The set of numbers which are solutions of the equation $x^{27} + 26x^{18} - 17x^{11} + 7x^3 - 10 = 0$.

(f)   The set of circles through the origin $(0, 0)$.

1.38. Use Theorem 1.5 to prove Corollary 1.6: If A, B and C are finite sets, then so is $A \cup B \cup C$ and
$$n(A \cup B \cup C) = n(A) + n(B) + n(C) - n(A \cap B) - n(A \cap C) - n(B \cap C) + n(A \cap B \cap C)$$

1.39. A survey of 100 students produced the following statistics:

32 study mathematics,

20 study physics,

45 study biology,

15 study mathematics and biology,

7 study mathematics and physics,

10 study physics and biology,

30 do not study any of the three subjects.

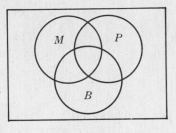

Fig. 1-19

(a) Find the number of students studying all three subjects.

(b) Fill in the number of students in each of the eight regions of the Venn diagram (Fig. 1-19), where M, P and B denote the sets of students studying mathematics, physics and biology respectively.

(c) Find the number of students taking exactly one of the three subjects.

CLASSES OF SETS

1.40. Find the power set $\mathcal{P}(A)$ of $A = \{1, 2, 3, 4, 5\}$.

1.41. Given $A = [\{a, b\}, \{c\}, \{d, e, f\}]$.

(a) State whether each of the following is true or false:

(i) $a \in A$, (ii) $\{c\} \subset A$, (iii) $\{d, e, f\} \in A$, (iv) $\{\{a, b\}\} \subset A$, (v) $\emptyset \subset A$.

(b) Find the power set of A.

1.42. Suppose A is a finite set and $n(A) = m$. Prove that $\mathcal{P}(A)$ has 2^m elements.

ARGUMENTS AND VENN DIAGRAMS

1.43. Employ Venn diagrams to find the conclusion to the following set of assumptions (adopted from a book on logic by Lewis Carroll, the author of *Alice in Wonderland*).

S_1: Babies are illogical.

S_2: Nobody is despised who can manage a crocodile.

S_3: Illogical people are despised.

1.44. Consider the following assumptions:

S_1: I planted all my expensive trees last year.

S_2: All my fruit trees are in my orchard.

S_3: No tree in the orchard was planted last year.

Determine whether or not each of the following is a valid conclusion:

(a) The fruit trees were planted last year.

(b) No expensive tree is in the orchard.

(c) No fruit tree is expensive.

INDUCTION

1.45. Prove: $2 + 4 + 6 + \cdots + 2n = n(n+1)$

1.46.   Prove: $1 + 4 + 7 + \cdots + (3n - 2) = \frac{1}{2}n(3n - 1)$

1.47.   Prove: $\dfrac{1}{1 \cdot 3} + \dfrac{1}{3 \cdot 5} + \dfrac{1}{5 \cdot 7} + \cdots + \dfrac{1}{(2n - 1)(2n + 1)} = \dfrac{1}{2n + 1}$

1.48.   Prove: $1^2 + 2^2 + 3^2 + \cdots + n^2 = \dfrac{n(n + 1)(2n + 1)}{6}$

MISCELLANEOUS PROBLEM

1.49.   By the *symmetric difference* of sets A and B, denoted by $A \triangle B$, we mean the set

$$A \triangle B = (A \cup B) \setminus (A \cap B)$$

i.e. $A \triangle B$ consists of those elements which belong to either A or B but not both. (See Fig. 1-20, which shows a Venn diagram of $A \triangle B$.)

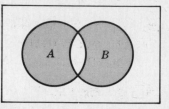

$A \triangle B$ is shaded.

Fig. 1-20

   (*a*)   Given: $A = \{1, 2, 3, 4, 5, 6\}$ $C = \{1, 3, 5, 7, 9\}$

                      $B = \{4, 5, 6, 7, 8, 9\}$ $D = \{2, 3, 5, 7, 8\}$

       Find: (i) $A \triangle B$ (iii) $C \triangle D$ (v) $A \cap (B \triangle D)$

                   (ii) $B \triangle C$ (iv) $A \triangle D$ (vi) $(A \cap B) \triangle (A \cap D)$

   (*b*)   Prove the following properties of the symmetric difference:

       (i) $A \triangle (B \triangle C) = (A \triangle B) \triangle C$ (associative law)

       (ii) $A \triangle B = B \triangle A$ (commutative law)

       (iii)   If $A \triangle B = A \triangle C$ then $B = C$ (cancellation law)

       (iv) $A \cap (B \triangle C) = (A \cap B) \triangle (A \cap C)$ (distributive law)

COMPUTER PROGRAMMING PROBLEMS

1.50.   The first card of a deck contains five integers, the elements of a set A, and the second card contains seven integers, the elements of a set B. Write a program which prints (*a*) $A \cap B$, (*b*) $A \cup B$, (*c*) $A \setminus B$.

       Test the program with the following input:

       (i)   First card: 1, 3, −5, 8, −4
             Second card: 7, −4, 2, 3, 8, −6, 5

       (ii)   First card: 1, 3, −5, 8, −4
             Second card: 7, −2, 3, 2, −6, 9, −4

   In Problems 1.51–1.53, A and B are one-dimensional arrays with M and N elements respectively. We view A and B as sets of elements.

1.51.   Let X be a number. Write a subprogram BLONG such that BLONG(X,A,M) has the value 1 or −1 according as X does or does not belong to A.

1.52.   Write a subprogram SUBSET such that SUBSET(A,M,B,N) has the value 1 or −1 according as A is or is not a subset of B. (*Hint:* Use the subprogram BLONG from Problem 1.51.)

1.53.   Write a subprogram EQUAL such that EQUAL(A,M,B,N) has the value 1 or −1 according as A is or is not equal to B (as sets). (*Hint:* Use the subprogram SUBSET from Problem 1.52.)

Answers to Supplementary Problems

1.23. $\{1,2\} = \{2,1\} = \{1,2,1\} = B = C,$ $\{1,3\} = \{3,1,3\} = A = D$

1.24. $A = \{a,e,i,o,u\},$ $B = D = \{l,i,t,e\},$ $C = \{a,b,c,d,e\}$

1.25. (a) C and $E,$ (b) D and $E,$ (c) A, B and $D,$ (d) None

1.26. (a) $\emptyset \subset A,$ (b) $D \subset E,$ (c) $A \not\subset B,$ (d) $D \not\subset A,$ (e) $B \subset C,$ (f) $D \subset C,$ (g) $C \not\subset D,$ (h) $B \not\subset D$

1.27.
(a) $A \cup B = \{a,b,c,d,e,f,g\}$ (d) $A \cap (B \cup D) = \{a,b,d,e\}$ (g) $(A \cup D) \setminus C = \{a,d,f\}$
(b) $B \cap C = \{b,g\}$ (e) $B \setminus (C \cup D) = \{a\}$ (h) $B \cap C \cap D = \{g\}$
(c) $C \setminus D = \{b,c\}$ (f) $(A \cap D) \cup B = \{a,b,d,e,f,g\}$ (i) $(C \setminus A) \setminus D = \emptyset$

1.28.   See Fig. 1-21.

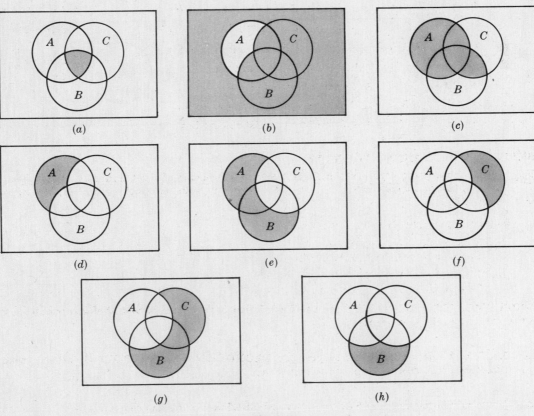

Fig. 1-21

1.29.   See Fig. 1-22.

1.33. $A \cup B = (A^c \cap B^c)^c$

1.34. (b) Let $A = \{1\}, B = \{2\}, C = \{3\}.$ Then $A \cup (B \setminus C) = \{1,2\}$ but $(A \cup B) \setminus (A \cup C) = \{2\}.$

1.35. (a) $A \cap (A \cup B) = A$ (b) $(A \cup B) \cap (A^c \cup B) \cap (A \cup B^c) \cap (A^c \cup B^c) = \emptyset$

1.37. (a) Infinite, (b) finite, (c) infinite, (d) finite, (e) finite, (f) infinite

1.39. (a) 5, (b) see Fig. 1-23, (c) 48

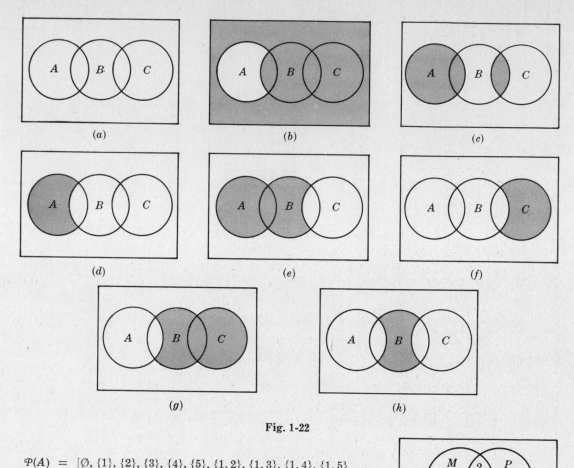

Fig. 1-22

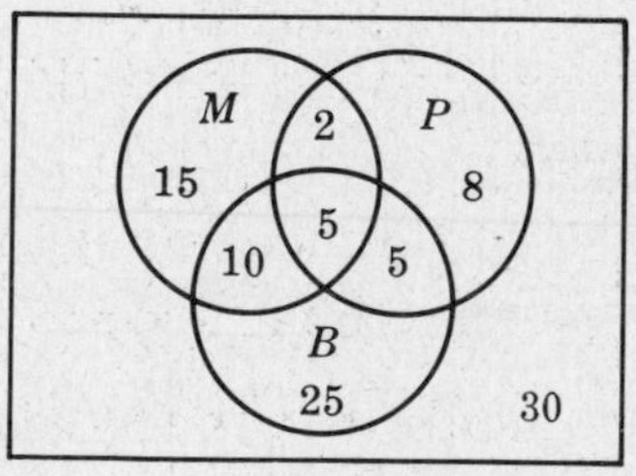

1.40. $\mathcal{P}(A) = [\varnothing, \{1\}, \{2\}, \{3\}, \{4\}, \{5\}, \{1,2\}, \{1,3\}, \{1,4\}, \{1,5\},$
$\{2,3\}, \{2,4\}, \{2,5\}, \{3,4\}, \{3,5\}, \{4,5\}, \{1,2,3\},$
$\{1,2,4\}, \{1,2,5\}, \{2,3,4\}, \{2,3,5\}, \{3,4,5\}, \{1,3,4\},$
$\{1,3,5\}, \{1,4,5\}, \{2,4,5\}, \{1,2,3,4\}, \{1,2,3,5\},$
$\{1,2,4,5\}, \{1,3,4,5\}, \{2,3,4,5\}, A]$

There are $2^5 = 32$ sets in $\mathcal{P}(A)$.

Fig. 1-23

1.41. (a) (i) F, (ii) F, (iii) T, (iv) T, (v) T

(b) Note $n(A) = 3$; hence $\mathcal{P}(A)$ has $2^3 = 8$ elements:

$$\mathcal{P}(A) = \{A, [\{a,b\}, \{c\}], [\{a,b\}, \{d,e,f\}], [\{c\}, \{d,e,f\}],$$
$$[\{a,b\}], [\{c\}], [\{d,e,f\}], \varnothing\}$$

1.42. Let X be an arbitrary member of $\mathcal{P}(A)$. For each $a \in A$, there are two possibilities: $a \in X$ or $a \notin X$. But there are m elements in A; hence there are

$$\overbrace{2 \cdot 2 \cdot 2 \cdots \cdots 2}^{m \text{ times}} = 2^m$$

different sets X. That is, $\mathcal{P}(A)$ has 2^m members.

Chapter 2

Relations

2.1 INTRODUCTION

In the first chapter we saw that the order of the elements of a set is irrelevant, i.e. $\{3,5\} = \{5,3\}$. In this chapter we will consider *ordered pairs* (a, b) of elements; here a is designated as the first element and b as the second element. Thus $(a, b) \neq (b, a)$ unless $a = b$. Furthermore,

$$(a, b) = (c, d)$$

if and only if $a = c$ and $b = d$.

The reader is familiar with many *relations* which are used in mathematics: "less than", "is parallel to", "is congruent to", "is a subset of", etc. In a certain sense, these relations consider the existence or nonexistence of a certain connection between pairs of objects taken in a definite order. Formally, we define a relation simply in terms of ordered pairs of objects. We then discuss relations which are similar to the relation of equality, and show their connection to partitions of sets. Although matrices are covered in Chapter 4, we have included their connection with relations here for completeness. However, such sections can be ignored at a first reading by those with no previous knowledge of matrix theory.

2.2 PRODUCT SETS

Consider two arbitrary sets A and B. The set of all ordered pairs (a, b) where $a \in A$ and $b \in B$ is called the *product*, or *cartesian product*, of A and B. A short designation of this product is $A \times B$, which is read "A cross B". By definition,

$$A \times B = \{(a, b) : a \in A \text{ and } b \in B\}$$

One frequently writes A^2 instead of $A \times A$.

EXAMPLE 2.1. **R** denotes the set of real numbers and so $\mathbf{R}^2 = \mathbf{R} \times \mathbf{R}$ is the set of ordered pairs of real numbers. The reader is familiar with the geometrical representation of $\mathbf{R}^2$ as points in the plane as in Fig. 2-1. Here each point P represents an ordered pair (a, b) of real numbers and vice versa; the vertical line through P meets the x axis at a, and the horizontal line through P meets the y axis at b. $\mathbf{R}^2$ is frequently called the *cartesian plane*.

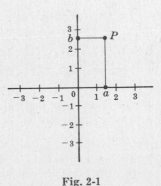

Fig. 2-1

EXAMPLE 2.2. Let $A = \{1, 2\}$ and $B = \{a, b, c\}$. Then

$$A \times B = \{(1, a), (1, b), (1, c), (2, a), (2, b), (2, c)\}$$
$$B \times A = \{(a, 1), (a, 2), (b, 1), (b, 2), (c, 1), (c, 2)\}$$

Also

$$A \times A = \{(1, 1), (1, 2), (2, 1), (2, 2)\}$$

There are two things worth noting in the above example. First of all $A \times B \neq B \times A$. The cartesian product deals with ordered pairs, so naturally the order in which the sets are considered is important. Secondly,

$$n(A \times B) = 6 = 2 \cdot 3 = n(A) \cdot n(B)$$

where $n(A) =$ number of elements in A. In fact, $n(A \times B) = n(A) \cdot n(B)$ for any finite sets A and B. This follows from the observation that, for an ordered pair (a, b) in $A \times B$, there are $n(A)$ possibilities for a, and for each of these there are $n(B)$ possibilities for b.

The idea of a product of sets can be extended to any finite number of sets. For any sets $A_1, A_2, \ldots, A_n$, the set of all ordered n-tuples $(a_1, a_2, \ldots, a_n)$ where $a_1 \in A_1$, $a_2 \in A_2$, $\ldots$, $a_n \in A_n$ is called the *product* of the sets $A_1, \ldots, A_n$ and is denoted by

$$A_1 \times A_2 \times \cdots \times A_n \qquad \text{or} \qquad \prod_{i=1}^{n} A_i$$

Just as we write A^2 instead of $A \times A$, so we write A^n instead of $A \times A \times \cdots \times A$ where there are n factors all equal to A. For example, $\mathbf{R}^3 = \mathbf{R} \times \mathbf{R} \times \mathbf{R}$ denotes the usual three-dimensional space.

2.3 RELATIONS

Let A and B be sets. A *binary relation*, R, from A to B is a subset of $A \times B$. If $(x, y) \in R$, we say that x is *R-related* to y and denote this by

$$x \, R \, y$$

If $(x, y) \notin R$, we write $x \, \not{R} \, y$ and say that x is not R-related to y. If R is a relation from A to A, i.e. is a subset of $A \times A$, then we say that R is a relation *on* A.

Since we will deal mainly with binary relations, the word "relation" will mean binary relation unless otherwise specified.

The *domain* of a relation R is the set of all first elements of the ordered pairs which belong to R, and the *range* of R is the set of second elements.

EXAMPLE 2.3.

(a) Let $A = \{1, 2, 3\}$, and $R = \{(1, 2), (1, 3), (3, 2)\}$. Then R is a relation on A since it is a subset of $A \times A$. With respect to this relation,

$$1 \, R \, 2, \quad 1 \, R \, 3, \quad 3 \, R \, 2, \quad \text{but} \quad 1 \, \not{R} \, 1, \quad 2 \, \not{R} \, 1, \quad 2 \, \not{R} \, 2, \quad 2 \, \not{R} \, 3, \quad 3 \, \not{R} \, 1, \quad 3 \, \not{R} \, 3$$

The domain of R is $\{1, 3\}$ and the range of R is $\{2, 3\}$.

(b) Let $A = \{\text{eggs}, \text{milk}, \text{corn}\}$ and $B = \{\text{cows}, \text{goats}, \text{hens}\}$. We can define a relation R from A to B by $(a, b) \in R$ if a is produced by b. In other words,

$$R = \{(\text{eggs, hens}), (\text{milk, cows}), (\text{milk, goats})\}$$

With respect to this relation,

$$\text{eggs } R \text{ hens}, \quad \text{milk } R \text{ cows, etc.}$$

(c) Suppose we say that two countries are adjacent if they have some part of their boundaries in common. Then "is adjacent to" is a relation R on the countries of the earth. Thus

$$(\text{Italy, Switzerland}) \in R \qquad \text{but} \qquad (\text{Canada, Mexico}) \notin R$$

(d) A familiar relation on the set $\mathbf{Z}$ of integers is "m divides n". A common notation for this relation is to write $m \mid n$ when m divides n. Thus $6 \mid 30$ but $7 \nmid 25$.

(e) Among the basic relations considered in geometry are "is congruent to" and "is similar to", which are relations on the set of geometric figures in the plane. The relation "is parallel to" is a relation on the set of lines in the plane.

(f) Let A be any set. An important relation on A is that of *equality*,

$$\{(a, a) : \ a \in A\}$$

which is usually denoted by " $=$ ". This relation is also called the *identity relation* in A and is sometimes denoted by $\triangle_A$.

(g) Let A be any set. Then $A \times A$ and $\emptyset$ are subsets of $A \times A$ and hence are relations on A called the *universal relation* and *empty relation* respectively.

2.4 PICTORIAL REPRESENTATIONS OF RELATIONS

We first consider a relation S on the set $\mathbf{R}$ of real numbers, i.e. S is a subset of $\mathbf{R}^2 = \mathbf{R} \times \mathbf{R}$. Since $\mathbf{R}^2$ can be represented by the set of points in the plane, we can picture S by emphasizing those points in the plane which belong to S. The pictorial representation of the relation is sometimes called the *graph* of the relation.

Frequently, the relation S consists of all ordered pairs of real numbers which satisfy some given equation

$$E(x, y) \ = \ 0$$

We usually identify the relation with the equation, i.e. we speak of the relation $E(x, y) = 0$.

EXAMPLE 2.4.   Consider the relation S defined by the equation

$$x^2 + y^2 \ = \ 25$$

That is, S consists of all ordered pairs (x_0, y_0) which satisfy the given equation. The graph of the equation is a circle having its center at the origin and radius 5.   See Fig. 2-2.

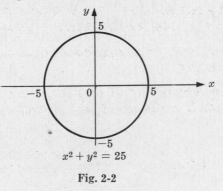

$$x^2 + y^2 = 25$$

Fig. 2-2

Suppose now A and B are finite sets.   We list three different ways of picturing a relation R from A to B.

(i) Represent the elements of A as points on some horizontal axis, and the elements of B as points on some vertical axis. The elements of $A \times B$ are then represented by the points of intersection of the vertical lines through the points of A and the horizontal lines through the points of B; this representation is called a *coordinate diagram* of $A \times B$. The relation R is pictured by adding emphasis to those points of $A \times B$ which belong to R.

(ii) Form a rectangular array whose rows are labeled by the elements of A and whose columns are labeled by the elements of B.   Put a 1 or 0 in each position of the array according as $a \in A$ is or is not related to $b \in B$. This array is called the *matrix of the relation*.

(iii) Write down the elements of A and the elements of B in two disjoint disks, and then draw an arrow from $a \in A$ to $b \in B$ whenever a is related to b. This picture will be called the *arrow diagram* of the relation.

In Fig. 2-3 we picture the first two relations of Example 2.3 by the above three ways.

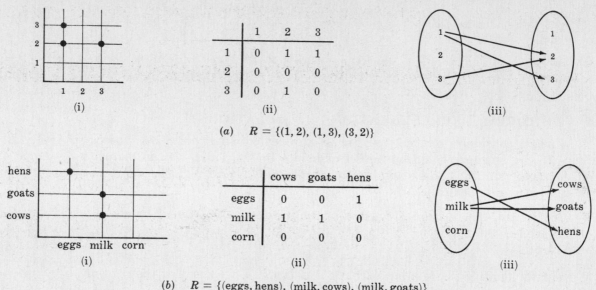

(a) $R = \{(1, 2), (1, 3), (3, 2)\}$

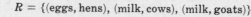

(b) $R = \{(\text{eggs}, \text{hens}), (\text{milk}, \text{cows}), (\text{milk}, \text{goats})\}$

Fig. 2-3

There is another way of picturing a relation when it is from a finite set into itself. We write down the elements of the set and then draw an arrow from an element x to an element y whenever x is related to y. This diagram is called the *directed graph* of the relation. Figure 2-4 gives the directed graph of a relation R on the set $A = \{1, 2, 3, 4\}$. Observe that there is an arrow from 2 to itself since 2 is related to 2.

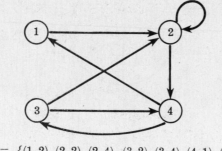

$R = \{(1, 2), (2, 2), (2, 4), (3, 2), (3, 4), (4, 1), (4, 3)\}$

Fig. 2-4

2.5 INVERSE RELATIONS

Let R be a relation from A to B. The *inverse* of R, denoted by R^{-1}, is the relation from B to A which consists of those ordered pairs which when reversed belong to R:

$$R^{-1} = \{(b, a) : (a, b) \in R\}$$

In other words, $bR^{-1}a$ if and only if aRb.

EXAMPLE 2.5.

(a) Let R be the following relation on $A = \{1, 2, 3\}$:

$$R = \{(1, 2), (1, 3), (2, 3)\}$$

Then $R^{-1} = \{(2, 1), (3, 1), (3, 2)\}$

(b) The inverses of the relations defined by

$$\text{``}x \text{ is the husband of } y\text{''} \quad \text{ and } \quad \text{``}x \text{ is taller than } y\text{''}$$

are respectively

$$\text{``}x \text{ is the wife of } y\text{''} \quad \text{ and } \quad \text{``}x \text{ is shorter than } y\text{''}$$

Clearly, if R is any relation then $(R^{-1})^{-1} = R$. Also, the domain of R^{-1} equals the range of R, and the range of R^{-1} equals the domain of R. Furthermore, if M_R is the matrix of a relation R between finite sets, then the matrix $M_{R^{-1}}$ of R^{-1} is the transpose of the matrix of R:

$$M_{R^{-1}} = M_R^T$$

For example, the matrices of the relations in Example 2.5a are as follows:

$$M_R = \begin{pmatrix} 0 & 1 & 1 \\ 0 & 0 & 1 \\ 0 & 0 & 0 \end{pmatrix} \quad \text{ and } \quad M_{R^{-1}} = \begin{pmatrix} 0 & 0 & 0 \\ 1 & 0 & 0 \\ 1 & 1 & 0 \end{pmatrix} = M_R^T$$

2.6 COMPOSITION OF RELATIONS

Let A, B, and C be sets, and let R be a relation from A to B and let S be a relation from B to C. That is, R is a subset of $A \times B$ and S is a subset of $B \times C$. Then R and S give rise to a relation from A to C denoted by $R \circ S$ and defined by:

$$a(R \circ S)c \quad \text{if for some } b \in B \text{ we have } aRb \text{ and } bSc$$

That is,

$$R \circ S = \{(a, c) : \text{ there exists } b \in B \text{ for which } (a, b) \in R \text{ and } (b, c) \in S\}$$

The relation $R \circ S$ is called the *composition* of R and S; it is sometimes denoted simply by RS.

The arrow diagrams give us a geometrical interpretation of $R \circ S$ as seen in the following example.

EXAMPLE 2.6. Let $A = \{1, 2, 3, 4\}$, $B = \{a, b, c, d\}$, $C = \{x, y, z\}$ and let

$$R = \{(1, a), (2, d), (3, a), (3, b), (3, d)\}$$

$$S = \{(b, x), (b, z), (c, y), (d, z)\}$$

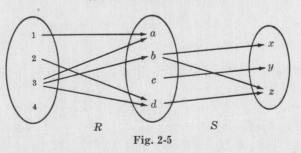

Fig. 2-5

Consider the arrow diagrams of R and S as in Fig. 2-5. Observe that there is an arrow from 2 to d which is followed by an arrow from d to z. We can view these two arrows as a "path" which "connects" the element $2 \in A$ to the element $z \in C$. Thus

$$2(R \circ S)z \quad \text{since} \quad 2Rd \text{ and } dSz$$

Similarly there are paths from 3 to x and from 3 to z. Hence

$$3(R \circ S)x \quad \text{and} \quad 3(R \circ S)z$$

No other element of A is connected to an element of C. Accordingly,

$$R \circ S = \{(2, z), (3, x), (3, z)\}$$

There is another way of finding $R \circ S$. Let M_R and M_S denote respectively the matrices of the relations R and S.   Then

$$
M_R = \begin{array}{c} 1 \\ 2 \\ 3 \\ 4 \end{array}
\begin{pmatrix}
a & b & c & d \\
1 & 0 & 0 & 0 \\
0 & 0 & 0 & 1 \\
1 & 1 & 0 & 1 \\
0 & 0 & 0 & 0
\end{pmatrix}
\quad \text{and} \quad
M_S = \begin{array}{c} a \\ b \\ c \\ d \end{array}
\begin{pmatrix}
x & y & z \\
0 & 0 & 0 \\
1 & 0 & 1 \\
0 & 1 & 0 \\
0 & 0 & 1
\end{pmatrix}
$$

Multiplying M_R and M_S we obtain the matrix

$$
M = M_R M_S = \begin{array}{c} 1 \\ 2 \\ 3 \\ 4 \end{array}
\begin{pmatrix}
x & y & z \\
0 & 0 & 0 \\
0 & 0 & 1 \\
1 & 0 & 2 \\
0 & 0 & 0
\end{pmatrix}
$$

The nonzero entries in this matrix tell us which elements are related by $R \circ S$. Thus $M = M_R M_S$ and $M_{R \circ S}$ have the same nonzero entries.

Our first theorem tells us that the composition of relations is associative.

Theorem 2.1:   Let A, B, C and D be sets.   Suppose R is a relation from A to B, S is a relation from B to C and T is a relation from C to D.   Then

$$(R \circ S) \circ T = R \circ (S \circ T)$$

We prove this theorem in Problem 2.10.

2.7 PROPERTIES OF RELATIONS

Let R be a relation on a set A.   We list four types of relations:

(1) R is *reflexive* if $a\,R\,a$ for every a in A.

(2) R is *symmetric* if $a\,R\,b$ implies $b\,R\,a$.

(3) R is *anti-symmetric* if $a\,R\,b$ and $b\,R\,a$ implies $a = b$.

(4) R is *transitive* if $a\,R\,b$ and $b\,R\,c$ implies $a\,R\,c$.

Observe that these properties are only defined for relations on a set.

EXAMPLE 2.7.

(a) Consider the relation $\subset$ of set inclusion on any collection C of sets.   Note that:

  (1) $A \subset A$ for any set A in C, so $\subset$ is reflexive;

  (2) $A \subset B$ does not imply $B \subset A$, so $\subset$ is not symmetric;

  (3) if $A \subset B$ and $B \subset A$ then $A = B$, so $\subset$ is anti-symmetric;

  (4) if $A \subset B$ and $B \subset C$ then $A \subset C$, so $\subset$ is transitive.

  (We assume that C has more than one set.)

(b) Consider the relation $=$ of equality on any set A. Note that $=$ satisfies all four of the above properties. That is,

  (1) $a = a$ for any element $a \in A$,

  (2) if $a = b$ then $b = a$,

(3) if $a = b$ and $b = a$ then $a = b$,

(4) if $a = b$ and $b = c$ then $a = c$.

This example shows that symmetric and anti-symmetric relations are not the negations of each other.

(c) Consider the relation $R = \{(1, 1), (1, 2), (2, 1), (2, 3)\}$ on $A = \{1, 2, 3\}$.   Then:

    (1) 2 is in A but $2 \not{R} 2$, so R is not reflexive;

    (2) $2 R 3$ but $3 \not{R} 2$, so R is not symmetric;

    (3) $1 R 2$ and $2 R 1$, but $1 \neq 2$, so R is not anti-symmetric;

    (4) $1 R 2$ and $2 R 3$ but $1 \not{R} 3$, so R is not transitive.

(d) Consider the relation $\perp$ of perpendicularity on the set L of lines in the Euclidean plane. If line a is perpendicular to line b then b is perpendicular to a, i.e. if $a \perp b$ then $b \perp a$; hence $\perp$ is symmetric. However, $\perp$ is neither reflexive, anti-symmetric nor transitive.

2.8 PARTITIONS

Let S be any nonempty set. A *partition* of S is a subdivision of S into nonoverlapping, nonempty subsets. Precisely, a *partition* of S is a collection $\{A_i\}$ of nonempty subsets of S such that:

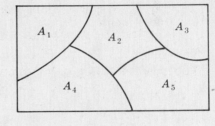

(i)   Each a in S belongs to one of the A_i.

(ii) The sets of $\{A_i\}$ are mutually disjoint; that is, if

$$A_i \neq A_j \text{ then } A_i \cap A_j = \emptyset$$

Fig. 2-6

The subsets in a partition are called *cells*. Figure 2-6 is a Venn diagram of a partition of the rectangular set of points into five cells.

EXAMPLE 2.8.   Consider the following collections of subsets of $S = \{1, 2, \ldots, 8, 9\}$:

(i) $[\{1, 3, 5\}, \{2, 6\} \{4, 8, 9\}]$

(ii) $[\{1, 3, 5\}, \{2, 4, 6, 8\}, \{5, 7, 9\}]$

(iii) $[\{1, 3, 5\}, \{2, 4, 6, 8\}, \{7, 9\}]$

Then (i) is not a partition of S since 7 in S does not belong to any of the subsets. Furthermore, (ii) is not a partition of S since $\{1, 3, 5\}$ and $\{5, 7, 9\}$ are not disjoint. On the other hand, (iii) is a partition of S.

2.9 EQUIVALENCE RELATIONS

A relation R on a set S is called an *equivalence relation* if R is reflexive, symmetric and transitive (see Section 2.8). The general idea behind an equivalence relation is that it is a classification of objects which are in some way "alike." Example 2.7(b) shows that equality is itself an equivalence relation. We list other equivalence relations in Example 2.9.

EXAMPLE 2.9.

(a) Consider the set L of lines and the set T of triangles in the Euclidean plane. The relation "is parallel to or equal to" is an equivalence relation on L, and congruence and similarity are equivalence relations on T.

(b) The classification of animals by species, that is, the relation "is of the same species as" is an equivalence relation on the set of animals.

(c) The relation $\subset$ of set inclusion is not an equivalence relation. It is reflexive and transitive, but it is not symmetric since $A \subset B$ does not imply $B \subset A$.

(d) Let m be a fixed positive integer. Two integers a and b are said to be *congruent modulo* m, written

$$a \equiv b \pmod{m}$$

if m divides $a - b$. For example, for $m = 4$ we have $11 \equiv 3 \pmod 4$ since 4 divides $11 - 3$, and $22 \equiv 6 \pmod 4$ since 4 divides $22 - 6$. We prove in Problem 2.16 that congruence modulo m is an equivalence relation.

2.10 EQUIVALENCE RELATIONS AND PARTITIONS

Let R be an equivalence relation in a set A and, for each $a \in A$, let $[a]$, called the *equivalence class* of A, be the set of elements to which a is related:

$$[a] = \{x : (a, x) \in R\}$$

The collection of equivalence classes of A, denoted by A / R, is called the *quotient* of A by R:

$$A / R = \{[a] : a \in A\}$$

The fundamental property of a quotient set is contained in the following theorem.

Theorem 2.2:   Let R be an equivalence relation in a set A. Then the quotient set A / R is a partition of A. Specifically,

   (i) $a \in [a]$, for every $a \in A$;

   (ii) $[a] = [b]$ if and only if $(a, b) \in R$;

   (iii) if $[a] \neq [b]$, then $[a]$ and $[b]$ are disjoint.

We prove Theorem 2.2 in Problem 2.20.

EXAMPLE 2.10.

(a) Let $S = \{1, 2, 3\}$. The relation

$$R = \{(1, 1), (1, 2), (2, 1), (3, 3)\}$$

is an equivalence relation on S. Under the relation R,

$$[1] = \{1, 2\}, \qquad [2] = \{1, 2\} \qquad \text{and} \qquad [3] = \{3\}$$

Observe that $[1] = [2]$ and $\{[1], [3]\}$ is a partition of S.

(b) Let R_5 be the relation in $\mathbf{Z}$, the set of integers, defined by

$$x \equiv y \pmod 5$$

which reads "x is congruent to y modulo 5" and which mean that the difference $x - y$ is divisible by 5. Then R_5 is an equivalence relation in $\mathbf{Z}$. There are exactly five distinct equivalence classes in $\mathbf{Z}/R_5$:

$$A_0 = \{\ldots, -10, -5, 0, 5, 10, \ldots\}$$
$$A_1 = \{\ldots, -9, -4, 1, 6, 11, \ldots\}$$
$$A_2 = \{\ldots, -8, -3, 2, 7, 12, \ldots\}$$
$$A_3 = \{\ldots, -7, -2, 3, 8, 13, \ldots\}$$
$$A_4 = \{\ldots, -6, -1, 4, 9, 14, \ldots\}$$

Observe that each integer x, which is uniquely expressible in the form $x = 5q + r$ where $0 \le r < 5$, is a member of the equivalence class A_r, where r is the remainder. Note that the equivalence classes are pairwise disjoint and that

$$\mathbf{Z} = A_0 \cup A_1 \cup A_2 \cup A_3 \cup A_4$$

2.11 PARTIAL ORDERING RELATIONS

We define another important class of relations. A relation R on a set S is called a *partial ordering* on S if R is reflexive, anti-symmetric and transitive. We shall study partial orderings in more detail in Chapter 10, so here we will simply give some examples.

EXAMPLE 2.11.

(a) The relation $\subset$ of set inclusion is a partial ordering on any collection of sets since set inclusion has the three desired properties. That is, (i) $A \subset A$, for any set A; (ii) if $A \subset B$ and $B \subset A$ then $A = B$; and (iii) if $A \subset B$ and $B \subset C$ then $A \subset C$.

(b) The relation $\leqq$ on the real numbers $\mathbf{R}$ is reflexive, anti-symmetric and transitive. Thus $\leqq$ is a partial ordering.

(c) The relation "a divides b" is a partial ordering on the set $\mathbf{N}$ of positive integers. However, "a divides b" is not a partial ordering on the set $\mathbf{Z}$ of integers since $a \mid b$ and $b \mid a$ does not imply $a = b$. For example, $3 \mid -3$ and $-3 \mid 3$ but $3 \neq -3$.

2.12 *n*-ARY RELATIONS

So far we have only discussed binary relations. An *n-ary relation* is a set of ordered *n*-tuples. If S is a set, a subset of S^n is called an *n*-ary relation on S. In particular, a subset of S^3 is called a *ternary relation* on S.

EXAMPLE 2.12.

(a) Let L be a line in the plane. Then "betweenness" is a ternary relation on the points of L.

(b) The equation $x^2 + y^2 + z^2 = 1$ determines a ternary relation T on the set $\mathbf{R}$ of real numbers. A triple (x, y, z) belongs to T if and only if (x, y, z) is the triple of coordinates of a point on the sphere with radius 1 and center at the origin $(0, 0, 0)$.

Solved Problems

PRODUCT SETS

2.1. Given $A = \{1, 2, 3\}$ and $B = \{a, b\}$. Find (a) $A \times B$, (b) $B \times A$, (c) $B \times B$.

(a) $A \times B$ consists of all ordered pairs (x, y) where $x \in A$ and $y \in B$. Hence

$$A \times B = \{(1, a), (1, b), (2, a), (2, b), (3, a), (3, b)\}$$

(b) $B \times A$ consists of all ordered pairs (y, x) where $y \in B$ and $x \in A$. Hence

$$B \times A = \{(a, 1), (a, 2), (a, 3), (b, 1), (b, 2), (b, 3)\}$$

(c) $B \times B$ consists of all ordered pairs (x, y) where $x, y \in B$. Hence

$$B \times B = \{(a, a), (a, b), (b, a), (b, b)\}$$

As expected, the number of elements in the product set is equal to the product of the numbers of the elements in each set.

2.2.   Given $A = \{1, 2\}$, $B = \{x, y, z\}$ and $C = \{3, 4\}$.   Find $A \times B \times C$.

$A \times B \times C$ consists of all ordered triplets (a, b, c) where $a \in A$, $b \in B$, $c \in C$. These elements of $A \times B \times C$ can be systematically obtained by a so-called tree diagram (Fig. 2-7).

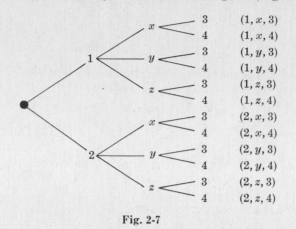

Fig. 2-7

The elements of $A \times B \times C$ are precisely the 12 ordered triplets to the right of the tree diagram.

Observe that $n(A) = 2$, $n(B) = 3$, and $n(C) = 2$ and, as expected,

$$n(A \times B \times C) \ = \ 12 \ = \ n(A) \cdot n(B) \cdot n(C)$$

2.3.   Given $(2x, x + y) = (6, 2)$.   Find x and y.

Two ordered pairs are equal if and only if the corresponding components are equal. Hence we obtain the equations

$$2x \ = \ 6 \qquad \text{and} \qquad x + y \ = \ 2$$

The two equations yield $x = 3$ and $y = -1$.

2.4.   Given $A = \{1, 2\}$, $B = \{a, b, c\}$ and $C = \{c, d\}$.   Find

$$(A \times B) \cap (A \times C) \ \text{ and } \ A \times (B \cap C)$$

We have

$$A \times B \ = \ \{(1, a), (1, b), (1, c), (2, a), (2, b), (2, c)\}$$
$$A \times C \ = \ \{(1, c), (1, d), (2, c), (2, d)\}$$

Hence

$$(A \times B) \cap (A \times C) \ = \ \{(1, c), (2, c)\}$$

Since $B \cap C = \{c\}$,

$$A \times (B \cap C) \ = \ \{(1, c,), (2, c)\}$$

Observe that $(A \times B) \cap (A \times C) = A \times (B \cap C)$.   This is true for any sets A, B and C (see Problem 2.25).

RELATIONS AND THEIR GRAPHS

2.5.   Let $M = \{a, b, c, d\}$ and let R be the relation on M consisting of those points which are displayed on the coordinate diagram of $M \times M$ on the right.

(a) Find all the elements in M which are related to b, that is, $\{x : (x, b) \in R\}$.

(b) Find all those elements in M to which d is related, that is, $\{x : (d, x) \in R\}$.

(c) Find the inverse relation R^{-1}.

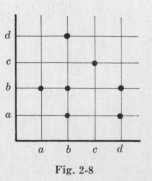

Fig. 2-8

(a) The horizontal line through b contains all points of R in which b appears as the second element: (a, b), (b, b) and (d, b). Hence the desired set is $\{a, b, d\}$.

(b) The vertical line through d contains all the points of R in which d appears as the first element: (d, a) and (d, b). Hence $\{a, b\}$ is the desired set.

(c) First write R as a set of ordered pairs, and then write the pairs in reverse order:

$$R = \{(a, b), (b, a), (b, b), (b, d), (c, c), (d, a), (d, b)\}$$
$$R^{-1} = \{(b, a), (a, b), (b, b), (d, b), (c, c), (a, d), (b, d)\}$$

2.6. Given $A = \{1, 2, 3, 4\}$ and $B = \{x, y, z\}$. Consider the following relation from A to B:

$$R = \{(1, y), (1, z), (3, y), (4, x), (4, z)\}$$

(a) Plot R on a coordinate diagram of $A \times B$.

(b) Determine the matrix of the relation.

(c) Draw the arrow diagram of R.

(d) Find the inverse relation R^{-1} of R.

(e) Determine the domain and range of R.

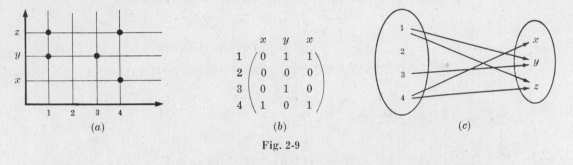

Fig. 2-9

(a) See Fig. 2-9(a).

(b) See Fig. 2-9(b). Observe that the rows of the matrix are labeled by the elements of A and the columns by the elements of B. Also observe that the entry in the matrix corresponding to $a \in A$ and $b \in B$ is 1 if a is related to b and 0 otherwise.

(c) See Fig. 2-9(c). Observe that there is an arrow from $a \in A$ to $b \in B$ iff a is related to b, i.e. iff $(a, b) \in R$.

(d) Reverse the ordered pairs of R to obtain R^{-1}:

$$R^{-1} = \{(y, 1), (z, 1), (y, 3), (x, 4), (z, 4)\}$$

Observe that by reversing the arrows in Fig. 2-9(c) we obtain the arrow diagram of R^{-1}.

(e) The domain of R consists of the first elements of the ordered pairs of R, and the range of R consists of the second elements. Thus,

domain of $R = \{1, 3, 4\}$ and   range of $R = \{x, y, z\}$

2.7. Let $A = \{1, 2, 3, 4, 6\}$, and let R be the relation on A defined by "x divides y", written $x \mid y$. (Note $x \mid y$ iff there exists an integer z such that $xz = y$.)

(a) Write R as a set of ordered pairs.

(b) Plot R on a coordinate diagram of $A \times A$, and draw its directed graph.

(c) Find the inverse relation R^{-1} of R. Can R^{-1} be described in words?

(a) Find those numbers in A divisible by 1, 2, 3, 4 and then 6. These are:

$$1\,|\,1,\ 1\,|\,2,\ 1\,|\,3,\ 1\,|\,4,\ 1\,|\,6,\ 2\,|\,2,\ 2\,|\,4,\ 2\,|\,6,\ 3\,|\,3,\ 3\,|\,6, 4\,|\,4,\ 6\,|\,6$$

Hence

$$R \;=\; \{(1,1),\,(1,2),\,(1,3),\,(1,4),\,(1,6),$$
$$(2,2),\,(2,4),\,(2,6),\,(3,3),\,(3,6),\,(4,4),\,(6,6)\}$$

(b) See Fig. 2-10.

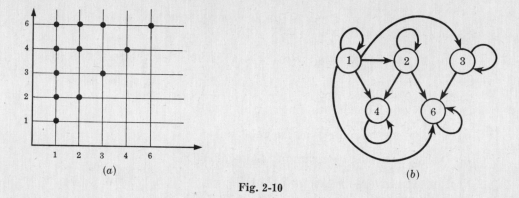

(a)   (b)

Fig. 2-10

(c) Reverse the ordered pairs of R to obtain R^{-1}:

$$R^{-1} \;=\; \{(1,1),\,(2,1),\,(3,1),\,(4,1),\,(6,1),\,(2,2),\,(4,2),$$
$$(6,2),\,(3,3),\,(6,3),\,(4,4),\,(6,6)\}$$

R^{-1} can be described by the statement "x is a multiple of y".

2.8. Let R and S be the following relations on $A = \{1,2,3\}$:

$$R \;=\; \{(1,1),\,(1,2),\,(2,3),\,(3,1),\,(3,3)\}$$
$$S \;=\; \{1,2),\,(1,3),\,(2,1),\,(3,3)\}$$

Find $R \cap S$, $R \cup S$ and R^c.

Treat R and S simply as sets, and take the usual intersection and union. For R^c, use the fact that $A \times A$ is the universal relation on A.

$$R \cap S \;=\; \{(1,2),\,(3,3)\}$$
$$R \cup S \;=\; \{(1,1),\,(1,2),\,(1,3),\,(2,1),\,(2,3),\,(3,1),\,(3,3)\}$$
$$R^c \;=\; \{(1,3),\,(2,1),\,(2,2),\,(3,2)\}$$

2.9. Let $A = \{1,2,3\}$, $B = \{a,b,c\}$ and $C = \{x,y,z\}$. Consider the following relations R and S from A to B and from B to C respectively.

$$R \;=\; \{(1,b),\,(2,a),\,(2,c)\} \quad \text{and} \quad S \;=\; \{(a,y),\,(b,x),\,(c,y),\,(c,z)\}$$

(a) Find the composition relation $R \circ S$.

(b) Find the matrices M_R, M_S, and $M_{R \circ S}$ of the respective relations R, S and $R \circ S$, and compare $M_{R \circ S}$ to the product $M_R M_S$.

(a) Draw the arrow diagram of the relations R and S as in Fig. 2-11. Observe that 1 in A is "connected" to x in C by the path $1 \to b \to x$; hence $(1,x)$ belongs to $R \circ S$. Similarly, $(2,y)$ and $(2,z)$ belong to $R \circ S$. We have

$$R \circ S \;=\; \{(1,x),\,(2,y),\,(2,z)\}$$

(See Example 2.6.)

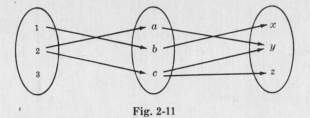

Fig. 2-11

(b) The matrices of M_R, M_S and $M_{R \circ S}$ follow:

$$M_R = \begin{array}{c} 1 \\ 2 \\ 3 \end{array} \begin{pmatrix} 0 & 1 & 0 \\ 1 & 0 & 1 \\ 0 & 0 & 0 \end{pmatrix} \quad \begin{array}{c} a\ \ b\ \ c \end{array}$$

$$M_S = \begin{array}{c} a \\ b \\ c \end{array} \begin{pmatrix} 0 & 1 & 0 \\ 1 & 0 & 0 \\ 0 & 1 & 1 \end{pmatrix} \quad \begin{array}{c} x\ \ y\ \ z \end{array}$$

$$M_{R \circ S} = \begin{array}{c} 1 \\ 2 \\ 3 \end{array} \begin{pmatrix} 1 & 0 & 0 \\ 0 & 1 & 1 \\ 0 & 0 & 0 \end{pmatrix} \quad \begin{array}{c} x\ \ y\ \ z \end{array}$$

Multiplying M_R and M_S we obtain

$$M_R M_S = \begin{pmatrix} 1 & 0 & 0 \\ 0 & 2 & 1 \\ 0 & 0 & 0 \end{pmatrix}$$

Observe that $M_{R \circ S}$ and $M_R M_S$ have the same zero entries.

2.10. Prove Theorem 2.1: Let A, B, C and D be sets. Suppose R is a relation from A to B, S is a relation from B to C and T is a relation from C to D. Then $(R \circ S) \circ T = R \circ (S \circ T)$.

We need to show that each ordered pair in $(R \circ S) \circ T$ belongs to $R \circ (S \circ T)$, i.e. that $(R \circ S) \circ T \subset R \circ (S \circ T)$, and vice versa.

Suppose (a, d) belongs to $(R \circ S) \circ T$. Then there exists a c in C such that $(a, c) \in R \circ S$ and $(c, d) \in T$. Since $(a, c) \in R \circ S$, there exists a b in B such that $(a, b) \in R$ and $(b, c) \in S$. Since $(b, c) \in S$ and $(c, d) \in T$, we have $(b, d) \in S \circ T$; and since $(a, b) \in R$ and $(b, d) \in S \circ T$, we have $(a, d) \in R \circ (S \circ T)$. Thus $(R \circ S) \circ T \subset R \circ (S \circ T)$. Similarly $R \circ (S \circ T) \subset (R \circ S) \circ T$. Both inclusion relations prove $(R \circ S) \circ T = R \circ (S \circ T)$.

PROPERTIES OF RELATIONS

2.11. Determine when a relation R on a set A is (a) not reflexive, (b) not symmetric, (c) not transitive, (d) not anti-symmetric.

(a) There exists $a \in A$ such that (a, a) does not belong to R.

(b) There exists (a, b) in R such that (b, a) does not belong to R.

(c) There exist (a, b) and (b, c) in R such that (a, c) does not belong to R.

(d) There exist distinct elements a and b such that (a, b) and (b, a) belong to R.

2.12. Consider the following five relations on the set $A = \{1, 2, 3\}$.

$$R = \{(1,1), (1,2), (1,3), (3,3)\}$$
$$S = \{(1,1), (1,2), (2,1), (2,2), (3,3)\}$$
$$T = \{(1,1), (1,2), (2,2), (2,3)\}$$
$$\varnothing = \text{empty relation}$$
$$A \times A = \text{universal relation}$$

Determine whether or not each of the above relations on A is (a) reflexive, (b) symmetric, (c) transitive, (d) anti-symmetric.

(a) R is not reflexive since $2 \in A$ but $(2,2) \notin R$. T is not reflexive since $(3,3) \notin T$ and, similarly, $\emptyset$ is not reflexive. S and $A \times A$ are reflexive.

(b) R is not symmetric since $(1,2) \in R$ but $(2,1) \notin R$, and similarly T is not symmetric. S, $\emptyset$ and $A \times A$ are symmetric.

(c) T is not transitive since $(1,2)$ and $(2,3)$ belong to T, but $(1,3)$ does not belong to T. The other four relations are transitive.

(d) S is not anti-symmetric since $1 \neq 2$, and $(1,2)$ and $(2,1)$ both belong to S. Similarly, $A \times A$ is not anti-symmetric. The other three relations are anti-symmetric.

2.13. Given $A = \{1,2,3,4\}$. Consider the following relation in A:

$$R = \{(1,1),(2,2),(2,3),(3,2),(4,2),(4,4)\}$$

(a) Plot R on a coordinate diagram of $A \times A$, and draw its **directed** graph.

(b) Is R (i) reflexive, (ii) symmetric, (iii) transitive, (iv) anti-symmetric?

(a) See Fig. 2-12.

(b) (i) R is not reflexive because $3 \in A$ but $3 \not{R} 3$, i.e. $(3,3) \notin R$.

(ii) R is not symmetric because $4 R 2$ but $2 \not{R} 4$, i.e. $(4,2) \in R$ but $(2,4) \notin R$.

(iii) R is not transitive because $4 R 2$ and $2 R 3$ but $4 \not{R} 3$, i.e. $(4,2) \in R$ and $(2,3) \in R$ but $(4,3) \notin R$.

(iv) R is not anti-symmetric because $2 R 3$ and $3 R 2$ but $2 \neq 3$.

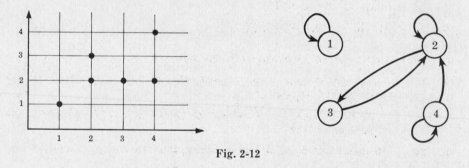

Fig. 2-12

2.14. Give examples of relations R on $A = \{1,2,3\}$ having the stated property.

(a) R is both symmetric and anti-symmetric.

(b) R is neither symmetric nor anti-symmetric.

(c) R is transitive but $R \cup R^{-1}$ is not transitive.

There are several possible examples for each answer. One possible set of examples follows:

(a) $R = \{(1,1),(2,2)\}$

(b) $R = \{(1,2),(2,1),(2,3)\}$

(c) $R = \{(1,2)\}$

2.15. Let R and S be relations on a set A. Prove:

(a) If R and S are transitive then $R \cap S$ is transitive.

(b) If R is anti-symmetric, then $R \cap S$ is anti-symmetric.

(a) Suppose (a,b) and (b,c) are in $R \cap S$. Then (a,b) and (b,c) are in both R and S. Since both relations are transitive, $(a,c) \in R$ and $(a,c) \in S$. Thus $(a,c) \in R \cap S$, and so $R \cap S$ is transitive.

(b) Suppose (a,b) and (b,a) are both in $R \cap S$. Then, in particular, (a,b) and (b,a) are both in R. Since R is anti-symmetric, $a = b$. Hence $R \cap S$ is anti-symmetric.

EQUIVALENCE RELATIONS AND PARTITIONS

2.16. Consider the set **Z** of integers and an integer $m > 1$. We say that x is congruent to y modulo m, written

$$x \equiv y \pmod{m}$$

if $x - y$ is divisible by m. Show that this defines an equivalence relation on **Z**.

For any x in **Z**, we have $x \equiv x \pmod{m}$ because $x - x = 0$ is divisible by m. Hence the relation is reflexive.

Suppose $x \equiv y \pmod{m}$, so $x - y$ is divisible by m. Then $-(x - y) = y - x$ is also divisible by m, so $y \equiv x \pmod{m}$. Thus the relation is symmetric.

Now suppose $x \equiv y \pmod{m}$ and $y \equiv z \pmod{m}$, so $x - y$ and $y - z$ are each divisible by m. Then the sum

$$(x - y) + (y - z) = x - z$$

is also divisible by m; hence $x \equiv z \pmod{m}$. Thus the relation is transitive.

We have shown that the relation of congruence modulo m on **Z** is reflexive, symmetric and transitive; hence it is an equivalence relation.

2.17. Given $X = \{1, 2, 3, 4, 5, 6, 7, 8, 9\}$. Determine which of the following are partitions of X:

(a) $[\{1, 3, 6\}, \{2, 8\}, \{5, 7, 9\}]$

(b) $[\{1, 5, 7\}, \{2, 4, 8, 9\}, \{3, 5, 6\}]$

(c) $[\{2, 4, 5, 8\}, \{1, 9\}, \{3, 6, 7\}]$

(d) $[\{1, 2, 7\}, \{3, 5\}, \{4, 6, 8, 9\}, \{3, 5\}]$

(a) No; because $4 \in X$ does not belong to any cell. In other words, X is not the union of the cells.

(b) No; because $5 \in X$ belongs to two distinct cells, $\{1, 5, 7\}$ and $\{3, 5, 6\}$. In other words, the two distinct cells are not disjoint.

(c) Yes; because each element of X belongs to exactly one cell. In other words, the cells are disjoint and their union is X.

(d) Yes. Although 3 and 5 appear in two places, the cells are not distinct.

2.18. Find all the partitions of $X = \{a, b, c, d\}$.

Note first that each partition of X contains either 1, 2, 3, or 4 distinct sets. The partitions are as follows:

(1) $[\{a, b, c, d\}]$

(2) $[\{a\}, \{b, c, d\}]$, $[\{b\}, \{a, c, d\}]$, $[\{c\}, \{a, b, d\}]$, $[\{d\}, \{a, b, c\}]$,

         $[\{a, b\}, \{c, d\}]$, $[\{a, c\}, \{b, d\}]$, $[\{a, d\}, \{b, c\}]$

(3) $[\{a\}, \{b\}, \{c, d\}]$, $[\{a\}, \{c\}, \{b, d\}]$, $[\{a\}, \{d\}, \{b, c\}]$,

         $[\{b\}, \{c\}, \{a, d\}]$, $[\{b\}, \{d\}, \{a, c\}]$, $[\{c\}, \{d\}, \{a, b\}]$

(4) $[\{a\}, \{b\}, \{c\}, \{d\}]$

There are fifteen different partitions of X.

2.19. Let R be the following equivalence relation on the set $A = \{1, 2, 3, 4, 5, 6\}$:

$$R = \{(1, 1), (1, 5), (2, 2), (2, 3), (2, 6), (3, 2), (3, 3),$$
$$(3, 6), (4, 4), (5, 1), (5, 5), (6, 2), (6, 3), (6, 6)\}$$

Find the partition of A *induced by* R, i.e. find the equivalence classes of R.

Those elements related to 1 are 1 and 5 hence

$$[1] \;=\; \{1, 5\}$$

We pick an element which does not belong to [1], say 2. Those elements related to 2 are 2, 3 and 6; hence

$$[2] \;=\; \{2, 3, 6\}$$

The only element which does not belong to [1] or [2] is 4. The only element related to 4 is 4. Thus

$$[4] \;=\; \{4\}$$

Accordingly, $[\{1, 5\}, \{2, 3, 6\}, \{4\}]$

is the partition of A induced by R.

2.20. Prove Theorem 2.2: Let R be an equivalence relation in a set A. Then the quotient set A/R is a partition of A. Specifically,

(i) $a \in [a]$, for every $a \in A$;

(ii) $[a] = [b]$ if and only if $(a, b) \in R$;

(iii) if $[a] \neq [b]$, then $[a]$ and $[b]$ are disjoint.

Proof of (i). Since R is reflexive, $(a, a) \in R$ for every $a \in A$ and therefore $a \in [a]$.

Proof of (ii). Suppose $(a, b) \in R$. We want to show that $[a] = [b]$. Let $x \in [b]$; then $(b, x) \in R$. But by hypothesis $(a, b) \in R$ and so, by transitivity, $(a, x) \in R$. Accordingly $x \in [a]$. Thus $[b] \subset [a]$. To prove that $[a] \subset [b]$, we observe that $(a, b) \in R$ implies, by symmetry, that $(b, a) \in R$. Then by a similar argument, we obtain $[a] \subset [b]$. Consequently, $[a] = [b]$.

On the other hand, if $[a] = [b]$, then, by (i), $b \in [b] = [a]$; hence $(a, b) \in R$.

Proof of (iii). We prove the equivalent contrapositive statement:

$$\text{if} \quad [a] \cap [b] \neq \emptyset \quad \text{then} \quad [a] = [b]$$

If $[a] \cap [b] \neq \emptyset$, then there exists an element $x \in A$ with $x \in [a] \cap [b]$. Hence $(a, x) \in R$ and $(b, x) \in R$. By symmetry, $(x, b) \in R$ and by transitivity, $(a, b) \in R$. Consequently by (ii), $[a] = [b]$.

Supplementary Problems

PRODUCT SETS

2.21. Let $A = \{a, b, c, d\}$. Find the ordered pairs corresponding to the points P_1, P_2, P_3 and P_4 which appear in the adjacent coordinate diagram of $A \times A$ (Fig. 2-13).

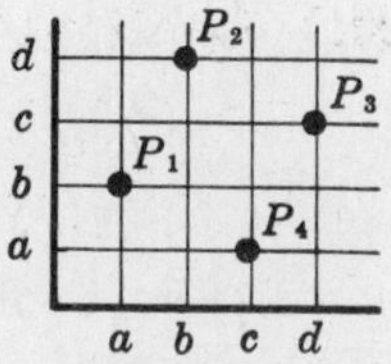

Fig. 2-13

2.22. Find x and y if: (a) $(x + 2, 4) = (5, 2x + y)$, (b) $(y - 2, 2x + 1) = (x - 1, y + 2)$.

2.23. Let $W = \{\text{Mark, Eric, Paul}\}$ and let $V = \{\text{Eric, David}\}$, Find: (a) $W \times V$, (b) $V \times W$, (c) $V \times V$.

2.24. Let $S = \{a, b, c\}$, $T = \{b, c, d\}$ and $W = \{a, d\}$. Construct the tree diagram of $S \times T \times W$ and then find $S \times T \times W$.

2.25. Prove: (a) $A \times (B \cap C) = (A \times B) \cap (A \times C)$, (b) $A \times (B \cup C) = (A \times B) \cup (A \times B)$.

RELATIONS

2.26. Let R be the relation on $A = \{2, 3, 4, 5, 6\}$ defined by "x is relatively prime to y", i.e. the only positive divisor of x and y is 1.

    (a) Write R as a set of ordered pairs.

    (b) Draw the directed graph of R.

2.27. Let R be the relation on $C = \{1, 2, 3, 4, 5\}$ given by the set of points displayed in the coordinate diagram of $C \times C$ in Fig. 2-14.

    (1) State whether each is true or false: (a) $1\,R\,4$, (b) $2\,R\,5$, (c) $3\,\not{R}\,1$, (d) $5\,\not{R}\,3$.

    (2) Find the elements in each of the following subsets of C:

        (a) $\{x : 3\,R\,x\}$ (c) $\{x : (x, 2) \notin R\}$

        (b) $\{x : (4, x) \in R\}$ (d) $\{x : x\,R\,5\}$

    (3) Find (a) the domain of R, (b) the range of R, (c) R^{-1}.

    (4) Draw the directed graph of R.

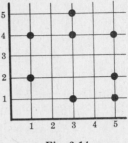

Fig. 2-14

2.28. Let R be the following relation on $A = \{1, 2, 3, 4\}$:

$$R = \{(1, 3), (1, 4), (3, 2), (3, 3), (3, 4)\}$$

    (a) Find the matrix M_R of R.   (d) Draw the directed graph of R.

    (b) Find the domain and range of R.   (e) Find the composition relation $R \circ R$.

    (c) Find R^{-1}.

2.29. Let R and S be the following relations on $B = \{a, b, c, d\}$:

$$R = \{(a, a), (a, c), (c, b), (c, d), (d, b)\}$$
$$S = \{(b, a), (c, c), (c, d), (d, a)\}$$

Find the following composition relations on B:

    (a) $R \circ S$, (b) $S \circ R$, (c) $R \circ R$, (d) $S \circ S$.

2.30. Let R be the relation on the positive integers **N** defined by the equation $x + 3y = 12$; that is,

$$R = \{(x, y) : x + 3y = 12\}$$

    (a) Write R as a set of ordered pairs.

    (b) Find (i) domain of R, (ii) range of R, and (iii) R^{-1}.

    (c) Find the composition relation $R \circ R$.

REFLEXIVE, SYMMETRIC, ANTI-SYMMETRIC AND TRANSITIVE RELATIONS

2.31. Let $W = \{1, 2, 3, 4\}$. Consider the following relations on W:

$$R_1 = \{(1, 1), (2, 1)\} \qquad\qquad R_4 = \{(1, 1), (2, 2), (3, 3)\}$$
$$R_2 = \{(1, 1), (2, 3), (4, 1)\} \qquad R_5 = \{(1, 3), (2, 4)\}$$
$$R_3 = \{(3, 4)\}$$

Determine which relations are (a) reflexive, (b) symmetric, (c) anti-symmetric, (d) transitive.

2.32. Each of the following defines a relation on the positive integers **N**:

    (1) "x is greater than y"   (3) $x + y = 10$

    (2) "xy is the square of an integer"   (4) $x + 4y = 10$

Determine which relations are (a) reflexive, (b) symmetric, (c) anti-symmetric, (d) transitive.

2.33. Let $P(A)$ be the collection of all subsets of $A = \{a, b, c\}$. Each of the following defines a relation on $P(A)$:

(1) "x is a proper subset of y"   (3) $x \cup y = A$

(2) "x is disjoint from y"

Determine which relations are (a) reflexive, (b) symmetric, (c) anti-symmetric, (d) transitive.

2.34. Let R and S be relations on a set A. Assuming A has at least three elements, state whether each of the following statements is true or false. If it is false, give a counter-example on the set $A = \{1, 2, 3\}$:

(a) If R and S are symmetric then $R \cap S$ is symmetric.

(b) If R and S are symmetric then $R \cup S$ is symmetric.

(c) If R and S are reflexive then $R \cap S$ is reflexive.

(d) If R and S are reflexive then $R \cup S$ is reflexive.

(e) If R and S are transitive then $R \cup S$ is transitive.

(f) If R and S are anti-symmetric then $R \cup S$ is anti-symmetric.

(g) If R is anti-symmetric, then R^{-1} is anti-symmetric.

(h) If R is reflexive then $R \cap R^{-1}$ is not empty.

(i) If R is symmetric then $R \cap R^{-1}$ is not empty.

PARTITIONS AND EQUIVALENCE RELATIONS

2.35. Let $W = \{1, 2, 3, 4, 5, 6\}$. Determine which of the following are partitions of W:

(a) $[\{1, 3, 5\}, \{2, 4\}, \{3, 6\}]$ (c) $[\{1, 5\}, \{2\}, \{4\}, \{1, 5\}, \{3, 6\}]$

(b) $[\{1, 5\}, \{2\}, \{3, 6\}]$ (d) $[\{1, 2, 3, 4, 5, 6\}]$

2.36. Find all partitions of $V = \{1, 2, 3\}$.

2.37. Let $[A_1, A_2, \ldots, A_m]$ and $[B_1, B_2, \ldots, B_n]$ be partitions of a set X. Show that the collection of sets
$$[A_i \cap B_j : \ i = 1, \ldots, m, j = 1, \ldots, n] \setminus \emptyset$$
is also a partition (called the *cross partition*) of X. (Observe that we have deleted the empty set $\emptyset$.)

2.38. Prove that if R is an equivalence relation on a set A, then R^{-1} is also an equivalence relation on A.

2.39. Let $S = \{1, 2, 3, \ldots, 19, 20\}$. Let R be the equivalence relation on S defined by $x \equiv y$ (mod 5), that is, $x - y$ is divisible by 5. Find the partition of S induced by R, i.e. the quotient set S/R.

2.40. Let $\mathbf{N} = \{1, 2, 3, \ldots\}$ and let $\simeq$ be the relation on $\mathbf{N} \times \mathbf{N}$ defined by
$$(a, b) \simeq (c, d) \qquad \text{if} \qquad ad = bc$$
Prove that $\simeq$ is an equivalence relation.

2.41. Let $A = \{1, 2, 3, \ldots, 9\}$ and let $\sim$ be the relation on $A \times A$ defined by
$$(a, b) \sim (c, d) \qquad \text{if} \qquad a + d = b + c$$

(a) Prove that $\sim$ is an equivalence relation.

(b) Find $[(2, 5)]$, i.e. the equivalence class of $(2, 5)$.

COMPUTER PROGRAMMING PROBLEMS

Suppose a relation R on the set $S = \{1, 2, 3, 4\}$ contains six elements. Suppose a deck contains six cards and each card contains a pair of integers which is an element of R.

2.42.   Write a program which prints (a) the domain of R, and (b) the range of R.

2.43.   Write a program which prints the composition relation $R \circ R = R^2$.

2.44.   Write a program which decides whether or not the relation R is:

(a) reflexive, (b) symmetric, (c) transitive, (d) anti-symmetric.

2.45.   Write a program which prints the 4×4 matrix M of the relation R.

Test the programs in Problems 2.41 to 2.45 with:

(i) $R = \{(1,1),\ (1,2),\ (1,3),\ (2,2),\ (2,3),\ (4,4)\}$

(ii) $R = \{(1,2),\ (2,1),\ (2,2),\ (3,4),\ (3,3),\ (4,3)\}$

(iii) $R = \{(1,4),\ (2,3),\ (4,3),\ (2,4),\ (3,1),\ (3,2)\}$

2.46.   Now suppose R is an equivalence relation on S. Write a program which prints the equivalence classes [1], [2], [3] and [4]. Test the program with the relation

$$R = \{(1,1),\ (2,2),\ (2,4),\ (3,3),\ (4,2),\ (4,4)\}$$

Answers to Supplementary Problems

2.21. $P_1 = (a,b),\ P_2 = (b,d),\ P_3 = (d,c),\ P_4 = (c,a)$

2.22.   (a) $x = 3,\ y = -2$, (b) $x = 2,\ y = 3$

2.23.   (a) $W \times V = \{(\text{Mark, Eric}),\ (\text{Mark, David}),\ (\text{Eric, Eric}),\ (\text{Eric, David}),\ (\text{Paul, Eric}),\ (\text{Paul, David})\}$.

(b) $V \times W = \{(\text{Eric, Mark}),\ (\text{David, Mark}),\ (\text{Eric, Eric}),\ (\text{David, Eric}),\ (\text{Eric, Paul}),\ (\text{David, Paul})\}$.

(c) $V \times V = \{(\text{Eric, Eric}),\ (\text{Eric, David}),\ (\text{David, Eric}),\ (\text{David, David})\}$.

2.24.   See Fig. 2-15.

$$S \times T \times W = \{(a,b,a),\ (a,b,d),\ (a,c,a),\ (a,c,d),\ (a,d,a),\ (a,d,d),$$
$$(b,b,a),\ (b,b,d),\ (b,c,a),(b,c,d),\ (b,d,a),\ (b,d,d),$$
$$(c,b,a),\ (c,b,d),\ (c,c,a),\ (c,c,d),\ (c,d,a),\ (c,d,d)\}$$

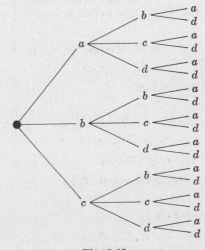

Fig. 2-15

2.26. $R = \{(2,3), (2,5), (3,2), (3,4), (3,5), (4,3), (4,5)$
                $(5,2), (5,3), (5,4), (5,6), (6,5)\}$

2.27.   (1)   (a) True,   (b) False,   (c) False,   (d) True

         (2)   (a) $\{1,4,5\}$,   (b) $\emptyset$,   (c) $\{2,3,4\}$,   (d) $\{3\}$

         (3)   (a) $\{1,3,5\}$,   (b) $\{1,2,4,5\}$,   (c) $R^{-1} = \{(1,3), (1,5), (2,1), (2,5), (4,1), (4,3), (4,5), (5,3)\}$

         (4)   See Fig. 2-16.

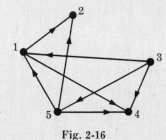

Fig. 2-16

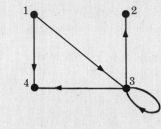

Fig. 2-17

2.28.   (a) $M_R = \begin{pmatrix} 0 & 0 & 1 & 1 \\ 0 & 0 & 0 & 0 \\ 0 & 1 & 1 & 1 \\ 0 & 0 & 0 & 0 \end{pmatrix}$ (b) Domain $= \{1,3\}$, range $= \{2,3,4\}$

         (c) $R^{-1} = \{(3,1), (4,1), (2,3), (3,3), (4,3)\}$

         (d)   See Fig. 2-17.

         (e) $R \circ R = \{(1,2), (1,3), (1,4), (3,2), (3,3), (3,4)\}$

2.29.   (a) $R \circ S = \{(a,c), (a,d), (c,a), (d,a)\}$

         (b) $S \circ R = \{(b,a), (b,c), (c,b), (c,d), (d,a), (d,c)\}$

         (c) $R \circ R = \{(a,a), (a,b), (a,c), (a,d), (c,b)\}$

         (d) $S \circ S = \{(c,c), (c,a), (c,d)\}$

2.30.   (a) $\{(9,1), (6,2), (3,3)\}$,   (b) (i) $\{9,6,3\}$,   (ii) $\{1,2,3\}$,   (iii) $\{(1,9), (2,6), (3,3)\}$,   (c) $\{(3, ?$

2.31.   (a) None,   (b) R_4,   (c) all,   (d) all except R_2

2.32.   (a) None,   (b) (2) and (3),   (c) (1) and (4),   (d) all except (3)

2.33.   (a) None,   (b) (2) and (3),   (c) (1),   (d) (1)

2.34.   All are true except   (e) $R = \{(1,2)\}$, $S = \{(2,3)\}$ and   (f) $R = \{(1,2)\}$, $S = \{(2,1)\}$.

2.35.   (c) and (d)

2.37.   There are five: $[\{1,2,3\}]$, $[\{1\}, \{2,3\}]$, $[\{2\}, \{1,3\}]$, $[\{3\}, \{1,2\}]$ and $[\{1\}, \{2\}, \{3\}]$

2.39. $[\{1,6,11,16\}, \{2,7,12,17\}, \{3,8,13,18\}, \{4,9,14,19\}, \{5,10,15,20\}]$

2.41.   (b) $\{(1,4), (2,5), (3,6), (4,7), (5,8), (6,9)\}$

Chapter 3

Functions

3.1 INTRODUCTION

One of the most important concepts in mathematics is that of a function. This concept appears and plays a major role in all branches of mathematics. The terms "map", "mapping", "transformation" and many others are sometimes used instead of "function"; the choice of which word to use in a given situation is usually determined by tradition and by the mathematical background of the person using the term.

In this chapter we will investigate elementary properties of functions.

3.2 FUNCTIONS

Suppose that to each element of a set A there is assigned a unique element of a set B; the collection of such assignments is called a *function* from A into B. The set A is called the *domain* of the function, and the set B is called the *codomain*.

We usually represent a function by a letter, e.g. f, g, etc. If f is a function from A to B, then we write

$$f : A \to B$$

which is read "f *takes* (or *maps*) A into B". If $a \in A$, then the unique element of B which the function f assigns to a is called the *image* of a under f, or the *value* of f at a, and is designated by

$$f(a)$$

(read f of a). The set of all such image values is called the *image* of f and is denoted by $\mathrm{Im}\,(f)$ or $f(A)$. Observe that $\mathrm{Im}\,(f)$ is a subset (perhaps a proper subset) of B.

If a function f can be expressed by a mathematical formula, then there are several ways in which the function f may be described. For example, suppose f is the function from $\mathbf{R}$ into $\mathbf{R}$ which sends each real number into its square; then f may be described by any of the following:

$$f(x) = x^2 \quad \text{or} \quad x \mapsto x^2 \quad \text{or} \quad y = x^2$$

Here the barred arrow $\mapsto$ is read "goes into". In the last notation, x is called the *independent variable* and y is called the *dependent variable* since the value of y will depend on the value that x takes.

Remark: Whenever a function f is given by a formula using the independent variable x, as above, we assume unless otherwise stated or implied that f is a function from $\mathbf{R}$ (or the largest subset of $\mathbf{R}$ for which f has meaning) into $\mathbf{R}$.

EXAMPLE 3.1.

(a) Consider the function $f(x) = x^3$, i.e. f assigns to each real number its cube. Then the image of 2 is 8, and so we may write $f(2) = 8$. (By convention, f is a function from $\mathbf{R}$ into $\mathbf{R}$.)

(b) Let f assign to each country in the world its capital city. Here the domain of f is the set of countries in the world; the codomain is the list of cities of the world. The image of France is Paris, that is, $f(\text{France}) = \text{Paris}$.

(c) Figure 3-1 defines a function f from $A = \{a, b, c, d\}$ into $B = \{r, s, t, u\}$ in the obvious way. Here

$$f(a) = s, \quad f(b) = u, \quad f(c) = r, \quad f(d) = s$$

The image of f consists of the image values; hence

$$\text{Im}\,(f) = \{r, s, u\}$$

Note that t does not belong to the image of f because t is not the image of any element under f.

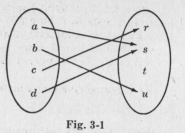

Fig. 3-1

(d) Let A be any set. The function from A into A which assigns to each element itself is called the *identity function* on A and is usually denoted by 1_A or simply 1. In other words

$$1_A(a) = a$$

for every element a in A.

3.3 GRAPH OF A FUNCTION

There is another point of view from which functions may be considered. First of all, every function $f : A \rightarrow B$ gives rise to a relation from A to B called the *graph of f* and defined by

$$\text{Graph of } f = \{(a, b) : a \in A \text{ and } b = f(a)\}$$

Two functions $f : A \rightarrow B$ and $g : A \rightarrow B$ are defined to be equal, written $f = g$, if $f(a) = g(a)$ for every $a \in A$; that is, if they have the same graph. Accordingly, we do not distinguish between a function and its graph.

The graph of a function $f : A \rightarrow B$ has the following basic property:

Each $a \in A$ belongs to a unique ordered pair (a, b) in the relation. $(\ast)$

On the other hand, suppose f is a relation from A to B satisfying $(\ast)$. Then f specifies an assignment of an element $b \in B$ to each $a \in A$; namely, if $(a, b) \in f$ then b is assigned to a. In other words, f is a function from A into B. Accordingly, the concepts of functions and of relations satisfying $(\ast)$ are one and the same. In fact, some texts define a function as a relation satisfying $(\ast)$.

Although we do not distinguish between a function and its graph, we will still use the terminology "graph of f" when referring to f as a set of ordered pairs. Moreover, since the graph of f is a relation, we can draw its picture as was done for relations in general, and this pictorial representation is itself sometimes called the graph of f. Also, the defining condition of a function, that each $a \in A$ belongs to a unique pair (a, b) in f, is equivalent to the geometrical condition of each vertical line intersecting the graph in exactly one point.

EXAMPLE 3.2.

(a) Let $f : A \rightarrow B$ be the function defined in Example 3.1(c). Then the graph of f is the following set of ordered pairs:

$$\{(a, s), (b, u), (c, r), (d, s)\}$$

(b) Consider the following relations on the set $A = \{1, 2, 3\}$:

$$f = \{(1, 3), (2, 3), (3, 1)\}$$

$$g = \{(1, 2), (3, 1)\}$$

$$h = \{(1, 3), (2, 1), (1, 2), (3, 1)\}$$

f is a function from A into A since each member of A appears as the first coordinate in exactly one ordered pair in f; here $f(1) = 3$, $f(2) = 3$ and $f(3) = 1$. g is not a function from A into A since $2 \in A$ is not the first coordinate of any pair in g and so g does not assign any image to 2. Also h is not a function from A into A since $1 \in A$ appears as the first coordinate of two distinct ordered pairs in h, $(1, 3)$ and $(1, 2)$. If h is to be a function it cannot assign both 3 and 2 to the element $1 \in A$.

(c) By a real polynomial function, we mean a function $f : \mathbf{R} \to \mathbf{R}$ of the form

$$f(x) = a_n x^n + a_{n-1} x^{n-1} + \cdots + a_1 x + a_0$$

where the a_i are real numbers. Since $\mathbf{R}$ is an infinite set, it would be impossible to plot each point of the graph. However, the graph of such a function can be approximated by first plotting some of its points and then drawing a smooth curve through these points. The points are usually obtained from a table where various values are assigned to x and the corresponding values of $f(x)$ computed. Figure 3-2 illustrates this technique using the function $f(x) = x^2 - 2x - 3$.

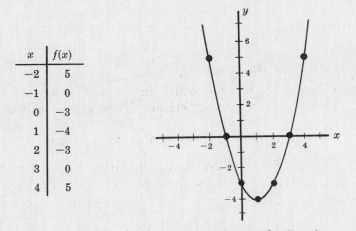

x	$f(x)$
-2	5
-1	0
0	-3
1	-4
2	-3
3	0
4	5

Graph of $f(x) = x^2 - 2x - 3$

Fig. 3-2

(d) (Composition function.) Consider functions $f : A \to B$ and $g : B \to C$; that is, where the codomain of f is the domain of g. Then we may define a new function from A to C, called the *composition* of f and g and written $g \circ f$, as follows:

$$(g \circ f)(a) \equiv g(f(a))$$

That is, we find the image of a under f and then find the image of $f(a)$ under g. This definition is not really new. If we view f and g as relations, then this function is the same as the composition of f and g as relations (see Section 2.6) except that here we use the functional notation $g \circ f$ for the composition of f and g instead of the notation $f \circ g$ which was used for relations.

If $f : A \to B$ is any function then

$$f \circ 1_A = f \quad \text{and} \quad 1_B \circ f = f$$

where 1_A and 1_B are the identity function on A and B respectively.

3.4 ONE-TO-ONE, ONTO AND INVERTIBLE FUNCTIONS

A function $f : A \to B$ is said to be *one-to-one* (written 1-1) if different elements in the domain A have distinct images. Another way of saying the same thing is that f is *one-to-one* if $f(a) = f(a')$ implies $a = a'$.

A function $f : A \to B$ is said to be an *onto* function if each element of B is the image of some element of A. In other words, $f : A \to B$ is onto if the image of f is the entire codomain, i.e. if $f(A) = B$. In such a case we say that f is a function from A *onto* B or that f maps A *onto* B.

A function $f: A \to B$ is *invertible* if its inverse relation f^{-1} is a function from B to A. In general, the inverse relation f^{-1} is not a function. The following theorem is a simple criterion which tells us when it is.

Theorem 3.1:   A function $f: A \to B$ is invertible if and only if f is both one-to-one and onto.

If $f: A \to B$ is one-to-one and onto, then f is called a *one-to-one correspondence* between A and B. This terminology comes from the fact that each element of A will then correspond to a unique element of B and vice versa.

Some texts use the term *injective* for a one-to-one function, *surjective* for an onto function, and *bijective* for a one-to-one correspondence.

EXAMPLE 3.3.   Consider the functions $f_1: A \to B$, $f_2: B \to C$, $f_3: C \to D$ and $f_4: D \to E$ defined by the diagram of Fig. 3-3. Now f_1 is one-to-one since no element of B is the image of more than one element of A. Similarly, f_2 is one-to-one. However, neither f_3 nor f_4 is one-to-one since $f_3(r) = f_3(u)$ and $f_4(v) = f_4(w)$.

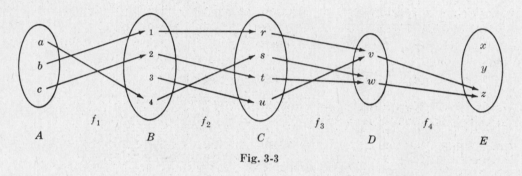

Fig. 3-3

As far as being onto is concerned, f_2 and f_3 are both onto functions since every element of C is the image under f_2 of some element of B and every element of D is the image under f_3 of some element of C, i.e. $f_2(B) = C$ and $f_3(C) = D$. On the other hand, f_1 is not onto since $3 \in B$ is not the image under f_1 of any element of A, and f_4 is not onto since $x \in E$ is not the image under f_4 of any element of D.

Thus f_1 is one-to-one but not onto, f_3 is onto but not one-to-one and f_4 is neither one-to-one nor onto. However, f_2 is both one-to-one and onto, i.e. is a one-to-one correspondence between A and B. Hence f_2 is invertible and f_2^{-1} is a function from C to B.

Since functions may be identified with their graphs, and since graphs may be plotted, we might wonder whether the concepts of being one-to-one and onto have geometrical meaning. We show that the answer is yes.

To say that a function $f: A \to B$ is one-to-one means that there are no two distinct pairs (a_1, b) and (a_2, b) in the graph of f; hence each horizontal line can intersect the graph of f in at most one point. On the other hand, to say that f is an onto function means that for every $b \in B$ there must be at least one $a \in A$ such that (a, b) belongs to the graph of f; hence each horizontal line must intersect the graph of f at least once. Accordingly, if f is both one-to-one and onto, i.e. invertible, then each horizontal line will intersect the graph of f in exactly one point.

EXAMPLE 3.4.   Consider the following four functions from **R** into **R**:

$$f_1(x) = x^2, \quad f_2(x) = 2^x, \quad f_3(x) = x^3 - 2x^2 - 5x + 6, \quad f_4(x) = x^3$$

The graphs of these functions appear in Fig. 3-4. Observe that there are horizontal lines which intersect the graph of f_1 twice and there are horizontal lines which do not intersect the graph of f_1 at all; hence f_1 is neither one-to-one nor onto. Similarly, f_2 is one-to-one but not onto, f_3 is onto but not one-to-one and f_4 is both one-to-one and onto. The inverse of f_4 is the cube root function, i.e.

$$f_4^{-1}(x) = \sqrt[3]{x}$$

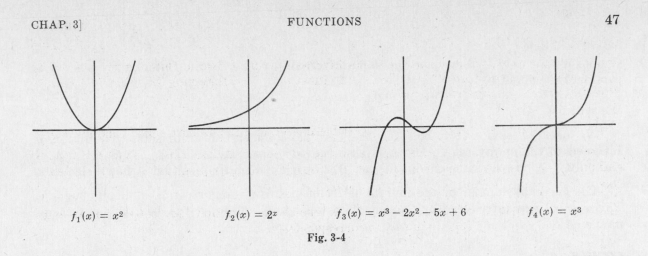

$$f_1(x) = x^2 \qquad f_2(x) = 2^x \qquad f_3(x) = x^3 - 2x^2 - 5x + 6 \qquad f_4(x) = x^3$$

Fig. 3-4

3.5 INDEXED CLASSES OF SETS

We next define a special kind of function, called an indexing function. Let I be any nonempty set, and let S be a class of sets. An indexing function from I to S is a function $f : I \to S$. For any $i \in I$, we denote the image $f(i)$ by A_i. Thus the indexing function f is usually denoted by

$$\{A_i : i \in I\} \qquad \text{or} \qquad \{A_i\}_{i \in I} \quad \text{or simply} \qquad \{A_i\}$$

The set I is called the *indexing set*, and the elements of I are called *indices*. If f is one-to-one and onto, we say that S is indexed by I.

We may define the concepts of union and intersection of an indexed class of sets by

$$\cup_{i \in I} A_i = \{x : x \in A_i \text{ for some } i \in I\}$$

and

$$\cap_{i \in I} A_i = \{x : x \in A_i \text{ for all } i \in I\}$$

In the case that I is a finite set, this is just the same as our previous definitions of union and intersection. If I is $\mathbf{N}$, we may denote the union and intersection by

$$A_1 \cup A_2 \cup \cdots \qquad \text{and} \qquad A_1 \cap A_2 \cap \cdots$$

respectively.

EXAMPLE 3.5. Let I be the set $\mathbf{Z}$ of integers. For each integer n we assign the following subset of $\mathbf{R}$:

$$A_n = \{x : x \leq n\}$$

(In other words, A_n is the infinite interval $(-\infty, n]$.) For any real number a, there exist integers n_1 and n_2 such that $n_1 < a < n_2$; so $a \in A_{n_2}$ but $a \notin A_{n_1}$. Hence

$$a \in \cup_n A_n \qquad \text{but} \qquad a \notin \cap_n A_n$$

Accordingly,

$$\cup_n A_n = \mathbf{R} \qquad \text{but} \qquad \cap_n A_n = \emptyset$$

3.6 CARDINALITY

Two sets A and B are said to have the *same cardinality* if there exists a one-to-one correspondence $f : A \to B$. A set A is *finite* if A is empty or if A has the same cardinality as the set $\{1, 2, \ldots, n\}$ for some positive integer n. A set is *infinite* if it is not finite. Familiar examples of infinite sets are the natural numbers $\mathbf{N}$, the integers $\mathbf{Z}$, the rational numbers $\mathbf{Q}$ and the real numbers $\mathbf{R}$.

We now introduce the idea of "cardinal numbers". We will consider cardinal numbers simply as symbols assigned to sets in such a way that two sets are assigned the same symbol if and only if they have the same cardinality. The cardinal number of a set A is commonly denoted by

$$|A|, \quad n(A), \quad \#(A) \quad \text{or} \quad \text{card}(A)$$

We will use $|A|$.

We use the obvious symbols for the cardinal numbers of finite sets. That is, 0 is assigned to the empty set $\varnothing$, and n is assigned to the set $\{1, 2, \ldots, n\}$. Thus $|A| = n$ if and only if A has the same cardinality as $\{1, 2, \ldots, n\}$ which implies that A has n elements.

The cardinal number of the infinite set $\mathbf{N}$ of positive integers is $\aleph_0$ ("aleph-naught"). This symbol was introduced by Cantor. Thus $|A| = \aleph_0$ if and only if A has the same cardinality as $\mathbf{N}$

EXAMPLE 3.6.

(a) $|\{x, y, z\}| = 3$ and $|\{1, 3, 5, 7, 9\}| = 5$.

(b) Let $E = \{2, 4, 6, \ldots\}$, the set of even positive integers. The function $f: \mathbf{N} \to E$ defined by $f(n) = 2n$ is a one-to-one correspondence between the positive integers $\mathbf{N}$ and E. Thus E has the same cardinality as $\mathbf{N}$ and so we may write

$$|E| = \aleph_0$$

A set with cardinality $\aleph_0$ is said to be *denumerable* or *countably infinite*. A set which is finite or denumerable is said to be *countable*. One can show that the set $\mathbf{Q}$ of rational numbers is countable. In fact, we have the following theorem which we will use subsequently.

Theorem 3.2: A countable union of countable sets is countable.

In other words, if $A_1, A_2, \ldots$ are each countable sets then the union

$$A_1 \cup A_2 \cup A_3 \cup \cdots$$

is also a countable set.

An important example of an infinite set which is uncountable, i.e. not countable, is given by the following theorem.

Theorem 3.3: The set I of all real numbers between 0 and 1 is uncountable.

Solved Problems

FUNCTIONS

3.1. State whether or not each diagram in Fig. 3-5 defines a function from $A = \{a, b, c\}$ into $B = \{x, y\, z\}$.

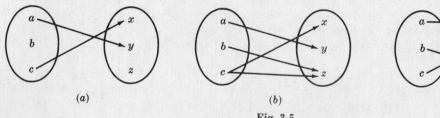

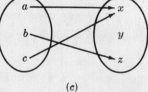

    (a)   (b)   (c)

Fig. 3-5

(a) No. There is nothing assigned to the element $b \in A$.

(b) No. Two elements, x and z, are assigned to $c \in A$.

(c) Yes.

3.2.   Let $A = \{1, 2, 3, 4, 5\}$ and let $f : A \to A$ be the function defined in Fig. 3-6.

    (*a*) Find $f(A)$, i.e. the image of f.

    (*b*) Find the graph of f, i.e. write f as a set of ordered pairs.

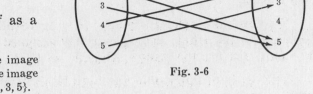

Fig. 3-6

    (*a*) The image $f(A)$ of f consists of all the image values. Now only 2, 3 and 5 appear as the image of any elements of A; hence $f(A) = \{2, 3, 5\}$.

    (*b*) The ordered pairs $(a, f(a))$, where $a \in A$, form the graph of f. Now $f(1) = 3$, $f(2) = 5$, $f(3) = 5$, $f(4) = 2$ and $f(5) = 3$; hence $f = \{(1, 3), (2, 5), (3, 5), (4, 2), (5, 3)\}$.

3.3.   Let $X = \{1, 2, 3, 4\}$. Determine whether or not each relation is a function from X into X.

    (*a*) $f = \{(2, 3), (1, 4), (2, 1), (3, 2), (4, 4)\}$

    (*b*) $g = \{(3, 1), (4, 2), (1, 1)\}$

    (*c*) $h = \{(2, 1), (3, 4), (1, 4), (2, 1), (4, 4)\}$

    Recall that a subset f of $X \times X$ is a function $f : X \to X$ if and only if each $a \in X$ appears as the first coordinate in exactly one ordered pair in f.

    (*a*) No. Two different ordered pairs $(2, 3)$ and $(2, 1)$ in f have the same number 2 as their first coordinate.

    (*b*) No. The element $2 \in X$ does not appear as the first coordinate in any ordered pair in g.

    (*c*) Yes. Although $2 \in X$ appears as the first coordinate in two ordered pairs in h, these two ordered pairs are equal.

3.4.   Let $W = \{a, b, c, d\}$. Determine whether the set of points in each coordinate diagram of $W \times W$ (Fig. 3-7) is a function from W into W.

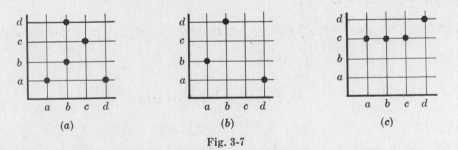

Fig. 3-7

    (*a*) No. The vertical line through b contains two points of the set, i.e. two different ordered pairs (b, b) and (b, d) contain the same first element b.

    (*b*) No. The vertical line through c contains no point of the set, i.e. $c \in W$ does not appear as the first element in any ordered pair.

    (*c*) Yes. Each vertical line contains exactly one point of the set.

3.5.   Let A be the set of students in a school. Determine which of the following assignments defines a function on A.

    (*a*) To each student assign his age.

    (*b*) To each student assign his teacher.

    (*c*) To each student assign his sex.

    (*d*) To each student assign his spouse.

A collection of assignments is a function on A if and only if each element a in A is assigned exactly one element.   Thus:

(a) Yes, because each student has one and only one age.

(b) Yes, if each student has only one teacher; no, if any student has more than one teacher.

(c) Yes.

(d) No, if any student is not married.

3.6. (a) Rewrite each of the following functions from **R** into **R** using a formula:

    (i)   To each number let f assign its cube.

    (ii) To each number let g assign the number 5.

    (iii) To each positive number let h assign its square, and to each nonpositive number let h assign the number 4.

(b) Find:

    (i) $f(4), f(-2), f(0)$;   (ii) $g(4), g(-2), g(0)$;   (iii) $h(4), h(-2), h(0)$.

(a) (i)   Since f assigns to any number x; its cube x^3, f can be defined by $f(x) = x^3$.

    (ii) Since g assigns 5 to any number x, we can define g by $g(x) = 5$.

    (iii) Two different rules are used to define h as follows:

$$h(x) \;=\; \begin{cases} x^2 & \text{if } x > 0 \\ 4 & \text{if } x \le 0 \end{cases}$$

(b) (i)   Now $f(x) = x^3$ for every number x; hence $f(4) = 4^3 = 64$, $f(-2) = (-2)^3 = -8$, $f(0) = 0^3 = 0$.

    (ii) Since $g(x) = 5$ for every number x, $g(4) = 5$, $g(-2) = 5$ and $g(0) = 5$.

    (iii) If $x > 0$, then $h(x) = x^2$; hence $h(4) = 4^2 = 16$. On the other hand, if $x \le 0$, then $h(x) = 4$; thus $h(-2) = 4$ and $h(0) = 4$.

3.7. Let $f : \mathbf{R} \to \mathbf{R}$ be defined by $f(x) = x^3$.

(a) Find: (i) $f(3)$, (ii) $f(-2)$, (iii) $f(y)$, (iv) $f(y+1)$, (v) $f(x+h)$, (vi) $[f(x+h) - f(x)]/h$

(b) Plot f on a coordinate diagram of $\mathbf{R} \times \mathbf{R}$.

(a) (i) $f(3) = 3^3 = 27$, (ii) $f(-2) = (-2)^3 = -8$, (iii) $f(y) = (y)^3 = y^3$

    (iv) $f(y+1) = (y+1)^3 = y^3 + 3y^2 + 3y + 1$, (v) $f(x+h) = (x+h)^3 = x^3 + 3x^2h + 3xh^2 + h^3$

    (vi) $[f(x+h) - f(x)]/h = (x^3 + 3x^2h + 3xh^2 + h^3 - x^3)/h = (3x^2h + 3xh^2 + h^3)/h = 3x^2 + 3xh + h^2$

(b) Since f is a polynomial function, it can be plotted by first plotting some of its points and then drawing a smooth curve through these points (Fig. 3-8).

x	$f(x)$
-3	-27
-2	-8
-1	-1
0	0
1	1
2	8
3	27

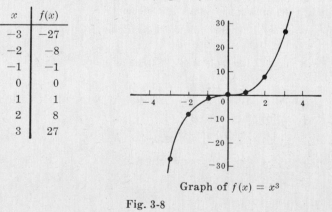

Graph of $f(x) = x^3$

Fig. 3-8

3.8. Sketch the graphs of

(a) $f(x) = 3x - 2$, (b) $g(x) = x^2 + x - 6$, (c) $h(x) = x^3 - 3x^2 - x + 3$.

In (a) the function is linear; only two points (three as a check) are needed to sketch its graph. Set up a table with three values of x, say, $x = -2, 0, 2$ and find the corresponding values of $f(x)$:

$$f(-2) = 3(-2) - 2 = -8 \qquad f(0) = 3(0) - 2 = -2 \qquad f(2) = 3(2) - 2 = 4$$

Draw the line through these points as in Fig. 3-9.

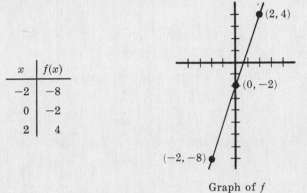

x	$f(x)$
-2	-8
0	-2
2	4

Graph of f

Fig. 3-9

In (b) and (c), set up a table of values for x and then find the corresponding values of the function. Plot the points in a coordinate diagram, and then draw a smooth continuous curve through the points as in Fig. 3-10.

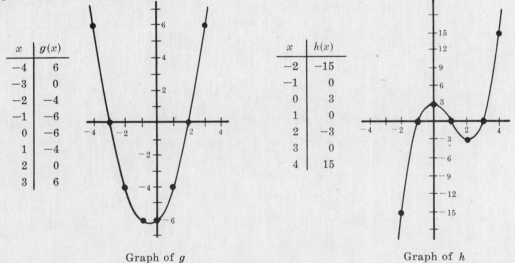

x	$g(x)$
-4	6
-3	0
-2	-4
-1	-6
0	-6
1	-4
2	0
3	6

x	$h(x)$
-2	-15
-1	0
0	3
1	0
2	-3
3	0
4	15

Graph of g Graph of h

Fig. 3-10

3.9. Let the functions $f : A \to B$ and $g : B \to C$ be defined by Fig. 3-11. Find the composition function $g \circ f : A \to C$.

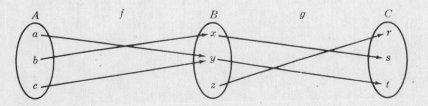

Fig. 3-11

We use the definition of the composition function to compute:

$$(g \circ f)(a) \;=\; g(f(a)) \;=\; g(y) \;=\; t$$
$$(g \circ f)(b) \;=\; g(f(b)) \;=\; g(x) \;=\; s$$
$$(g \circ f)(c) \;=\; g(f(c)) \;=\; g(y) \;=\; t$$

Note that we arrive at the same answer if we "follow the arrows" in the diagram:

$$a \to y \to t, \quad b \to x \to s, \quad c \to y \to t$$

3.10. Let the functions f and g be defined by $f(x) = 2x+1$ and $g(x) = x^2 - 2$. Find the formula defining the composition function $g \circ f$.

Compute $g \circ f$ as follows: $(g \circ f)(x) \;=\; g(f(x)) \;=\; g(2x+1) \;=\; (2x+1)^2 - 2 \;=\; 4x^2 + 4x - 1$.

Observe that the same answer can be found by writing

$$y \;=\; f(x) \;=\; 2x+1 \quad \text{and} \quad z \;=\; g(y) \;=\; y^2 - 2$$

and then eliminating y from both equations:

$$z \;=\; y^2 - 2 \;=\; (2x+1)^2 - 2 \;=\; 4x^2 + 4x - 1$$

ONE-TO-ONE, ONTO AND INVERTIBLE FUNCTIONS

3.11. Let $A = \{a, b, c, d, e\}$, and let B be the set of letters in the alphabet. Let the functions f, g and h from A into B be defined as follows:

	f		g		h
(a)	$a \to r$	(b)	$a \to z$	(c)	$a \to a$
	$b \to a$		$b \to y$		$b \to c$
	$c \to s$		$c \to x$		$c \to e$
	$d \to r$		$d \to y$		$d \to r$
	$e \to e$		$e \to z$		$e \to s$

Are any of these functions one-to-one?

Recall that a function is one-to-one if it assigns distinct image values to distinct elements in the domain.

(a) No.   For f assigns r to both a and d.

(b) No.   For g assigns z to both a and e.

(c) Yes.   For h assigns distinct images to different elements in the domain.

3.12. Determine if each function is one-to-one.

(a) To each person on the earth assign the number which corresponds to his age.

(b) To each country in the world assign the latitude and longitude of its capital.

(c) To each book written by only one author assign the author.

(d) To each country in the world which has a prime minister assign its prime minister.

(a) No.   Many people in the world have the same age.

(b) Yes.

(c) No.   There are different books with the same author.

(d) Yes.   Different countries in the world have different prime ministers.

3.13. Let the functions $f : A \to B$, $g : B \to C$ and $h : C \to D$ be defined by Fig. 3-12. (a) Determine if each function is onto. (b) Find the composition function $h \circ g \circ f$.

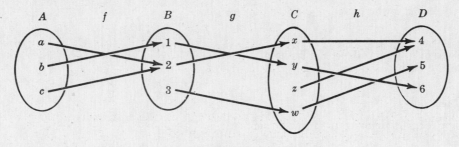

Fig. 3-12

(a) The function $f : A \to B$ is not onto since $3 \in B$ is not the image of any element in A.

    The function $g : B \to C$ is not onto since $z \in C$ is not the image of any element in B.

    The function $h : C \to D$ is onto since each element in D is the image of some element of C.

(b) Now $a \to 2 \to x \to 4$, $b \to 1 \to y \to 6$, $c \to 2 \to x \to 4$. Hence $h \circ g \circ f = \{(a, 4), (b, 6), (c, 4)\}$.

3.14. Consider functions $f : A \to B$ and $g : B \to C$. Prove the following:

(a) If f and g are one-to-one, then the composition function $g \circ f$ is one-to-one.

(b) If f and g are onto functions, then $g \circ f$ is an onto function.

(a) Suppose $(g \circ f)(x) = (g \circ f)(y)$; then $g(f(x)) = g(f(y))$. Hence $f(x) = f(y)$ because g is one-to-one. Furthermore, $x = y$ since f is one-to-one. Accordingly, $g \circ f$ is one-to-one.

(b) Let c be any arbitrary element of C. Since g is onto, there exists a $b \in B$ such that $g(b) = c$. Since f is onto, there exists an $a \in A$ such that $f(a) = b$. But then

$$(g \circ f)(a) \;=\; g(f(a)) \;=\; g(b) \;=\; c$$

Hence each $c \in C$ is the image of some element $a \in A$. Accordingly, $g \circ f$ is an onto function.

3.15. Let $W = \{1, 2, 3, 4, 5\}$ and let $f : W \to W$, $g : W \to W$ and $h : W \to W$ be defined by the diagrams in Fig. 3-13. Determine whether each function is invertible, and, if it is, find its inverse function.

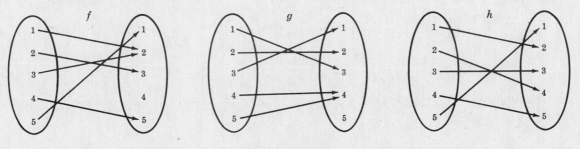

Fig. 3-13

    In order for a function to be invertible, the function must be both one-to-one and onto. Only h is one-to-one and onto, so only h is invertible. To find h^{-1}, the inverse of h, reverse the ordered pairs which belong to h. Note

$$h \;=\; \{(1, 2), (2, 4), (3, 3), (4, 5), (5, 1)\}$$

hence
$$h^{-1} \;=\; \{(2, 1), (4, 2), (3, 3), (5, 4), (1, 5)\}$$

Observe that h^{-1} can be obtained by reversing the arrows in the diagram for h.

3.16. Let $f : \mathbf{R} \to \mathbf{R}$ be defined by $f(x) = 2x - 3$. Now f is one-to-one and onto; hence f has an inverse function f^{-1}. Find a formula for f^{-1}.

Let y be the image of x under the function f:

$$y = f(x) = 2x - 3$$

Consequently, x will be the image of y under the inverse function f^{-1}. Solve for x in terms of y in the above equation:

$$x = (y + 3)/2$$

Then

$$f^{-1}(y) = (y + 3)/2$$

is a formula defining the inverse function.

INDEXED SETS

3.17. For each positive integer n in $\mathbf{N}$, let D_n be the following subset of $\mathbf{N}$:

$$D_n = \{n, 2n, 3n, 4n, \ldots\} = \{\text{multiples of } n\}$$

Find: (a) $D_3 \cap D_5$ (d) $\cup \{D_n : n \in \mathbf{N}\}$

       (b) $D_4 \cup D_8$ (e) $\cap \{D_n : n \in \mathbf{N}\}$

       (c) $D_3 \cap D_6$ (f) $\cup \{D_p : p \text{ is a prime number}\}$

(a) $D_3 \cap D_5$ consists of multiples of 3 and also multiples of 5, and so consists of multiples of 15. That is, $D_3 \cap D_5 = D_{15}$.

(b) $D_8 \subset D_4$ because every multiple of 8 is also a multiple of 4; hence $D_4 \cup D_8 = D_4$

(c) $D_6 \subset D_3$ because every multiple of 6 is also a multiple of 3; hence $D_3 \cap D_6 = D_6$.

(d) $\cup \{D_n : n \in \mathbf{N}\} = \mathbf{N}$

(e) $\cap \{D_n : n \in \mathbf{N}\} = \emptyset$

(f) $\cup_p D_p = \{2, 3, \ldots\} = \mathbf{N} \setminus \{1\}$ because every positive integer except 1 is a multiple of a prime number.

3.18. Prove the following generalization of DeMorgan's law: For any class of sets $\{A_i\}$, we have $(\cup_i A_i)^c = \cap_i A_i^c$.

We have:

$$x \in (\cup_i A_i)^c \text{ iff } x \notin \cup_i A_i$$
$$\text{iff } \forall_i \in I, \; x \notin A_i$$
$$\text{iff } \forall_i \in I, \; x \in A_i^c$$
$$\text{iff } x \in \cap_i A_i^c$$

Therefore, $(\cup_i A_i)^c = \cap_i A_i^c$. (Here we have used the logical notations iff for "if and only if" and $\forall$ for "for all".)

CARDINALITY

3.19. Find the cardinal number of each set

    (a) $A = \{a, b, c, \ldots, y, z\}$ (d) $D = \{10, 20, 30, 40, \ldots\}$

    (b) $B = \{1, -3, 5, 11, -28\}$ (e) $E = \{6, 7, 8, 9, \ldots\}$

    (c) $C = \{x : x \in \mathbf{N}, x^2 = 5\}$

(a) $|A| = 26$ since there are 26 letters in the English alphabet.

(b) $|B| = 5$.

(c) $|C| = 0$ since there is no positive integer whose square is 5, i.e. since C is empty.

(d) $|D| = \aleph_0$ because $f : \mathbf{N} \to D$, defined by $f(n) = 10n$, is a one-to-one correspondence between $\mathbf{N}$ and D.

(e) $|E| = \aleph_0$ because $g : \mathbf{N} \to E$ defined by $g(n) = n + 5$ is a one-to-one correspondence between $\mathbf{N}$ and E.

3.20. Show that the set $\mathbf{Z}$ of integers has cardinality $\aleph_0$.

The following diagram shows a one-to-one correspondence between $\mathbf{N}$ and $\mathbf{Z}$:

$$
\begin{array}{ccccccccc}
\mathbf{N} = & 1 & 2 & 3 & 4 & 5 & 6 & 7 & 8 & \cdots \\
& \downarrow & \downarrow & \downarrow & \downarrow & \downarrow & \downarrow & \downarrow & \downarrow & \\
\mathbf{Z} = & 0 & 1 & -1 & 2 & -2 & 3 & -3 & 4 & \cdots
\end{array}
$$

That is, the following function $f : \mathbf{N} \to \mathbf{Z}$ is one-to-one and onto:

$$
f(n) = \begin{cases} n/2 & \text{if } n \text{ is even} \\ (1-n)/2 & \text{if } n \text{ is odd} \end{cases}
$$

Accordingly, $|\mathbf{Z}| = |\mathbf{N}| = \aleph_0$.

3.21. Let $A_1, A_2, \ldots$ be a countable number of finite sets. Prove that the union $S = \cup_i A_i$ is countable.

Essentially, we list the elements of A_1, then we list the elements of A_2 which do not belong to A_1, then we list the elements of A_3 which do not belong to A_1 or A_2, i.e. which have not already been listed, and so on. Since the A_i are finite, we can always list the elements of each set. This process is done formally as follows.

First we define sets $B_1, B_2, \ldots$ where B_i contains the elements of A_i which do not belong to preceding sets, i.e. we define

$$
B_1 = A_1 \qquad \text{and} \qquad B_k = A_k \setminus (A_1 \cup A_2 \cup \cdots \cup A_{k-1})
$$

Then the B_i are disjoint and $S = \cup_i B_i$. Let $b_{i1}, b_{i2}, \ldots, b_{im_i}$ be the elements of B_i. Then $S = \{b_{ij}\}$. Let $f : S \to \mathbf{N}$ be defined as follows:

$$
f(b_{ij}) = m_1 + m_2 + \cdots + m_{i-1} + j
$$

If S is finite, then S is countable. If S is infinite then f is a one-to-one correspondence between S and $\mathbf{N}$. Thus S is countable.

3.22. Prove Theorem 3.2: A countable union of countable sets is countable.

Suppose A_1, A_2, A_3, $\ldots$ are a countable number of countable sets. In particular, suppose $a_{i1}, a_{i2}, a_{i3}, \ldots$ are the elements of A_i. Define sets $B_2, B_3, B_4, \ldots$ as follows:

$$
B_k = \{a_{ij} : i + j = k\}
$$

Observe that each B_k is finite and

$$
S = \cup_i A_i = \cup_k B_k
$$

By the preceding problem $\cup_k B_k$ is countable. Hence $S = \cup_i A_i$ is countable and the theorem is proved.

3.23. Prove Theorem 3.3: The set I of all real numbers between 0 and 1 inclusive is uncountable.

The set I is clearly infinite, since it contains 1, 1/2, 1/3, ... Suppose I is denumerable. Then there exists a one-to-one correspondence $f: \mathbf{N} \to I$. Let $f(1) = a_1$, $f(2) = a_2$, ...; that is, $I = \{a_1, a_2, a_3, \ldots\}$. We list the elements a_1, a_2, ... in a column and express each in its decimal expansion:

$$a_1 = 0.x_{11}x_{12}x_{13}x_{14}\ldots$$
$$a_2 = 0.x_{21}x_{22}x_{23}x_{24}\ldots$$
$$a_3 = 0.x_{31}x_{32}x_{33}x_{34}\ldots$$
$$a_4 = 0.x_{41}x_{42}x_{43}x_{44}\ldots$$
$$\cdots\cdots\cdots\cdots\cdots\cdots$$

where $x_{ij} \in \{0, 1, 2, \ldots, 9\}$. (For those numbers which can be expressed in two different decimal expansions, e.g. $0.2000000\ldots = 0.1999999\ldots$, we choose the expansion which ends with nines.)

Let $b = 0.y_1y_2y_3y_4\ldots$ be the real number obtained as follows:

$$y_i = \begin{cases} 1 & \text{if } x_{ii} \neq 1 \\ 2 & \text{if } x_{ii} = 1 \end{cases}$$

Now $b \in I$. But

$$b \neq a_1 \quad \text{because} \quad y_1 \neq x_{11}$$
$$b \neq a_2 \quad \text{because} \quad y_2 \neq x_{22}$$
$$b \neq a_3 \quad \text{because} \quad y_3 \neq x_{33}$$
$$\cdots\cdots\cdots\cdots\cdots\cdots$$

Therefore b does not belong to $I = \{a_1, a_2, \ldots\}$. This contradicts the fact that $b \in I$. Hence the assumption that I is denumerable must be false, so I is uncountable.

MISCELLANEOUS PROBLEMS

3.24. Determine which of the graphs in Fig. 3-14 are functions from **R** into **R**:

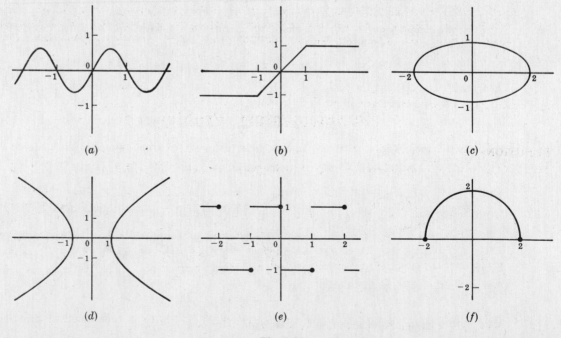

Fig. 3-14

Geometrically speaking, a set of points on a coordinate diagram is a function if and only if every vertical line contains exactly one point of the set. (a) Yes. (b) Yes. (c) No. (d) No. (e) Yes. (f) No; however the graph does define a function from D into **R** where $D = \{x: -2 \leqq x \leqq 2\}$.

3.25. Find the domain D of each of the following real-valued functions of a real variable:

(a) $f(x) = 1/(x - 2)$ (c) $f(x) = \sqrt{25 - x^2}$

(b) $f(x) = x^2 - 3x - 4$ (d) $f(x) = x^2$ where $0 \leqq x \leqq 2$

     When a real-valued function of a real variable is given by a formula $f(x)$, then the domain D consists of the largest subset of **R** for which $f(x)$ has meaning and is real, unless otherwise specified.

(a) f is not defined for $x - 2 = 0$, i.e. for $x = 2$; hence $D = \mathbf{R} \setminus \{2\}$.

(b) f is defined for every real number; hence $D = \mathbf{R}$.

(c) f is not defined when $25 - x^2$ is negative; hence

$$D = [-5, 5] = \{x : -5 \leqq x \leqq 5\}$$

(d) Although the formula for f is meaningful for every real number, the domain of f is explicitly given as $D = \{x : 0 \leqq x \leqq 2\}$.

3.26. An inequality relation is defined for cardinal numbers as follows: For any sets A and B, we define $|A| \leqq |B|$ if there exists a function $f : A \to B$ which is one-to-one. We also write $|A| < |B|$ if $|A| \leqq |B|$ but $|A| \neq |B|$. Prove Cantor's Theorem: For any set A, we have $|A| < |P(A)|$ (where $P(A)$ is the collection of subsets of A, i.e. the power set of A).

     The function $g : A \to P(A)$ which sends each a in A into the set consisting of a alone, i.e. defined by $g(a) = \{a\}$, is clearly one-to-one. Hence $|A| \leqq |P(A)|$.

     If we show that $|A| \neq |P(A)|$, then the theorem will follow. Suppose the contrary; that is, suppose $|A| = |P(A)|$ and that $f : A \to P(A)$ is a function which is both one-to-one and onto. Let $a \in A$ be called a "bad" element if $a \notin f(a)$, and let B be the set of "bad" elements. In other words,

$$B = \{x : x \in A, x \notin f(x)\}$$

     Now B is a subset of A. Since $f : A \to P(A)$ is onto, there exists an element $b \in A$ such that $f(b) = B$. Is b a "bad" or "good" element? If $b \in B$ then, by definition of B, $b \notin f(b) = B$, which is impossible. Likewise, if $b \notin B$ then $b \in f(b) = B$ which is also impossible. Thus the original assumption that $|A| = |P(A)|$ has led to a contradiction. Hence the assumption is false and so the theorem is true.

Supplementary Problems

FUNCTIONS

3.27. State whether each diagram of Fig. 3-15 defines a function from $\{1, 2, 3\}$ into $\{4, 5, 6\}$.

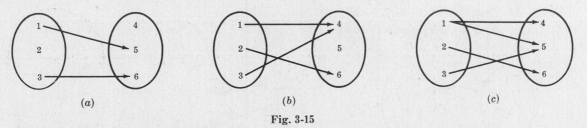

        (a)   (b)   (c)

Fig. 3-15

3.28. Define each function from **R** into **R** by a formula:

(a) To each number let f assign its square plus 3.

(b) To each number let g assign its cube plus twice the number.

(c) To each number greater than or equal to 3 let h assign the number squared, and to each number less than 3 let h assign the number -2.

3.29. Determine the number of different functions from $\{a, b\}$ into $\{1, 2, 3\}$.

3.30. Let $g : R \to R$ be defined by $\quad g(x) = \begin{cases} x^2 - 3x & \text{if } x \geqq 2 \\ x + 2 & \text{if } x < 2 \end{cases}$. Find $g(5)$, $g(0)$ and $g(-2)$.

3.31. Let $W = \{a, b, c, d\}$. Determine whether each set of ordered pairs is a function from W into W.

    (a) $\{(b, a), (c, d), (d, a), (c, d), (a, d)\}$ (c) $\{(a, b), (b, b), (c, b), (d, b)\}$

    (b) $\{(d, d), (c, a), (a, b), (d, b)\}$ (d) $\{(a, a), (b, a), (a, b), (c, d)\}$

3.32. Let the function g assign to each name in the set {Betty, Martin, David, Alan, Rebecca} the number of different letters needed to spell the name. Find the graph of g, i.e. write g as a set of ordered pairs.

3.33. Let $W = \{1, 2, 3, 4\}$ and let $g : W \to W$ be defined by Fig. 3-16. (a) Write g as a set of ordered pairs. (b) Find the image of g. (c) Write the composition function $g \circ g$ as a set of ordered pairs.

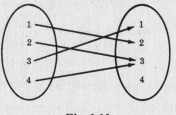

Fig. 3-16

3.34. Let $V = \{1, 2, 3, 4\}$. Determine whether the set of points in each coordinate diagram of $V \times V$ (Fig. 3-17) is a function from V into V.

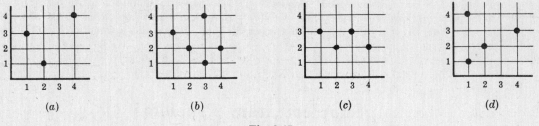

      (a)   (b)   (c)   (d)

Fig. 3-17

3.35. The diagrams in Fig. 3-18 define functions f, g and h which map the set $\{1, 2, 3, 4\}$ into itself.

    (a) Find the images of f, g and h.

    (b) Find the composition functions (1) $f \circ g$, (2) $h \circ f$, (3) g^2, i.e. $g \circ g$.

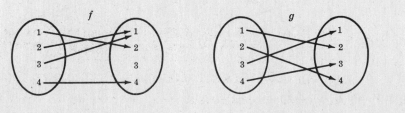
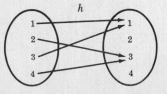

Fig. 3-18

3.36. Consider the functions $f(x) = x^2 + 3x + 1$ and $g(x) = 2x - 3$. Find formulas defining the composition functions (a) $f \circ g$, (b) $g \circ f$.

ONE-TO-ONE, ONTO AND INVERTIBLE FUNCTIONS

3.37. Which functions, if any, in Problem 3.35 are (a) one-to-one, (b) onto functions?

3.38. Let $f : R \to R$ be defined by $f(x) = 3x - 7$. Find a formula for the inverse function $f^{-1} : R \to R$.

3.39. Let $g : R \to R$ be defined by $g(x) = x^3 + 2$. Find a formula for the inverse function $g^{-1} : R \to R$.

3.40. Prove: If $f : A \to B$ and $g : B \to A$ satisfy $g \circ f = 1_A$, then f is one-to-one and g is onto.

3.41. Prove Theorem 3.1: A function $f : A \to B$ is invertible if and only if f is both one-to-one and onto.

3.42. Prove: If $f : A \to B$ is invertible with inverse function $f^{-1} : B \to A$, then $f^{-1} \circ f = 1_A$ and $f \circ f^{-1} = 1_B$.

INDEXED SETS

3.43. For each positive integer n in **N**, let D_n be the following subset of **N**:

$$D_n = \{n, 2n, 3n, 4n, \ldots\} = \{\text{multiples of } n\}$$

   (a) Find: (1) $D_2 \cap D_7$, (2) $D_6 \cap D_8$, (3) $D_3 \cup D_{12}$, (4) $D_3 \cap D_{12}$.

   (b) Prove that $\cap(D_i : i \in J) = \emptyset$ where J is an infinite subset of **N**.

3.44. For each positive integer n in **N**, let A_n be the following subset of the real numbers **R**:

$$A_n = (0, 1/n) = \{x : 0 < x < 1/n\}$$

Find: (a) $A_5 \cup A_8$ (d) $\cap(A_i : i \in J)$

          (b) $A_3 \cap A_7$ (e) $\cup(A_i : i \in K)$

          (c) $\cup(A_i : i \in J)$ (f) $\cap(A_i : i \in K)$

where J is a finite subset of **N** and K is an infinite subset of **N**.

3.45. Consider an indexed class of sets $\{A_i : i \in I\}$, a set B and an index i_0 in I. Prove:

   (a) $B \cap (\cup_i A_i) = \cup_i (B \cap A_i)$

   (b) $\cap(A_i : i \in I) \subset A_{i_0} \subset \cup(A_i : i \in I)$

CARDINAL NUMBERS

3.46. Find the cardinal number of each set:

   (a) {Monday, Tuesday, ..., Sunday}

   (b) $\{x : x$ is a letter in the word "BASEBALL"$\}$

   (c) $\{x : x^2 = 9, 2x = 8\}$

   (d) The power set $P(A)$ of $A = \{1, 3, 7, 11\}$

   (e) Collection of functions from $A = \{a, b, c\}$ into $B = \{1, 2, 3, 4\}$

   (f) Set of relations on $A = \{a, b, c\}$

3.47. Figure 3-19 lists the elements of $\mathbf{N} \times \mathbf{N}$. Show that this listing is equivalent to the function $f : \mathbf{N} \times \mathbf{N} \to \mathbf{N}$ defined by

$$f(r, s) = 1 + 2 + \cdots + (r + s - 2) + s = \tfrac{1}{2}(r + s - 2)(r + s - 1) + s$$

and show that f is one-to-one and onto; hence $|\mathbf{N} \times \mathbf{N}| = \aleph_0$.

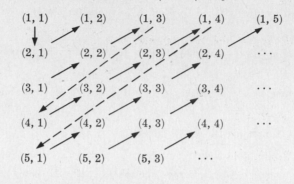

Fig. 3-19

3.48. Prove that: (*a*) $|A \times B| = |B \times A|$. (*b*) If $A \subset B$ then $|A| \leqq |B|$.
(*c*) If $|A| = |B|$ then $|P(A)| = |P(B)|$.

3.49. Prove that:

(*a*) Every infinite set A has a denumerable subset D.

(*b*) Each subset of a denumerable set is finite or denumerable.

(*c*) If A and B are denumerable, then $A \times B$ is denumerable.

(*d*) The set $\mathbf{Q}$ of rational numbers is denumerable.

MISCELLANEOUS PROBLEMS

3.50. Find the domain D of each of the following real-valued functions of a real variable (see Problem 3.25):

(*a*) $f(x) = 1/(x + 3)$ (*c*) $f(x) = \sqrt{16 - x^2}$

(*b*) $f(x) = 1/(x - 3)$ where $x > 0$ (*d*) $f(x) = \log(x + 3)$

3.51. Sketch the graph of each function:

(*a*) $f(x) = \tfrac{1}{2}x - 1$, (*b*) $g(x) = x^3 - 3x + 2$, (*c*) $h(x) = \begin{cases} 0 & \text{if } x = 0 \\ 1/x & \text{if } x \neq 0 \end{cases}$

COMPUTER PROGRAMMING PROBLEMS

A relation f on the set $D = \{1, 2, 3, 4, 5, 6\}$ contains six elements. Suppose a deck has six cards and each card contains a pair of integers which is an element of f.

3.52. Write a program which decides whether or not f is a function. Test the program with the relation:

(i) $f = \{(1,3,) (4,3,) (2,4,) (6,3), (3,6), (5,6)\}$

(ii) $f = \{(1,3), (2,4), (6,3), (4,4), (2,2), (5,3)\}$

3.53. Assuming f is a function, write a program which (*a*) prints Im (f), (*b*) prints $f \circ f = f^2$, (*c*) decides whether or not f is one-to-one. Test the program with the relation:

(i) $f = \{(1,4), (4,3), (2,3), (6,5), (3,4), (5,6)\}$

(ii) $f = \{(1,5), (2,3), (6,4), (4,1), (3,2), (5,1)\}$

Answers to Supplementary Problems

3.27. (a) No, (b) Yes, (c) No

3.28. (a) $f(x) = x^2 + 3$, (b) $g(x) = x^3 + 2x$, (c) $h(x) = \begin{cases} x^2 & \text{if } x \geqq 3 \\ -2 & \text{if } x < 3 \end{cases}$

3.29. Nine

3.30. $g(5) = 10$, $g(0) = 2$, $g(-2) = 0$

3.31. (a) Yes, (b) No, (c) Yes, (d) No

3.32. $g = \{(\text{Betty}, 4), (\text{Martin}, 6), (\text{David}, 4), (\text{Alan}, 3), (\text{Rebecca}, 5)\}$

3.33. (a) $g = \{(1, 2), (2, 3), (3, 1), (4, 3)\}$, (b) $\{1, 2, 3\}$, (c) $g \circ g = \{(1, 3), (2, 1), (3, 2), (4, 1)\}$

3.34. (a) No, (b) No, (c) Yes, (d) No

3.35. (a) $\text{Im}(f) = \{1, 2, 4\}$, $\text{Im}(g) = \{1, 2, 3, 4\}$, $\text{Im}(h) = \{1, 3\}$

(b)

x	$(f \circ g)(x)$	$(h \circ f)(x)$	$g^2(x)$
1	1	3	4
2	4	1	3
3	2	1	2
4	1	3	1

3.36. (a) $(f \circ g)(x) = 4x^2 - 6x + 1$ (b) $(g \circ f)(x) = 2x^2 + 6x - 1$

3.37. (a) Only g is one-to-one. (b) Only g is an onto function.

3.38. $f^{-1}(x) = (x + 7)/3$

3.39. $g^{-1}(x) = \sqrt[3]{x - 2}$

3.43. (1) D_{14}, (2) D_{24}, (3) D_3, (4) D_{12}

3.44. (a) A_5 (d) A_s where s is largest integer in J

(b) A_7 (e) A_r where r is smallest integer in K

(c) A_r where r is smallest integer in J (f) $\emptyset$

3.46. (a) 7, (b) 5, (c) 0, (d) 16, (e) $4^3 = 64$, (f) $2^9 = 512$

3.50. (a) $\mathbf{R} \setminus \{-3\}$, (b) $\{x : x > 0, x \neq 3\}$, (c) $-4 \leqq x \leqq 4$, (d) $x > -3$

3.51. *(a)*

x	$f(x)$
-2	-2
0	-1
2	0

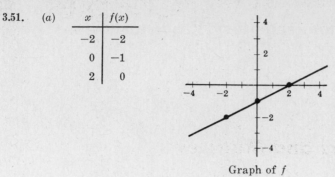

Graph of f

Fig. 3-20

(b)

x	$g(x)$
-3	-16
-2	0
-1	4
0	2
1	0
2	4
3	20

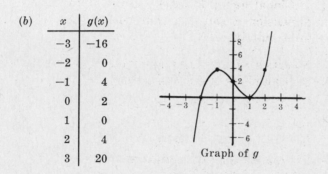

Graph of g

Fig. 3-21

(c)

x	$h(x)$
4	$\frac{1}{4}$
2	$\frac{1}{2}$
1	1
$\frac{1}{2}$	2
$\frac{1}{4}$	4
0	0
$-\frac{1}{4}$	-4
$-\frac{1}{2}$	-2
-1	-1
-2	$-\frac{1}{2}$
-4	$-\frac{1}{4}$

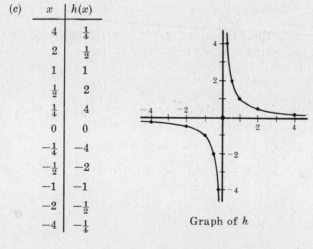

Graph of h

Fig. 3-22

Chapter 4

Vectors and Matrices

4.1 INTRODUCTION

Data is frequently arranged in arrays, i.e. sets whose elements are indexed by one or more subscripts. Formally, a one-dimensional array is called a *vector* and a two-dimensional array is called a *matrix*. (The dimension denotes the number of subscripts.) Also, vectors may be viewed as a special type of matrix.

In this chapter we study basic properties of vectors and matrices whose entries are real numbers. In such a context, the real numbers are called *scalars*.

4.2 VECTORS

By a vector u we simply mean a list (or n-tuple) of numbers:

$$u = (u_1, u_2, \ldots, u_n)$$

The numbers u_i are called the components of u. If all the $u_i = 0$, then u is called the *zero vector*. Two vectors u and v are *equal*, written $u = v$, if they have the same number of components and if corresponding components are equal.

The *sum* of two vectors u and v with the same number of components, written $u + v$, is the vector obtained by adding corresponding components from u and v:

$$u + v = (u_1, u_2, \ldots, u_n) + (v_1, v_2, \ldots, v_n)$$
$$= (u_1 + v_1, u_2 + v_2, \ldots, u_n + v_n)$$

The *product* of a scalar k and a vector u, written $k \cdot u$ or simply ku, is the vector obtained by multiplying each component of u by k:

$$k \cdot u = k(u_1, u_2, \ldots, u_n) = (ku_1, ku_2, \ldots, ku_n)$$

We also define

$$-u = -1 \cdot u \quad \text{and} \quad u - v = u + (-v)$$

and we let 0 denote the zero vector.

The *dot product* or *inner product* of vectors $u = (u_1, u_2, \ldots, u_n)$ and $v = (v_1, v_2, \ldots, v_n)$ is denoted and defined by

$$u \cdot v = u_1 v_1 + u_2 v_2 + \cdots + u_n v_n$$

and the *norm* (or *length*) of u is denoted and defined by

$$\|u\| = \sqrt{u \cdot u} = \sqrt{u_1^2 + u_2^2 + \cdots + u_n^2}$$

We note that $\|u\| = 0$ if and only if $u = 0$.

EXAMPLE 4.1.   Let $u = (2, 3, -4)$ and $v = (1, -5, 8)$.   Then

$$u + v = (2 + 1, 3 - 5, -4 + 8) = (3, -2, 4)$$

$$5u = (5 \cdot 2, 5 \cdot 3, 5 \cdot (-4)) = (10, 15, -20)$$

$$-v = -1 \cdot (1, -5, 8) = (-1, 5, -8)$$

$$2u - 3v = (4, 6, -8) + (-3, 15, -24) = (1, 21, -32)$$

$$u \cdot v = 2 \cdot 1 + 3 \cdot (-5) + (-4) \cdot 8 = 2 - 15 - 32 = -45$$

$$\|u\| = \sqrt{2^2 + 3^2 + (-4)^2} = \sqrt{4 + 9 + 16} = \sqrt{29}$$

Vectors under the operations of vector addition and scalar multiplication have various properties, e.g.

$$k(u + v) = ku + kv$$

where k is a scalar and u and v are vectors.   Since vectors may be viewed as a special case of matrices, Theorem 4.1 (see Section 4.4) contains a list of such properties.

4.3 MATRICES

By a matrix A we mean a rectangular array of numbers:

$$A = \begin{pmatrix} a_{11} & a_{12} & \dots & a_{1n} \\ a_{21} & a_{22} & \dots & a_{2n} \\ \dots\dots\dots\dots\dots\dots \\ a_{m1} & a_{m2} & \dots & a_{mn} \end{pmatrix}$$

The m horizontal n-tuples

$$(a_{11}, a_{12}, \dots, a_{1n}), \ (a_{21}, a_{22}, \dots, a_{2n}), \ \dots, \ (a_{m1}, a_{m2}, \dots, a_{mn})$$

are called the *rows* of A, and the n vertical m-tuples

$$\begin{pmatrix} a_{11} \\ a_{21} \\ \dots \\ a_{m1} \end{pmatrix}, \quad \begin{pmatrix} a_{12} \\ a_{22} \\ \dots \\ a_{m2} \end{pmatrix}, \quad \dots, \quad \begin{pmatrix} a_{1n} \\ a_{2n} \\ \dots \\ a_{mn} \end{pmatrix}$$

its *columns*.   Note that the element a_{ij}, called the *ij-entry*, appears in the ith row and the jth column.   We frequently denote such a matrix simply by $A = (a_{ij})$.

A matrix with m rows and n columns is said to be an m by n matrix, written $m \times n$. The pair of numbers m and n is called the *size* of the matrix.   Two matrices A and B are *equal*, written $A = B$, if they have the same size and if corresponding elements are equal.

A matrix with only one row is sometimes called a *row vector*, and a matrix with only one column is called a *column vector*.   A matrix whose entries are all zero is called a *zero matrix* and will usually be denoted by 0.

EXAMPLE 4.2.

(*a*) The rectangular array $\begin{pmatrix} 1 & -3 & 4 \\ 0 & 5 & -2 \end{pmatrix}$ is a 2×3 matrix.   Its rows are $(1, -3, 4)$ and $(0, 5, -2)$ and its

columns are $\begin{pmatrix} 1 \\ 0 \end{pmatrix}, \begin{pmatrix} -3 \\ 5 \end{pmatrix}$ and $\begin{pmatrix} 4 \\ -2 \end{pmatrix}$.

(*b*) The 2×4 zero matrix is

$$\begin{pmatrix} 0 & 0 & 0 & 0 \\ 0 & 0 & 0 & 0 \end{pmatrix}$$

(c) The statement

$$\begin{pmatrix} x+y & 2z+w \\ x-y & z-w \end{pmatrix} = \begin{pmatrix} 3 & 5 \\ 1 & 4 \end{pmatrix}$$

is equivalent to the system of equations

$$\begin{cases} x+y &= 3 \\ x-y &= 1 \\ 2z+w &= 5 \\ z-w &= 4 \end{cases}$$

The solution of the system of equations is $x=2$, $y=1$, $z=3$, $w=-1$.

4.4 MATRIX ADDITION AND SCALAR MULTIPLICATION

Let A and B be two matrices with the same size, i.e. the same number of rows and of columns. The sum of A and B, written $A+B$, is the matrix obtained by adding corresponding elements from A and B:

$$\begin{pmatrix} a_{11} & a_{12} & \ldots & a_{1n} \\ a_{21} & a_{22} & \ldots & a_{2n} \\ \hdotsfor{4} \\ a_{m1} & a_{m2} & \ldots & a_{mn} \end{pmatrix} + \begin{pmatrix} b_{11} & b_{12} & \ldots & b_{1n} \\ b_{21} & b_{22} & \ldots & b_{2n} \\ \hdotsfor{4} \\ b_{m1} & b_{m2} & \ldots & b_{mn} \end{pmatrix} = \begin{pmatrix} a_{11}+b_{11} & a_{12}+b_{12} & \ldots & a_{1n}+b_{1n} \\ a_{21}+b_{21} & a_{22}+b_{22} & \ldots & a_{2n}+b_{2n} \\ \hdotsfor{4} \\ a_{m1}+b_{m1} & a_{m2}+b_{m2} & \ldots, & a_{mn}+b_{mn} \end{pmatrix}$$

Observe that $A+B$ has the same size as A and B. The sum of two matrices with different sizes is not defined.

The product of a scalar k and a matrix A, written kA or Ak, is the matrix obtained by multiplying each element of A by k:

$$k\begin{pmatrix} a_{11} & a_{12} & \ldots & a_{1n} \\ a_{21} & a_{22} & \ldots & a_{2n} \\ \hdotsfor{4} \\ a_{m1} & a_{m2} & \ldots & a_{mn} \end{pmatrix} = \begin{pmatrix} ka_{11} & ka_{12} & \ldots & ka_{1n} \\ ka_{21} & ka_{22} & \ldots & ka_{2n} \\ \hdotsfor{4} \\ ka_{m1} & ka_{m2} & \ldots & ka_{mn} \end{pmatrix}$$

Note that A and kA have the same size. We also define

$$-A = (-1)A \quad \text{and} \quad A-B = A+(-B)$$

The matrix $-A$ is called the *negative* of the matrix A.

EXAMPLE 4.3.

(a)
$$\begin{pmatrix} 1 & -2 & 3 \\ 0 & 4 & 5 \end{pmatrix} + \begin{pmatrix} 3 & 0 & -6 \\ 2 & -3 & 1 \end{pmatrix} = \begin{pmatrix} 1+3 & -2+0 & 3+(-6) \\ 0+2 & 4+(-3) & 5+1 \end{pmatrix} = \begin{pmatrix} 4 & -2 & -3 \\ 2 & 1 & 6 \end{pmatrix}$$

(b)
$$3\begin{pmatrix} 1 & -2 & 0 \\ 4 & 3 & -5 \end{pmatrix} = \begin{pmatrix} 3\cdot 1 & 3\cdot(-2) & 3\cdot 0 \\ 3\cdot 4 & 3\cdot 3 & 3\cdot(-5) \end{pmatrix} = \begin{pmatrix} 3 & -6 & 0 \\ 12 & 9 & -15 \end{pmatrix}$$

(c)
$$2\begin{pmatrix} 3 & -1 \\ 4 & 6 \end{pmatrix} - 5\begin{pmatrix} 0 & 2 \\ 1 & -3 \end{pmatrix} = \begin{pmatrix} 6 & -2 \\ 8 & 12 \end{pmatrix} + \begin{pmatrix} 0 & -10 \\ -5 & 15 \end{pmatrix} = \begin{pmatrix} 6 & -12 \\ 3 & 27 \end{pmatrix}$$

Matrices under matrix addition and scalar multiplication have the following properties:

Theorem 4.1: Let A, B and C be matrices with the same size and let k and k' be scalars. Then

    (i) $(A+B)+C = A+(B+C)$, i.e. addition is associative

    (ii) $A+B = B+A$, i.e. addition is commutative

    (iii) $A+0 = 0+A = A$

(iv) $A + (-A) = (-A) + A = 0$

(v) $k(A + B) = kA + kB$

(vi) $(k + k')A = kA + k'A$

(vii) $(kk')A = k(k'A)$

(viii) $1A = A$

Since n-component vectors may be identified with either $1 \times n$ matrices or $n \times 1$ matrices, this theorem also holds for vectors under vector addition and scalar multiplication.

4.5 SUMMATION SYMBOL

Before we define matrix multiplication, it will be convenient to first introduce the *summation symbol* Σ (the Greek letter sigma).

Suppose $f(k)$ is an algebraic expression involving the variable k. Then the expression

$$\sum_{k=1}^{n} f(k) \quad \text{or equivalently} \quad \sum_{k=1}^{n} f(k)$$

has the following meaning. First we let $k = 1$ in $f(k)$, obtaining

$$f(1)$$

Then we let $k = 2$ in $f(k)$, obtaining $f(2)$, and add this to $f(1)$, obtaining

$$f(1) + f(2)$$

Next we let $k = 3$ in $f(k)$, obtaining $f(3)$, and add this to the previous sum, obtaining

$$f(1) + f(2) + f(3)$$

We continue this process until we obtain the sum

$$f(1) + f(2) + f(3) + \cdots + f(n-1) + f(n)$$

Observe that at each step we increase the value of k by 1 until k is equal to n. Naturally we may use another variable than k.

We also generalize our definition by allowing the sum to range from any integer n_1 to any integer n_2 where $n_1 \leq n_2$; that is, we define

$$\sum_{k=n_1}^{n_2} f(k) = f(n_1) + f(n_1 + 1) + f(n_1 + 2) + \cdots + f(n_2)$$

Thus we have, for example,

$$\sum_{k=1}^{5} x_k = x_1 + x_2 + x_3 + x_4 + x_5$$

$$\sum_{i=1}^{n} a_i b_i = a_1 b_1 + a_2 b_2 + \cdots + a_n b_n$$

$$\sum_{j=2}^{5} j^2 = 2^2 + 3^2 + 4^2 + 5^2 = 4 + 9 + 16 + 25 = 54$$

$$\sum_{i=0}^{n} a_i x^i = a_0 + a_1 x + a_2 x^2 + \cdots + a_n x^n$$

$$\sum_{k=1}^{p} a_{ik} b_{kj} = a_{i1} b_{1j} + a_{i2} b_{2j} + a_{i3} b_{3j} + \cdots + a_{ip} b_{pj}$$

4.6 MATRIX MULTIPLICATION

Now suppose A and B are two matrices such that the number of columns of A is equal to the number of rows of B, say A is an $m \times p$ matrix and B is a $p \times n$ matrix. Then the product of A and B, written AB, is the $m \times n$ matrix whose ij-entry is obtained by multiplying the elements of the ith row of A by the corresponding elements of the jth column of B and then adding:

$$\begin{pmatrix} a_{11} & \cdots & a_{1p} \\ \cdot & \cdots & \cdot \\ a_{i1} & \cdots & a_{ip} \\ \cdot & \cdots & \cdot \\ a_{m1} & \cdots & a_{mp} \end{pmatrix} \begin{pmatrix} b_{11} & \cdots & b_{1j} & \cdots & b_{1n} \\ \cdot & \cdots & \cdot & \cdots & \cdot \\ \cdot & \cdots & \cdot & \cdots & \cdot \\ \cdot & \cdots & \cdot & \cdots & \cdot \\ b_{p1} & \cdots & b_{pj} & \cdots & b_{pn} \end{pmatrix} = \begin{pmatrix} c_{11} & \cdots & c_{1n} \\ \cdot & \cdots & \cdot \\ \cdot & c_{ij} & \cdot \\ \cdot & \cdots & \cdot \\ c_{m1} & \cdots & c_{mn} \end{pmatrix}$$

where
$$c_{ij} = a_{i1}b_{1j} + a_{i2}b_{2j} + \cdots + a_{ip}b_{pj} = \sum_{k=1}^{p} a_{ik}b_{kj}$$

If the number of columns of A is not equal to the number of rows of B, say A is $m \times p$ and B is $q \times n$ where $p \neq q$, then the product AB is not defined.

EXAMPLE 4.4.

(a)
$$\begin{pmatrix} r & s \\ t & u \end{pmatrix} \begin{pmatrix} a_1 & a_2 & a_3 \\ b_1 & b_2 & b_3 \end{pmatrix} = \begin{pmatrix} ra_1 + sb_1 & ra_2 + sb_2 & ra_3 + sb_3 \\ ta_1 + ub_1 & ta_2 + ub_2 & ta_3 + ub_3 \end{pmatrix}$$

(b)
$$\begin{pmatrix} 1 & 2 \\ 3 & 4 \end{pmatrix} \begin{pmatrix} 1 & 1 \\ 0 & 2 \end{pmatrix} = \begin{pmatrix} 1 \cdot 1 + 2 \cdot 0 & 1 \cdot 1 + 2 \cdot 2 \\ 3 \cdot 1 + 4 \cdot 0 & 3 \cdot 1 + 4 \cdot 2 \end{pmatrix} = \begin{pmatrix} 1 & 5 \\ 3 & 11 \end{pmatrix}$$

$$\begin{pmatrix} 1 & 1 \\ 0 & 2 \end{pmatrix} \begin{pmatrix} 1 & 2 \\ 3 & 4 \end{pmatrix} = \begin{pmatrix} 1 \cdot 1 + 1 \cdot 3 & 1 \cdot 2 + 1 \cdot 4 \\ 0 \cdot 1 + 2 \cdot 3 & 0 \cdot 2 + 2 \cdot 4 \end{pmatrix} = \begin{pmatrix} 4 & 6 \\ 6 & 8 \end{pmatrix}$$

(c) A system of linear equations, such as

$$\begin{cases} x + 2y - 3z = 4 \\ 5x - 6y + 8z = 8 \end{cases}$$

is equivalent to the matrix equation

$$\begin{pmatrix} 1 & 2 & -3 \\ 5 & -6 & 8 \end{pmatrix} \begin{pmatrix} x \\ y \\ z \end{pmatrix} = \begin{pmatrix} 4 \\ 8 \end{pmatrix}$$

That is, any solution to the system of equations is also a solution to the matrix equation, and vice versa.

We see by the preceding example that matrices under the operation of matrix multiplication do not satisfy the commutative law, i.e. the products AB and BA of matrices need not be equal.

Matrix multiplication does, however, satisfy the following properties:

Theorem 4.2:   (i) $(AB)C = A(BC)$

(ii) $A(B + C) = AB + AC$

(iii) $(B + C)A = BA + CA$

(iv) $k(AB) = (kA)B = A(kB)$, where k is a scalar.

We assume that the sums and products in the above theorem are defined.

4.7 TRANSPOSE

The *transpose* of a matrix A, written A^T, is the matrix obtained by writing the rows of A, in order, as columns:

$$\begin{pmatrix} a_1 & a_2 & \ldots & a_n \\ b_1 & b_2 & \ldots & b_n \\ \cdots\cdots\cdots\cdots\cdots \\ c_1 & c_2 & \ldots & c_n \end{pmatrix}^T = \begin{pmatrix} a_1 & b_1 & \ldots & c_1 \\ a_2 & b_2 & \ldots & c_2 \\ \cdots\cdots\cdots\cdots\cdots \\ a_n & b_n & \ldots & c_n \end{pmatrix}$$

Note that if A is an $m \times n$ matrix, then A^T is an $n \times m$ matrix. Also, if $B = (b_{ij})$ is the transpose of $A = (a_{ij})$ then $b_{ij} = a_{ji}$ for all i and j.

EXAMPLE 4.5.

$$\begin{pmatrix} 1 & 2 & 3 \\ 4 & -5 & 6 \end{pmatrix}^T = \begin{pmatrix} 1 & 4 \\ 2 & -5 \\ 3 & 6 \end{pmatrix}$$

The transpose operation on matrices satisfies the following properties.

Theorem 4.3:   (i) $(A + B)^T = A^T + B^T$

(ii) $(kA)^T = kA^T$, for k a scalar

(iii) $(AB)^T = B^T A^T$

(iv) $(A^T)^T = A$

4.8 SQUARE MATRICES

A matrix with the same number of rows as columns is called a *square* matrix. A square matrix with n rows and n columns is said to be of *order n*, and is called an *n-square matrix*. The *main diagonal*, or simply *diagonal*, of a square matrix $A = (a_{ij})$ consists of the numbers $a_{11}, a_{22}, \ldots, a_{nn}$.

EXAMPLE 4.6.   The matrix $\begin{pmatrix} 1 & -2 & 0 \\ 0 & -4 & -1 \\ 5 & 3 & 2 \end{pmatrix}$ is a square matrix of order 3. The numbers along the main diagonal are 1, -4 and 2.

The n-square matrix with 1's along the main diagonal and 0's elsewhere, e.g.,

$$\begin{pmatrix} 1 & 0 & 0 & 0 \\ 0 & 1 & 0 & 0 \\ 0 & 0 & 1 & 0 \\ 0 & 0 & 0 & 1 \end{pmatrix}$$

is called the *unit matrix* and will be denoted by I. The unit matrix I plays the same role in matrix multiplication as the number 1 does in the usual multiplication of numbers. Specifically,

$$AI = IA = A$$

for any square matrix A.

We can form powers of a square matrix A by defining

$$A^2 = AA, \ A^3 = A^2A, \ \ldots \ \text{and} \ A^0 = I$$

We can also form polynomials in A. That is, for any polynomial

$$f(x) = a_0 + a_1x + a_2x^2 + \cdots + a_nx^n$$

we define $f(A)$ to be the matrix

$$f(A) = a_0 I + a_1 A + a_2 A^2 + \cdots + a_n A^n$$

In the case that $f(A)$ is the zero matrix, then A is said to be a *zero* or *root* of the polynomial $f(x)$.

EXAMPLE 4.7.   Let $A = \begin{pmatrix} 1 & 2 \\ 3 & -4 \end{pmatrix}$; then $A^2 = \begin{pmatrix} 7 & -6 \\ -9 & 22 \end{pmatrix}$.   If $f(x) = 2x^2 - 3x + 5$, then

$$f(A) = 2\begin{pmatrix} 7 & -6 \\ -9 & 22 \end{pmatrix} - 3\begin{pmatrix} 1 & 2 \\ 3 & -4 \end{pmatrix} + 5\begin{pmatrix} 1 & 0 \\ 0 & 1 \end{pmatrix} = \begin{pmatrix} 16 & -18 \\ -27 & 61 \end{pmatrix}$$

On the other hand, if $g(x) = x^2 + 3x - 10$ then

$$g(A) = \begin{pmatrix} 7 & -6 \\ -9 & 22 \end{pmatrix} + 3\begin{pmatrix} 1 & 2 \\ 3 & -4 \end{pmatrix} - 10\begin{pmatrix} 1 & 0 \\ 0 & 1 \end{pmatrix} = \begin{pmatrix} 0 & 0 \\ 0 & 0 \end{pmatrix}$$

Thus A is a zero of the polynomial $g(x)$.

4.9 INVERTIBLE MATRICES

A square matrix A is said to be *invertible* if there exists a matrix B with the property that

$$AB = BA = I, \text{ the identity matrix}$$

Such a matrix B is unique; it is called the *inverse* of A and is denoted by A^{-1}.   Observe that B is the inverse of A if and only if A is the inverse of B.   For example, suppose

$$A = \begin{pmatrix} 2 & 5 \\ 1 & 3 \end{pmatrix} \quad \text{and} \quad B = \begin{pmatrix} 3 & -5 \\ -1 & 2 \end{pmatrix}$$

Then

$$AB = \begin{pmatrix} 6-5 & -10+10 \\ 3-3 & -5+6 \end{pmatrix} = \begin{pmatrix} 1 & 0 \\ 0 & 1 \end{pmatrix} \quad \text{and} \quad BA = \begin{pmatrix} 6-5 & 15-15 \\ -2+2 & -5+6 \end{pmatrix} = \begin{pmatrix} 1 & 0 \\ 0 & 1 \end{pmatrix}$$

Thus A and B are inverses.

It is known that $AB = I$ if and only if $BA = I$; hence it is necessary to test only one product to determine whether two given matrices are inverses, as in the next example.

EXAMPLE 4.8.

$$\begin{pmatrix} 1 & 0 & 2 \\ 2 & -1 & 3 \\ 4 & 1 & 8 \end{pmatrix}\begin{pmatrix} -11 & 2 & 2 \\ -4 & 0 & 1 \\ 6 & -1 & -1 \end{pmatrix} = \begin{pmatrix} -11+0+12 & 2+0-2 & 2+0-2 \\ -22+4+18 & 4+0-3 & 4-1-3 \\ -44-4+48 & 8+0-8 & 8+1-8 \end{pmatrix} = \begin{pmatrix} 1 & 0 & 0 \\ 0 & 1 & 0 \\ 0 & 0 & 1 \end{pmatrix}$$

Thus the two matrices are invertible and are inverses of each other.

4.10 DETERMINANTS

To each n-square matrix $A = (a_{ij})$ we assign a specific number called the *determinant* of A, denoted by $\det(A)$ or $|A|$ or

$$\begin{vmatrix} a_{11} & a_{12} & \ldots & a_{1n} \\ a_{21} & a_{22} & \ldots & a_{2n} \\ \ldots & \ldots & \ldots & \ldots \\ a_{n1} & a_{n2} & \ldots & a_{nn} \end{vmatrix}$$

We emphasize that a square array of numbers enclosed by straight lines, called a *determinant of order n*, is not a matrix but denotes the number that the determinant function assigns to the enclosed array of numbers, i.e. the enclosed square matrix.

The determinants of order one, two and three are defined as follows:

$$|a_{11}| = a_{11}$$

$$\begin{vmatrix} a_{11} & a_{12} \\ a_{21} & a_{22} \end{vmatrix} = a_{11}a_{22} - a_{12}a_{21}$$

$$\begin{vmatrix} a_{11} & a_{12} & a_{13} \\ a_{21} & a_{22} & a_{23} \\ a_{31} & a_{32} & a_{33} \end{vmatrix} = a_{11}a_{22}a_{33} + a_{12}a_{23}a_{31} + a_{13}a_{21}a_{32} - a_{13}a_{22}a_{31} - a_{12}a_{21}a_{33} - a_{11}a_{23}a_{32}$$

The following diagram may help the reader remember the determinant of order two:

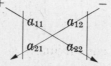

That is, the determinant equals the product of the elements along the plus-labeled arrow minus the product of the elements along the minus-labeled arrow. There is an analogous way to remember a determinant of order three. For notational convenience we have separated the plus-labeled and minus-labeled arrows:

We emphasize that there are no such diagrammatic tricks to remember determinants of higher order.

EXAMPLE 4.9.

(a) $\begin{vmatrix} 5 & 4 \\ 2 & 3 \end{vmatrix} = 5 \cdot 3 - 4 \cdot 2 = 15 - 8 = 7 \qquad \begin{vmatrix} 2 & 1 \\ -4 & 6 \end{vmatrix} = 2 \cdot 6 - 1 \cdot (-4) = 12 + 4 = 16$

(b) $\begin{vmatrix} 2 & 1 & 3 \\ 4 & 6 & -1 \\ 5 & 1 & 0 \end{vmatrix} = 2 \cdot 6 \cdot 0 + 1 \cdot (-1) \cdot 5 + 3 \cdot 4 \cdot 1 - 3 \cdot 6 \cdot 5 - 1 \cdot 4 \cdot 0 - 2 \cdot (-1) \cdot 1$

$$= 0 - 5 + 12 - 90 - 0 + 2 = -81$$

The general definition of a determinant of order n is as follows:

$$\det (A) = \sum \operatorname{sgn}(\sigma) a_{1j_1} a_{2j_2} \cdots a_{nj_n}$$

where the sum is taken over all permutations $\sigma = \{j_1, j_2, \ldots, j_n\}$ of $\{1, 2, \ldots, n\}$. Here sgn (σ) equals plus or minus one according as an even or an odd number of interchanges are required to change σ so that its numbers are in the usual order. We have included the general definition of the determinant function for completeness. The reader is referred to texts in matrix theory or linear algebra for techniques for computing determinants of order greater than three. Permutations are studied in Chapter 8.

An important property of the determinant function is that it is multiplicative. That is,

Theorem 4.4: For any two n-square matrices A and B, we have

$$\det(AB) = \det(A) \cdot \det(B)$$

4.11 INVERTIBLE MATRICES AND DETERMINANTS

We now calculate the inverse of a general 2×2 matrix $A = \begin{pmatrix} a & b \\ c & d \end{pmatrix}$. We seek scalars x, y, z and w such that

$$\begin{pmatrix} a & b \\ c & d \end{pmatrix}\begin{pmatrix} x & y \\ z & w \end{pmatrix} = \begin{pmatrix} 1 & 0 \\ 0 & 1 \end{pmatrix} \quad \text{or} \quad \begin{pmatrix} ax+bz & ay+bw \\ cx+dz & cy+dw \end{pmatrix} = \begin{pmatrix} 1 & 0 \\ 0 & 1 \end{pmatrix}$$

which reduces to solving the following two systems of linear equations in two unknowns:

$$\begin{cases} ax + bz = 1 \\ cx + dz = 0 \end{cases} \quad \begin{cases} ay + bw = 0 \\ cy + dw = 1 \end{cases}$$

If $|A| = ad - bc$ is not zero, then we can uniquely solve for the unknowns x, y, z and w, obtaining:

$$x = \frac{d}{ad-bc} = \frac{d}{|A|}, \; y = \frac{-b}{ad-bc} = \frac{-b}{|A|}, \; z = \frac{-c}{ad-bc} = \frac{-c}{|A|}, \; w = \frac{a}{ad-bc} = \frac{a}{|A|}$$

Accordingly,

$$A^{-1} = \begin{pmatrix} a & b \\ c & d \end{pmatrix}^{-1} = \begin{pmatrix} d/|A| & -b/|A| \\ -c/|A| & a/|A| \end{pmatrix} = \frac{1}{|A|}\begin{pmatrix} d & -b \\ -c & a \end{pmatrix}$$

In other words, we can obtain the inverse of a 2×2 matrix, with determinant nonzero, by (i) interchanging the elements on the main diagonal, (ii) taking the negative of the other elements, and (iii) dividing each element by the determinant of the original matrix.

For example, if

$$A = \begin{pmatrix} 2 & 3 \\ 4 & 5 \end{pmatrix} \quad \text{then} \quad |A| = -2, \quad \text{and so} \quad A^{-1} = \frac{1}{-2}\begin{pmatrix} 5 & -3 \\ -4 & 2 \end{pmatrix} = \begin{pmatrix} -\frac{5}{2} & \frac{3}{2} \\ 2 & -1 \end{pmatrix}$$

On the other hand, if $|A|$ is zero then we cannot solve for x, y, z and w, and A^{-1} would not exist. Although there is no simple formula for matrices of higher order, this result does hold in general. Namely,

Theorem 4.5: A matrix is invertible if and only if it has a nonzero determinant.

Solved Problems

VECTORS

4.1. Let $u = (2, -7, 1)$, $v = (-3, 0, 4)$ and $w = (0, 5, -8)$.

Find: (a) $u+v$, (b) $v+w$, (c) $-3u$, (d) $-w$

(a) Add corresponding components:

$$u + v = (2, -7, 1) + (-3, 0, 4) = (2-3, -7+0, 1+4) = (-1, -7, 5)$$

(b) Add corresponding components:

$$v + w = (-3, 0, 4) + (0, 5, -8) = (-3 + 0, 0 + 5, 4 - 8) = (-3, 5, -4)$$

(c) Multiply each component of u by the scalar -3:

$$-3u = -3(2, -7, 1) = (-6, 21, -3)$$

(d) Multiply each component of w by -1, i.e. change the sign of each component:

$$-w = -(0, 5, -8) = (0, -5, 8)$$

4.2. Let u, v and w be the row vectors of the preceding problem.

Find: (a) $3u - 4v$, (b) $2u + 3v - 5w$

First perform the scalar multiplication and then the vector addition.

(a) $3u - 4v = 3(2, -7, 1) - 4(-3, 0, 4) = (6, -21, 3) + (12, 0, -16) = (18, -21, -13)$

(b) $2u + 3v - 5w = 2(2, -7, 1) + 3(-3, 0, 4) - 5(0, 5, -8)$

$$= (4, -14, 2) + (-9, 0, 12) + (0, -25, 40)$$

$$= (4 - 9 + 0, -14 + 0 - 25, 2 + 12 + 40) = (-5, -39, 54)$$

4.3. Find x and y if $x(1, 1) + y(2, -1) = (1, 4)$.

First multiply by the scalars x and y and then add:

$$x(1, 1) + y(2, -1) = (x, x) + (2y, -y) = (x + 2y, x - y) = (1, 4)$$

Now set corresponding components equal to each other to obtain

$$x + 2y = 1$$

$$x - y = 4$$

Solve the system of equations to find $x = 3$ and $y = -1$.

4.4. Let u, v and w be the vectors in Problem 4.1.

Find: (a) $u \cdot v, u \cdot w, v \cdot w$, (b) $\|u\|, \|v\|, \|w\|$

(a) Multiply corresponding components and then add:

$$u \cdot v = 2 \cdot (-3) + (-7) \cdot 0 + 1 \cdot 4 = -6 + 0 + 4 = -2$$

$$u \cdot w = 0 - 35 - 8 = -43$$

$$v \cdot w = 0 + 0 - 32 = -32$$

(b) Take the square root of the sum of the squares of the components:

$$\|u\| = \sqrt{2^2 + (-7)^2 + 1^2} = \sqrt{4 + 49 + 1} = \sqrt{54} = 3\sqrt{6}$$

$$\|v\| = \sqrt{9 + 0 + 16} = \sqrt{25} = 5$$

$$\|w\| = \sqrt{0 + 25 + 64} = \sqrt{89}$$

MATRIX ADDITION AND SCALAR MULTIPLICATION

4.5. Compute:

(a) $\begin{pmatrix} 1 & 2 & 3 \\ 4 & 5 & 6 \end{pmatrix} + \begin{pmatrix} 1 & -1 & 2 \\ 0 & 3 & -5 \end{pmatrix}$ (b) $-2\begin{pmatrix} 1 & 7 \\ 2 & -3 \\ 0 & -1 \end{pmatrix}$ (c) $-\begin{pmatrix} 2 & -3 & 8 \\ 1 & -2 & -6 \end{pmatrix}$

(a) Add corresponding elements:

$$\begin{pmatrix} 1 & 2 & 3 \\ 4 & 5 & 6 \end{pmatrix} + \begin{pmatrix} 1 & -1 & 2 \\ 0 & 3 & -5 \end{pmatrix} = \begin{pmatrix} 1+1 & 2+(-1) & 3+2 \\ 4+0 & 5+3 & 6+(-5) \end{pmatrix}$$

$$= \begin{pmatrix} 2 & 1 & 5 \\ 4 & 8 & 1 \end{pmatrix}$$

(b) Multiply each element of the matrix by the scalar -2:

$$-2 \begin{pmatrix} 1 & 7 \\ 2 & -3 \\ 0 & -1 \end{pmatrix} = \begin{pmatrix} (-2) \cdot 1 & (-2) \cdot 7 \\ (-2) \cdot 2 & (-2) \cdot (-3) \\ (-2) \cdot 0 & (-2) \cdot (-1) \end{pmatrix} = \begin{pmatrix} -2 & -14 \\ -4 & 6 \\ 0 & 2 \end{pmatrix}$$

(c) Multiply each element of the matrix by -1, or equivalently change the sign of each element in the matrix:

$$-\begin{pmatrix} 2 & -3 & 8 \\ 1 & -2 & -6 \end{pmatrix} = \begin{pmatrix} -2 & 3 & -8 \\ -1 & 2 & 6 \end{pmatrix}$$

4.6. Compute: $3 \begin{pmatrix} 2 & -5 & 1 \\ 3 & 0 & -4 \end{pmatrix} - 2 \begin{pmatrix} 1 & -2 & -3 \\ 0 & -1 & 5 \end{pmatrix} + 4 \begin{pmatrix} 0 & 1 & -2 \\ 1 & -1 & -1 \end{pmatrix}$

First perform the scalar multiplication, and then the matrix addition:

$$3 \begin{pmatrix} 2 & -5 & 1 \\ 3 & 0 & -4 \end{pmatrix} - 2 \begin{pmatrix} 1 & -2 & -3 \\ 0 & -1 & 5 \end{pmatrix} + 4 \begin{pmatrix} 0 & 1 & -2 \\ 1 & -1 & -1 \end{pmatrix}$$

$$= \begin{pmatrix} 6 & -15 & 3 \\ 9 & 0 & -12 \end{pmatrix} + \begin{pmatrix} -2 & 4 & 6 \\ 0 & 2 & -10 \end{pmatrix} + \begin{pmatrix} 0 & 4 & -8 \\ 4 & -4 & -4 \end{pmatrix}$$

$$= \begin{pmatrix} 6+(-2)+0 & -15+4+4 & 3+6+(-8) \\ 9+0+4 & 0+2+(-4) & -12+(-10)+(-4) \end{pmatrix} = \begin{pmatrix} 4 & -7 & 1 \\ 13 & -2 & -26 \end{pmatrix}$$

4.7. Find x, y, z and w if:

$$3 \begin{pmatrix} x & y \\ z & w \end{pmatrix} = \begin{pmatrix} x & 6 \\ -1 & 2w \end{pmatrix} + \begin{pmatrix} 4 & x+y \\ z+w & 3 \end{pmatrix}$$

First write each side as a single matrix:

$$\begin{pmatrix} 3x & 3y \\ 3z & 3w \end{pmatrix} = \begin{pmatrix} x+4 & x+y+6 \\ z+w-1 & 2w+3 \end{pmatrix}$$

Set corresponding elements equal to each other to obtain the four linear equations

$$3x = x + 4 \qquad\qquad 2x = 4$$
$$3y = x + y + 6 \qquad\qquad 2y = 6 + x$$
$$3z = z + w - 1 \qquad \text{or} \qquad 2z = w - 1$$
$$3w = 2w + 3 \qquad\qquad w = 3$$

The solution of this system of equations is $x = 2$, $y = 4$, $z = 1$, $w = 3$.

MATRIX MULTIPLICATION

4.8. Let $(r \times s)$ denote an $r \times s$ matrix. Find the size of the following matrix products if the product is defined:

(a) $(2 \times 3)(3 \times 4)$ (d) $(5 \times 2)(2 \times 3)$

(b) $(4 \times 1)(1 \times 2)$ (e) $(4 \times 4)(3 \times 3)$

(c) $(1 \times 2)(3 \times 1)$ (f) $(2 \times 2)(2 \times 4)$

In each case the product is defined if the inner numbers are equal, and then the product will have the size of the outer numbers in the given order.

(a) The product is a 2×4 matrix.

(b) The product is a 4×2 matrix.

(c) The product is not defined since the inner numbers 2 and 3 are not equal.

(d) The product is a 5×3 matrix.

(e) The product is not defined since the inner numbers 4 and 3 are not equal.

(f) The product is a 2×4 matrix.

4.9.   Let $A = \begin{pmatrix} 1 & 3 \\ 2 & -1 \end{pmatrix}$ and $B = \begin{pmatrix} 2 & 0 & -4 \\ 3 & -2 & 6 \end{pmatrix}$. Find (a) AB, (b) BA.

(a) Now A is 2×2 and B is 2×3, so the product matrix AB is defined and is a 2×3 matrix. To obtain the elements in the first row of the product matrix AB, multiply the first row $(1, 3)$ of A by the columns $\begin{pmatrix} 2 \\ 3 \end{pmatrix}$, $\begin{pmatrix} 0 \\ -2 \end{pmatrix}$ and $\begin{pmatrix} -4 \\ 6 \end{pmatrix}$ of B, respectively:

$$\begin{pmatrix} 1 & 3 \\ 2 & -1 \end{pmatrix}\begin{pmatrix} 2 & 0 & -4 \\ 3 & -2 & 6 \end{pmatrix} = \begin{pmatrix} 1\cdot 2 + 3\cdot 3 & 1\cdot 0 + 3\cdot(-2) & 1\cdot(-4) + 3\cdot 6 \end{pmatrix}$$

$$= \begin{pmatrix} 11 & -6 & 14 \end{pmatrix}$$

To obtain the elements in the second row of the product matrix AB, multiply the second row $(2, -1)$ of A by the columns of B, respectively:

$$\begin{pmatrix} 1 & 3 \\ 2 & -1 \end{pmatrix}\begin{pmatrix} 2 & 0 & -4 \\ 3 & -2 & 6 \end{pmatrix} = \begin{pmatrix} 11 & -6 & 14 \\ 2\cdot 2 + (-1)\cdot 3 & 2\cdot 0 + (-1)\cdot(-2) & 2\cdot(-4) + (-1)\cdot 6 \end{pmatrix}$$

$$= \begin{pmatrix} 11 & -6 & 14 \\ 1 & 2 & -14 \end{pmatrix} = AB$$

(b) Now B is 2×3 and A is 2×2. Since the inner numbers 3 and 2 are not equal, the product BA is not defined.

4.10.   Let $A = \begin{pmatrix} 2 & -1 \\ 1 & 0 \\ -3 & 4 \end{pmatrix}$ and $B = \begin{pmatrix} 1 & -2 & -5 \\ 3 & 4 & 0 \end{pmatrix}$. Find AB.

Now A is 3×2 and B is 2×3, so the product AB is defined and is a 3×3 matrix. To obtain the first row of the product matrix AB, multiply the first row of A by each column of B, respectively:

$$\begin{pmatrix} 2 & -1 \\ 1 & 0 \\ -3 & 4 \end{pmatrix}\begin{pmatrix} 1 & -2 & -5 \\ 3 & 4 & 0 \end{pmatrix} = \begin{pmatrix} 2-3 & -4-4 & -10+0 \\ & & \\ & & \end{pmatrix} = \begin{pmatrix} -1 & -8 & -10 \\ & & \\ & & \end{pmatrix}$$

To obtain the second row of the product matrix AB, multiply the second row of A by each column of B, respectively:

$$\begin{pmatrix} 2 & -1 \\ 1 & 0 \\ -3 & 4 \end{pmatrix}\begin{pmatrix} 1 & -2 & -5 \\ 3 & 4 & 0 \end{pmatrix} = \begin{pmatrix} -1 & -8 & -10 \\ 1+0 & -2+0 & -5+0 \\ & & \end{pmatrix} = \begin{pmatrix} -1 & -8 & -10 \\ 1 & -2 & -5 \\ & & \end{pmatrix}$$

To obtain the third row of the product matrix AB, multiply the third row of A by each column of B, respectively:

$$\begin{pmatrix} 2 & -1 \\ 1 & 0 \\ -3 & 4 \end{pmatrix}\begin{pmatrix} 1 & -2 & -5 \\ 3 & 4 & 0 \end{pmatrix} = \begin{pmatrix} -1 & -8 & -10 \\ 1 & -2 & -5 \\ -3+12 & 6+16 & 15+0 \end{pmatrix} = \begin{pmatrix} -1 & -8 & -10 \\ 1 & -2 & -5 \\ 9 & 22 & 15 \end{pmatrix}$$

Thus $$AB = \begin{pmatrix} -1 & -8 & -10 \\ 1 & -2 & -5 \\ 9 & 22 & 15 \end{pmatrix}$$

4.11. Let $A = \begin{pmatrix} 2 & -1 & 0 \\ 1 & 0 & -3 \end{pmatrix}$ and $B = \begin{pmatrix} 1 & -4 & 0 & 1 \\ 2 & -1 & 3 & -1 \\ 4 & 0 & -2 & 0 \end{pmatrix}$.

(a) Determine the size of AB. (b) Let c_{ij} denote the element in the ith row and jth column of the product matrix AB, that is, $AB = (c_{ij})$. Find: c_{23}, c_{14}, c_{21} and c_{12}.

(a) Since A is 2×3 and B is 3×4, the product AB is a 2×4 matrix.

(b) Now c_{ij} is defined as the product of the ith row of A by the jth column of B. Hence:

$$c_{23} = (1,0,-3)\begin{pmatrix} 0 \\ 3 \\ -2 \end{pmatrix} = 1 \cdot 0 + 0 \cdot 3 + (-3) \cdot (-2) = 0 + 0 + 6 = 6$$

$$c_{14} = (2,-1,0)\begin{pmatrix} 1 \\ -1 \\ 0 \end{pmatrix} = 2 \cdot 1 + (-1) \cdot (-1) + 0 \cdot 0 = 2 + 1 + 0 = 3$$

$$c_{21} = (1,0,-3)\begin{pmatrix} 1 \\ 2 \\ 4 \end{pmatrix} = 1 \cdot 1 + 0 \cdot 2 + (-3) \cdot 4 = 1 + 0 - 12 = -11$$

$$c_{12} = (2,-1,0)\begin{pmatrix} -4 \\ -1 \\ 0 \end{pmatrix} = 2 \cdot (-4) + (-1) \cdot (-1) + 0 \cdot 0 = -8 + 1 + 0 = -7$$

4.12. Compute: (a) $\begin{pmatrix} 1 & 6 \\ -3 & 5 \end{pmatrix}\begin{pmatrix} 4 & 0 \\ 2 & -1 \end{pmatrix}$ (c) $\begin{pmatrix} 1 \\ -6 \end{pmatrix}\begin{pmatrix} 1 & 6 \\ -3 & 5 \end{pmatrix}$ (e) $(2,-1)\begin{pmatrix} 1 \\ -6 \end{pmatrix}$

(b) $\begin{pmatrix} 1 & 6 \\ -3 & 5 \end{pmatrix}\begin{pmatrix} 2 \\ -7 \end{pmatrix}$ (d) $\begin{pmatrix} 1 \\ 6 \end{pmatrix}(3,2)$

(a) The first factor is 2×2 and the second is 2×2, so the product is defined and is a 2×2 matrix:

$$\begin{pmatrix} 1 & 6 \\ -3 & 5 \end{pmatrix}\begin{pmatrix} 4 & 0 \\ 2 & -1 \end{pmatrix} = \begin{pmatrix} 1 \cdot 4 + 6 \cdot 2 & 1 \cdot 0 + 6 \cdot (-1) \\ (-3) \cdot 4 + 5 \cdot 2 & (-3) \cdot 0 + 5 \cdot (-1) \end{pmatrix} = \begin{pmatrix} 16 & -6 \\ -2 & -5 \end{pmatrix}$$

(b) The first factor is 2×2 and the second is 2×1, so the product is defined and is a 2×1 matrix:

$$\begin{pmatrix} 1 & 6 \\ -3 & 5 \end{pmatrix}\begin{pmatrix} 2 \\ -7 \end{pmatrix} = \begin{pmatrix} 1 \cdot 2 + 6 \cdot (-7) \\ (-3) \cdot 2 + 5 \cdot (-7) \end{pmatrix} = \begin{pmatrix} -40 \\ -41 \end{pmatrix}$$

(c) Now the first factor is 2×1 and the second is 2×2. Since the inner numbers 1 and 2 are distinct, the product is not defined.

(d) Here the first factor is 2×1 and the second is 1×2, so the product is defined and is a 2×2 matrix:

$$\binom{1}{6}(3,2) = \begin{pmatrix} 1\cdot 3 & 1\cdot 2 \\ 6\cdot 3 & 6\cdot 2 \end{pmatrix} = \begin{pmatrix} 3 & 2 \\ 18 & 12 \end{pmatrix}$$

(e) The first factor is 1×2 and the second is 2×1, so the product is defined and is a 1×1 matrix which we frequently write as a scalar.

$$(2,-1)\binom{1}{-6} = (2\cdot 1) + (-1)\cdot(-6) = (8) = 8$$

TRANSPOSE

4.13. Let A be an arbitrary matrix. Under what conditions is the product AA^T defined?

Suppose A is an $m \times n$ matrix; then A^T is $n \times m$. Thus the product AA^T is always defined. Observe that A^TA is also defined. Here AA^T is an $m \times m$ matrix, whereas A^TA is an $n \times n$ matrix.

4.14. Let $A = \begin{pmatrix} 1 & 2 & 0 \\ 3 & -1 & 4 \end{pmatrix}$. Find *(a)* AA^T, *(b)* A^TA.

To obtain A^T, rewrite the rows of A as columns: $A^T = \begin{pmatrix} 1 & 3 \\ 2 & -1 \\ 0 & 4 \end{pmatrix}$. Then

(a) $AA^T = \begin{pmatrix} 1 & 2 & 0 \\ 3 & -1 & 4 \end{pmatrix}\begin{pmatrix} 1 & 3 \\ 2 & -1 \\ 0 & 4 \end{pmatrix} = \begin{pmatrix} 1+4+0 & 3-2+0 \\ 3-2+0 & 9+1+16 \end{pmatrix} = \begin{pmatrix} 5 & 1 \\ 1 & 26 \end{pmatrix}$

(b) $A^TA = \begin{pmatrix} 1 & 3 \\ 2 & -1 \\ 0 & 4 \end{pmatrix}\begin{pmatrix} 1 & 2 & 0 \\ 3 & -1 & 4 \end{pmatrix} = \begin{pmatrix} 1+9 & 2-3 & 0+12 \\ 2-3 & 4+1 & 0-4 \\ 0+12 & 0-4 & 0+16 \end{pmatrix} = \begin{pmatrix} 10 & -1 & 12 \\ -1 & 5 & -4 \\ 12 & -4 & 16 \end{pmatrix}$

SQUARE MATRICES

4.15. Let $A = \begin{pmatrix} 1 & 2 \\ 4 & -3 \end{pmatrix}$.

Find: *(a)* A^2, *(b)* A^3, *(c)* $f(A)$, where $f(x) = 2x^3 - 4x + 5$. *(d)* Show that A is a zero of the polynomial $g(x) = x^2 + 2x - 11$.

(a) $A^2 = AA = \begin{pmatrix} 1 & 2 \\ 4 & -3 \end{pmatrix}\begin{pmatrix} 1 & 2 \\ 4 & -3 \end{pmatrix}$

$= \begin{pmatrix} 1\cdot 1 + 2\cdot 4 & 1\cdot 2 + 2\cdot(-3) \\ 4\cdot 1 + (-3)\cdot 4 & 4\cdot 2 + (-3)\cdot(-3) \end{pmatrix} = \begin{pmatrix} 9 & -4 \\ -8 & 17 \end{pmatrix}$

(b) $A^3 = AA^2 = \begin{pmatrix} 1 & 2 \\ 4 & -3 \end{pmatrix}\begin{pmatrix} 9 & -4 \\ -8 & 17 \end{pmatrix}$

$= \begin{pmatrix} 1\cdot 9 + 2\cdot(-8) & 1\cdot(-4) + 2\cdot 17 \\ 4\cdot 9 + (-3)\cdot(-8) & 4\cdot(-4) + (-3)\cdot 17 \end{pmatrix} = \begin{pmatrix} -7 & 30 \\ 60 & -67 \end{pmatrix}$

(c) To find $f(A)$, first substitute A for x and $5I$ for the constant 5 in the given polynomial $f(x) = 2x^3 - 4x + 5$:

$$f(A) = 2A^3 - 4A + 5I = 2\begin{pmatrix} -7 & 30 \\ 60 & -67 \end{pmatrix} - 4\begin{pmatrix} 1 & 2 \\ 4 & -3 \end{pmatrix} + 5\begin{pmatrix} 1 & 0 \\ 0 & 1 \end{pmatrix}$$

Then multiply each matrix by its respective scalar:

$$f(A) = \begin{pmatrix} -14 & 60 \\ 120 & -134 \end{pmatrix} + \begin{pmatrix} -4 & -8 \\ -16 & 12 \end{pmatrix} + \begin{pmatrix} 5 & 0 \\ 0 & 5 \end{pmatrix}$$

Lastly, add the corresponding elements in the matrices:

$$f(A) \;=\; \begin{pmatrix} -14 - 4 + 5 & 60 - 8 + 0 \\ 120 - 16 + 0 & -134 + 12 + 5 \end{pmatrix} \;=\; \begin{pmatrix} -13 & 52 \\ 104 & -117 \end{pmatrix}$$

(d) Now A is a zero of $g(x)$ if the matrix $g(A)$ is the zero matrix. Compute $g(A)$ as was done for $f(A)$, i.e. first substitute A for x and $11I$ for the constant 11 in $g(x) = x^2 + 2x - 11$:

$$g(A) \;=\; A^2 + 2A - 11I \;=\; \begin{pmatrix} 9 & -4 \\ -8 & 17 \end{pmatrix} + 2\begin{pmatrix} 1 & 2 \\ 4 & -3 \end{pmatrix} - 11\begin{pmatrix} 1 & 0 \\ 0 & 1 \end{pmatrix}$$

Then multiply each matrix by the scalar preceding it:

$$g(A) \;=\; \begin{pmatrix} 9 & -4 \\ -8 & 17 \end{pmatrix} + \begin{pmatrix} 2 & 4 \\ 8 & -6 \end{pmatrix} + \begin{pmatrix} -11 & 0 \\ 0 & -11 \end{pmatrix}$$

Lastly, add the corresponding elements in the matrices:

$$g(A) \;=\; \begin{pmatrix} 9 + 2 - 11 & -4 + 4 + 0 \\ -8 + 8 + 0 & 17 - 6 - 11 \end{pmatrix} \;=\; \begin{pmatrix} 0 & 0 \\ 0 & 0 \end{pmatrix}$$

Since $g(A) = 0$, A is a zero of the polynomial $g(x)$.

4.16. Compute the determinant of each matrix:

(a) $\begin{pmatrix} 3 & -2 \\ 4 & 5 \end{pmatrix}$　　(b) $\begin{pmatrix} -1 & 6 \\ 0 & 4 \end{pmatrix}$　　(c) $\begin{pmatrix} a - b & b \\ b & a + b \end{pmatrix}$　　(d) $\begin{pmatrix} a - b & a \\ a & a + b \end{pmatrix}$

(a) $\begin{vmatrix} 3 & -2 \\ 4 & 5 \end{vmatrix} \;=\; 3 \cdot 5 - (-2) \cdot 4 \;=\; 15 + 8 \;=\; 23$

(b) $\begin{vmatrix} -1 & 6 \\ 0 & 4 \end{vmatrix} \;=\; -1 \cdot 4 - 6 \cdot 0 \;=\; -4$

(c) $\begin{vmatrix} a - b & b \\ b & a + b \end{vmatrix} \;=\; (a - b)(a + b) - b \cdot b \;=\; a^2 - b^2 - b^2 \;=\; a^2 - 2b^2$

(d) $\begin{vmatrix} a - b & a \\ a & a + b \end{vmatrix} \;=\; (a - b)(a + b) - a \cdot a \;=\; a^2 - b^2 - a^2 \;=\; -b^2$

4.17. Find the determinant of each matrix:

(a) $\begin{pmatrix} 1 & 2 & 3 \\ 4 & -2 & 3 \\ 0 & 5 & -1 \end{pmatrix}$　　(b) $\begin{pmatrix} 4 & -1 & -2 \\ 0 & 2 & -3 \\ 5 & 2 & 1 \end{pmatrix}$　　(c) $\begin{pmatrix} 2 & -3 & 4 \\ 1 & 2 & -3 \\ -1 & -2 & 5 \end{pmatrix}$

(*Hint*: Use diagram on page 70.)

(a) $\begin{vmatrix} 1 & 2 & 3 \\ 4 & -2 & 3 \\ 0 & 5 & -1 \end{vmatrix} \;=\; 2 + 0 + 60 - 0 - 15 + 8 \;=\; 55$

(b) $\begin{vmatrix} 4 & -1 & -2 \\ 0 & 2 & -3 \\ 5 & 2 & 1 \end{vmatrix} \;=\; 8 + 15 + 0 + 20 + 24 + 0 \;=\; 67$

(c) $\begin{vmatrix} 2 & -3 & 4 \\ 1 & 2 & -3 \\ -1 & -2 & 5 \end{vmatrix} \;=\; 20 - 9 - 8 + 8 - 12 + 15 \;=\; 14$

4.18. Find the inverse of $\begin{pmatrix} 3 & 5 \\ 2 & 3 \end{pmatrix}$.

Method 1.

We seek scalars x, y, z and w for which

$$\begin{pmatrix} 3 & 5 \\ 2 & 3 \end{pmatrix}\begin{pmatrix} x & y \\ z & w \end{pmatrix} = \begin{pmatrix} 1 & 0 \\ 0 & 1 \end{pmatrix} \quad \text{or} \quad \begin{pmatrix} 3x + 5z & 3y + 5w \\ 2x + 3z & 2y + 3w \end{pmatrix} = \begin{pmatrix} 1 & 0 \\ 0 & 1 \end{pmatrix}$$

or which satisfy

$$\begin{cases} 3x + 5z = 1 \\ 2x + 3z = 0 \end{cases} \quad \text{and} \quad \begin{cases} 3y + 5w = 0 \\ 2y + 3w = 1 \end{cases}$$

To solve the first system, multiply the first equation by 2 and the second equation by -3 and then add:

$$\begin{array}{ll} 2 \times \text{first:} & 6x + 10z = 2 \\ -3 \times \text{second:} & -6x - 9z = 0 \\ \hline \text{Addition:} & z = 2 \end{array}$$

Substitute $z = 2$ into the first equation to obtain

$$3x + 5 \cdot 2 = 1 \quad \text{or} \quad 3x + 10 = 1 \quad \text{or} \quad 3x = -9 \quad \text{or} \quad x = -3$$

To solve the second system, multiply the first equation by 2 and the second equation by -3 and then add:

$$\begin{array}{ll} 2 \times \text{first:} & 6y + 10w = 0 \\ -3 \times \text{second:} & -6y - 9w = -3 \\ \hline \text{Addition:} & w = -3 \end{array}$$

Substitute $w = -3$ in the first equation to obtain

$$3y + 5 \cdot (-3) = 0 \quad \text{or} \quad 3y - 15 = 0 \quad \text{or} \quad 3y = 15 \quad \text{or} \quad y = 5$$

Thus the inverse of the given matrix is $\begin{pmatrix} -3 & 5 \\ 2 & -3 \end{pmatrix}$.

Method 2.

We use the general formula for the inverse of a 2×2 matrix. First find the determinant of the given matrix:

$$\begin{vmatrix} 3 & 5 \\ 2 & 3 \end{vmatrix} = 3 \cdot 3 - 2 \cdot 5 = 9 - 10 = -1$$

Now interchange the elements on the main diagonal of the given matrix and take the negative of the other elements to obtain

$$\begin{pmatrix} 3 & -5 \\ -2 & 3 \end{pmatrix}$$

Lastly, divide each element of this matrix by the determinant of the given matrix, that is, by -1:

$$\begin{pmatrix} -3 & 5 \\ 2 & -3 \end{pmatrix}$$

The above is the required inverse.

PROOFS

4.19. Prove Theorem 4.2(i): $(AB)C = A(BC)$.

Let $A = (a_{ij})$, $B = (b_{jk})$ and $C = (c_{kl})$. Furthermore, let $AB = S = (s_{ik})$ and $BC = T = (t_{jl})$. Then

$$s_{ik} = a_{i1}b_{1k} + a_{i2}b_{2k} + \cdots + a_{im}b_{mk} = \sum_{j=1}^{m} a_{ij}b_{jk}$$

$$t_{jl} = b_{j1}c_{1l} + b_{j2}c_{2l} + \cdots + b_{jn}c_{nl} = \sum_{k=1}^{n} b_{jk}c_{kl}$$

Now multiplying S by C, i.e. (AB) by C, the element in the ith row and lth column of the matrix $(AB)C$ is

$$s_{i1}C_{1l} + s_{i2}c_{2l} + \cdots + s_{in}c_{nl} \;=\; \sum_{k=1}^{n} s_{ik}c_{kl} \;=\; \sum_{k=1}^{n}\sum_{j=1}^{m} (a_{ij}b_{jk})c_{kl}$$

On the other hand, multiplying A by T, i.e. A by BC, the element in the ith row and lth column of the matrix $A(BC)$ is

$$a_{i1}t_{1l} + a_{i2}t_{2l} + \cdots + a_{im}t_{ml} \;=\; \sum_{j=1}^{m} a_{ik}t_{k} \;=\; \sum_{j=1}^{m}\sum_{k=1}^{n} a_{ij}(b_{jk}c_{kl})$$

Since the above sums are equal, the theorem is proven.

4.20. Prove Theorem 4.3(iii): $(AB)^T = B^T A^T$.

Let $A = (a_{ij})$ and $B = (b_{jk})$. Then the element in the ith row and jth column of the matrix AB is

$$a_{i1}b_{1j} + a_{i2}b_{2j} + \cdots + a_{im}b_{mj} \tag{1}$$

Thus (1) is the element which appears in the jth row and ith column of the transpose matrix $(AB)^T$.

On the other hand, the jth row of B^T consists of the elements from the jth column of B:

$$(b_{1j}\;\; b_{2j}\;\; \cdots \;\; b_{mj}) \tag{2}$$

Furthermore, the ith column of A^T consists of the elements from the ith row of A:

$$\begin{pmatrix} a_{i1} \\ a_{i2} \\ \cdot \\ \cdot \\ \cdot \\ a_{im} \end{pmatrix} \tag{3}$$

Consequently, the element appearing in the jth row and ith column of the matrix $B^T A^T$ is the product of (2) by (3) which gives (1). Thus $(AB)^T = B^T A^T$.

Supplementary Problems

VECTORS

4.21. Let $u = (2, -1, 0, -3)$, $v = (1, -1, -1, 3)$ and $w = (1, 3, -2, 2)$. Find: (a) $3u$, (b) $u + v$, (c) $2u - 3v$, (d) $5u - 3v - 4w$, (e) $-u + 2v - 2w$.

4.22. Find x and y if:

(a) $(x, x + y) = (y - 2, 6)$, (b) $x(3, 2) = 2(y, -1)$

4.23. Find x, y and z if: $x(1, 1, 0) + y(2, 0, -1) + z(0, 1, 1) = (-1, 3, 3)$.

4.24. Compute the following where u, v and w are the vectors in Problem 4.21:

(a) $u \cdot v$, (b) $u \cdot w$, (c) $w \cdot v$, (d) $||u||$, (e) $||v||$, (f) $||w||$

MATRIX OPERATIONS

For Problems 4.25–27, let

$$A = \begin{pmatrix} 1 & -1 & 2 \\ 0 & 3 & 4 \end{pmatrix}, \quad B = \begin{pmatrix} 4 & 0 & -3 \\ -1 & -2 & 3 \end{pmatrix}, \quad C = \begin{pmatrix} 2 & -3 & 0 & 1 \\ 5 & -1 & -4 & 2 \\ -1 & 0 & 0 & 3 \end{pmatrix} \quad \text{and} \quad D = \begin{pmatrix} 2 \\ -1 \\ 3 \end{pmatrix}$$

4.25. Find: (a) $A + B$, (b) $A + C$, (c) $3A - 4B$

4.26. Find: (a) AB, (b) AC, (c) AD, (d) BC, (e) BD, (f) CD

4.27. Find: (a) A^T, (b) $A^T C$, (c) $D^T A^T$, (d) $B^T A$, (e) $D^T D$, (f) DD^T

SQUARE MATRICES

4.28. Let $A = \begin{pmatrix} 2 & 2 \\ 3 & -1 \end{pmatrix}$. (a) Find A^2 and A^3. (b) If $f(x) = x^3 - 3x^2 - 2x + 4$, find $f(A)$. (c) If $g(x) = x^2 - x - 8$, find $g(A)$.

4.29. Let $B = \begin{pmatrix} 1 & 3 \\ 5 & 3 \end{pmatrix}$. (a) Find $f(B)$ where $f(x) = 2x^2 - 4x + 3$. (b) Find $g(B)$ where $g(x) = x^2 - 4x - 12$. (c) Find a nonzero column vector $u = \begin{pmatrix} x \\ y \end{pmatrix}$ such that $Bu = 6u$.

4.30. Matrices A and B are said to commute if $AB = BA$. Find all matrices $\begin{pmatrix} x & y \\ z & w \end{pmatrix}$ which commute with $\begin{pmatrix} 1 & 1 \\ 0 & 1 \end{pmatrix}$.

4.31. Let $A = \begin{pmatrix} 1 & 2 \\ 0 & 1 \end{pmatrix}$.   Find A^n.

DETERMINANTS

4.32. Find the determinant of each matrix:

(a) $\begin{pmatrix} 2 & 5 \\ 4 & 1 \end{pmatrix}$, (b) $\begin{pmatrix} 6 & 1 \\ 3 & -2 \end{pmatrix}$, (c) $\begin{pmatrix} 4 & -5 \\ 0 & 2 \end{pmatrix}$, (d) $\begin{pmatrix} 1 & 0 \\ 0 & 1 \end{pmatrix}$, (e) $\begin{pmatrix} -2 & 8 \\ -5 & -2 \end{pmatrix}$, (f) $\begin{pmatrix} 4 & 9 \\ 5 & -3 \end{pmatrix}$

4.33. Find the determinant of each matrix:

(a) $\begin{pmatrix} 2 & 1 & 1 \\ 0 & 5 & -2 \\ 1 & -3 & 4 \end{pmatrix}$, (b) $\begin{pmatrix} 3 & -2 & -4 \\ 2 & 5 & -1 \\ 0 & 6 & 1 \end{pmatrix}$, (c) $\begin{pmatrix} -2 & -1 & 4 \\ 6 & -3 & -2 \\ 4 & 1 & 2 \end{pmatrix}$, (d) $\begin{pmatrix} 7 & 6 & 5 \\ 1 & 2 & 1 \\ 3 & -2 & 1 \end{pmatrix}$

4.34. Find the inverse of each matrix: (a) $\begin{pmatrix} 3 & 2 \\ 7 & 5 \end{pmatrix}$, (b) $\begin{pmatrix} 2 & -3 \\ 1 & 3 \end{pmatrix}$

COMPUTER PROGRAMMING PROBLEMS

4.35. The first card of a deck contains 5 real numbers, the elements of a vector u, and the second card contains 5 real numbers, the elements of a vector v. Write a program which prints: (a) $2u - 3v$, (b) $u \cdot v$, (c) $\|u\|$. Test the program with the following data:

$$u = (3.0, 1.2, -4.0, 7.3, -2.8), \quad v = (5.4, -2.0, -6.3, 0.0, -5.8)$$

4.36. Let A, B and C be 2×2 matrices. Suppose A, B and C are punched into a deck of cards, e.g. the deck can contain six cards and each card can contain a row of a matrix. Write a program which prints: (a) $3A - 2B$, (b) AB, (c) $|A|$, (d) A^{-1}. Test the program with the following data:

$$A = \begin{pmatrix} 4.0 & 3.0 \\ 9.0 & 8.0 \end{pmatrix}, \quad B = \begin{pmatrix} 3.1 & -8.2 \\ 5.0 & 2.5 \end{pmatrix}, \quad C = \begin{pmatrix} -4.0 & 2.0 \\ 0.0 & 7.0 \end{pmatrix}$$

4.37. Let A and B be any two $n \times n$ matrices (real). Write a subroutine called PRDCT which computes the matrix product $C = AB$.

4.38. Let A be any $m \times k$ matrix and let B be any $k \times n$ matrix. Write a subroutine called PRDCT2 which computes the matrix $C = AB$.

4.39. Let A be any $n \times 2$ matrix and let X, Y be a pair of numbers. Write a subprogram called BLNG such that $\mathrm{BLNG}(A, N, X, Y)$ has the value 1 or -1 according as (X, Y) is or is not a row of A.

Answers to Supplementary Problems

4.21.   (a) $3u = (6, -3, 0, -9)$ (c) $2u - 3v = (1, 1, 3, -15)$ (e) $-u + 2v - 2w = (-2, -7, 2, 5)$
      (b) $u + v = (3, -2, -1, 0)$ (d) $5u - 3v - 4w = (3, -14, 11, -32)$

4.22.   (a) $x = 2, \; y = 4,$ (b) $x = -1, \; y = -3/2$

4.23. $x = 1, \; y = -1, \; z = 2$

4.24.   (a) -6, (b) -7, (c) 6, (d) $\sqrt{14}$, (e) $\sqrt{12} = 2\sqrt{3}$, (f) $\sqrt{18} = 3\sqrt{2}$

4.25.   (a) $\begin{pmatrix} 5 & -1 & -1 \\ -1 & 1 & 7 \end{pmatrix}$,   (b) Not defined,   (c) $\begin{pmatrix} -13 & -3 & 18 \\ 4 & 17 & 0 \end{pmatrix}$

4.26.   (a) Not defined   (c) $\begin{pmatrix} 9 \\ 9 \end{pmatrix}$ (e) $\begin{pmatrix} -1 \\ 9 \end{pmatrix}$

      (b) $\begin{pmatrix} -5 & -2 & 4 & 5 \\ 11 & -3 & -12 & 18 \end{pmatrix}$ (d) $\begin{pmatrix} 11 & -12 & 0 & -5 \\ -15 & 5 & 8 & 4 \end{pmatrix}$ (f) Not defined

4.27.   (a) $\begin{pmatrix} 1 & 0 \\ -1 & 3 \\ 2 & 4 \end{pmatrix}$,   (b) Not defined,   (c) $(9, 9)$ (d) $\begin{pmatrix} 4 & -7 & 4 \\ 0 & -6 & -8 \\ -3 & 12 & 6 \end{pmatrix}$,   (e) 14,   (f) $\begin{pmatrix} 4 & -2 & 6 \\ -2 & 1 & -3 \\ 6 & -3 & 9 \end{pmatrix}$

4.28.   (a) $A^2 = \begin{pmatrix} 10 & 2 \\ 3 & 7 \end{pmatrix}$, $A^3 = \begin{pmatrix} 26 & 18 \\ 27 & -1 \end{pmatrix}$,   (b) $f(A) = \begin{pmatrix} -4 & 8 \\ 12 & -16 \end{pmatrix}$,   (c) $g(A) = \begin{pmatrix} 0 & 0 \\ 0 & 0 \end{pmatrix}$

4.29.   (a) $f(B) = \begin{pmatrix} 31 & 12 \\ 20 & 39 \end{pmatrix}$,   (b) $g(B) = \begin{pmatrix} 0 & 0 \\ 0 & 0 \end{pmatrix}$,   (c) $u = \begin{pmatrix} 3 \\ 5 \end{pmatrix}$ or $\begin{pmatrix} 3k \\ 5k \end{pmatrix}$, $k \neq 0$.

4.30.   Only matrices of the form $\begin{pmatrix} a & b \\ 0 & a \end{pmatrix}$ commute with $\begin{pmatrix} 1 & 1 \\ 0 & 1 \end{pmatrix}$.

4.31. $A^n = \begin{pmatrix} 1 & 2n \\ 0 & 1 \end{pmatrix}$

4.32.   (a) -18, (b) -15, (c) 8, (d) 1, (e) 44, (f) -57

4.33.   (a) 21, (b) -11, (c) 102, (d) 0

4.34.   (a) $\begin{pmatrix} 5 & -2 \\ -7 & 3 \end{pmatrix}$,   (b) $\begin{pmatrix} \frac{1}{3} & \frac{1}{3} \\ -\frac{1}{9} & \frac{2}{9} \end{pmatrix}$

Chapter 5

Graph Theory

5.1 INTRODUCTION

The term "graph" in mathematics has different meanings. Previously we spoke of the "graph" of a function and a relation. We now use the word "graph" in a different context. However, the idea of a directed graph (which is discussed in Chapter 7) has already appeared in Chapter 2 on relations.

5.2 GRAPHS AND MULTIGRAPHS

A *graph* G consists of two things:

(i) A set V whose elements are called *vertices, points* or *nodes*.

(ii) A set E of unordered pairs of distinct vertices called *edges*.

We denote such a graph by $G(V, E)$ when we want to emphasize the two parts of G.

Vertices u and v are said to be *adjacent* if there is an edge $\{u, v\}$.

We picture graphs by diagrams in the plane in a natural way. That is, each vertex v in V is represented by a dot (or small circle) and each edge $e = \{v_1, v_2\}$ is represented by a curve which connects its *endpoints* v_1 and v_2. For example, Fig. 5-1(a) represents the graph $G(V, E)$ where (i) V consists of four vertices A, B, C, D; and (ii) E consists of five edges $e_1 = \{A, B\}$, $e_2 = \{B, C\}$, $e_3 = \{C, D\}$, $e_4 = \{A, C\}$, $e_5 = \{B, D\}$. In fact, we will usually denote a graph by drawing its diagram rather than explicitly listing its vertices and edges.

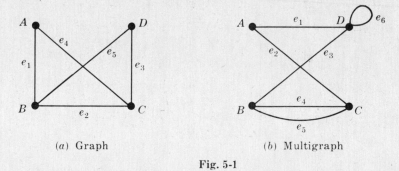

(a) Graph   (b) Multigraph

Fig. 5-1

The diagram of Fig. 5-1(b) is not a graph but a *multigraph*. The reason is that e_4 and e_5 are *multiple edges*, i.e. edges connecting the same endpoints, and e_6 is a *loop*, i.e. an edge whose endpoints are the same vertex. The definition of a graph does not permit such multiple edges or loops. In other words, we may define a graph to be a multigraph without multiple edges or loops.

Let $G(V, E)$ be a graph. Let V' be a subset of V and let E' be a subset of E whose endpoints belong to V'. Then $G(V', E')$ is a graph and is called a *subgraph* of $G(V, E)$. If E' contains all the edges of E whose endpoints lie in V', then $G(V', E')$ is called the subgraph *generated* by V'.

A multigraph is said to be *finite* if it has a finite number of vertices and a finite number of edges. Observe that a graph with a finite number of vertices must automatically have a finite number of edges and so must be finite. The finite graph with one vertex and no edges, i.e. a single point, is called the *trivial graph*. Unless otherwise specified, the multigraphs in this book shall be finite.

5.3 DEGREE

If v is an endpoint of an edge e, then we say that e is *incident* on v. The *degree* of a vertex v, written deg (v), is equal to the number of edges which are incident on v. Since each edge is counted twice in counting the degrees of the vertices of a graph, we have the following simple but important result.

Theorem 5.1: The sum of the degrees of the vertices of a graph is equal to twice the number of edges.

For example, in Fig. 5-1(a) we have

$$\deg(A) = 2, \quad \deg(B) = 3, \quad \deg(C) = 3, \quad \deg(D) = 2$$

The sum of the degrees equals ten which, as expected, is twice the number of edges. A vertex is said to be *even* or *odd* according as its degree is an even or an odd number. Thus A and D are even vertices whereas B and C are odd vertices.

Theorem 5.1 also holds for multigraphs where a loop is counted twice towards the degree of its endpoint. For example, in Fig. 5-1(b) we have deg $(D) = 4$ since the edge e_6 is counted twice; hence D is an even vertex.

A vertex of degree zero is called an *isolated* vertex.

5.4 CONNECTIVITY

A *walk* in a multigraph consists of an alternating sequence of vertices and edges of the form

$$v_0, \ e_1, \ v_1, \ e_2, \ v_2, \ \ldots, \ e_{n-1}, \ v_{n-1}, \ e_n, \ v_n$$

where each edge e_i is incident on v_{i-1} and v_i. The number n of edges is called the *length* of the walk. When there is no ambiguity we denote a walk by its sequence of edges $(e_1, e_2, \ldots, e_n)$ or by its sequence of vertices $(v_0, v_1, \ldots, v_n)$. The walk is said to be *closed* if $v_0 = v_n$. Otherwise, we say that the walk is from v_0 to v_n, or *between* v_0 and v_n, or *connects* v_0 to v_n.

A *trail* is a walk in which all edges are distinct. A *path* is a walk in which all vertices are distinct; hence a path must be a trail. A *cycle* is a closed walk such that all vertices are distinct except $v_0 = v_n$. A cycle of length k is called a *k-cycle*. In a graph, any cycle must have length three or more.

EXAMPLE 5.1. Consider the graph in Fig. 5-2, Then

$$(P_4, P_1, P_2, P_5, P_1, P_2, P_3, P_6)$$

is a walk from P_4 to P_6. It is not a trail since the edge $\{P_1, P_2\}$ is used twice. The sequence

$$(P_4, P_1, P_5, P_2, P_6)$$

is not a walk since there is no edge $\{P_2, P_6\}$. The sequence

$$(P_4, P_1, P_5, P_2, P_3, P_5, P_6)$$

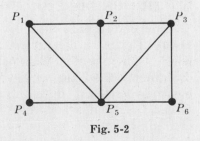

Fig. 5-2

is a trail since no edge is used twice; but it is not a path since the vertex P_5 is used twice. The sequence

$$(P_4, P_1, P_5, P_3, P_6)$$

is a path from P_4 to P_6. The shortest path (with respect to length) from P_4 to P_6 is (P_4, P_5, P_6) which has length 2.

By eliminating unnecessary edges, it is not difficult to see that any walk from a vertex u to a vertex v can be replaced by a path from u to v. We state this result formally.

Theorem 5.2:   There is a walk from a vertex u to a vertex v if and only if there is a path from u to v.

A graph is said to be *connected* if there is a path between any two of its vertices. The graph in Fig. 5-2 is connected, but the graph in Fig. 5-3(a) is not connected since, for example, there is no path between D and E.

Suppose $G(V, E)$ is a graph. A connected subgraph of $G(V, E)$ is called a *connected component* if it is not contained in any larger connected subgraph. It is intuitively clear that any graph can be partitioned into its connected components. For example, the graph in Figure 5-3(a) has three connected components.

The *distance* between vertices u and v of a connected graph G, written $d(u, v)$, is the length of the shortest path between u and v. The *diameter* of a connected graph G is the maximum distance between any two of its vertices. In Fig. 5-3(b), $d(A, F) = 2$ and the diameter of the graph is 3. (Although the edges $\{A, D\}$ and $\{B, C\}$ are pictured crossing in Fig. 5-3(b), they do not meet at a vertex.)

Let v be a vertex of a graph G. By $G - v$ we mean the graph obtained from G by deleting v and all edges incident on v. A vertex v in a connected graph G is called a *cut point* if $G - v$ is disconnected. The vertex D in Fig. 5-3(b) is a cut point.

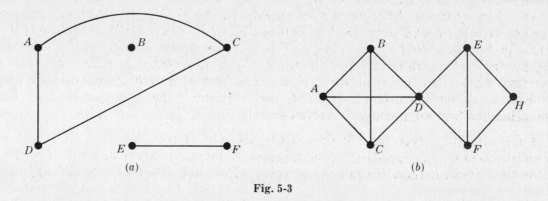

Fig. 5-3

5.5 THE BRIDGES OF KÖNIGSBERG, TRAVERSABLE MULTIGRAPHS

The eighteenth-century East Prussian town of Königsberg included two islands and seven bridges as shown in Fig. 5-4(a). Question: Beginning anywhere and ending anywhere, can a person walk through town crossing all seven bridges but not crossing any bridge twice? The people of Königsberg wrote to the celebrated Swiss mathematician L. Euler about this question. Euler proved in 1736 that such a walk is impossible. He replaced the islands and the two sides of the river by points and the bridges by curves, obtaining Fig. 5-4(b).

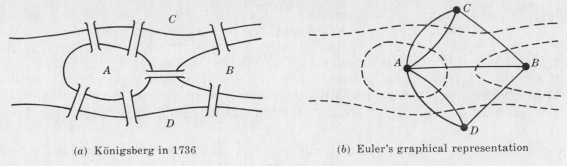

(a) Königsberg in 1736   (b) Euler's graphical representation

Fig. 5-4

Observe that Fig. 5-4(b) is a multigraph. A multigraph is said to be *traversable* if it "can be drawn without any breaks in the curve and without repeating any edge", that is, if there is a walk which includes all vertices and uses each edge exactly once. Such a walk must be a trail (since no edge is used twice) and will be called a *traversable trail*. Clearly a traversable multigraph must be finite and connected. Fig. 5-5(b) shows a traversable trail of the multigraph Fig. 5-5(a). To indicate the direction of the trail, the diagram misses touching vertices which are actually traversed. Now it is not difficult to see that the walk in Königsberg is possible if and only if the multigraph in Fig. 5-4(b) is traversable.

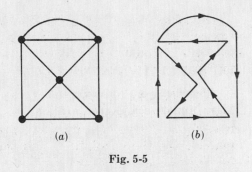

(a)   (b)

Fig. 5-5

We now show how Euler proved that the multigraph in Fig. 5-4(b) is not traversable and hence the walk in Königsberg is impossible. Recall first that a vertex is even or odd according as its degree is an even or an odd number. Suppose a multigraph is traversable and that a traversable trail does not begin or end at a vertex P. We claim that P is an even vertex. For whenever the traversable trail enters P by an edge, there must always be an edge not previously used by which the trail can leave P. Thus the edges in the trail incident with P must appear in pairs, and so P is an even vertex. Therefore if a vertex Q is odd, the traversable trail must begin or end at Q. Consequently, a multigraph with more than two odd vertices cannot be traversable. Observe that the multigraph corresponding to the Königsberg bridge problem has four odd vertices. Thus one cannot walk through Königsberg so that each bridge is crossed exactly once.

Euler actually proved the converse of the above statement, which is contained in the following theorem and corollary. (The theorem is proved in Problem 5.8.) A graph G is an *eulerian* graph if there exists a closed traversable trail, called an *eulerian* trail.

Theorem (Euler) 5.3:   A finite connected graph is eulerian if and only if each vertex has even degree.

Corollary 5.4:   Any finite connected graph with two odd vertices is traversable. A traversable trail may begin at either odd vertex and will end at the other odd vertex.

Another problem, closely related to that above, was first posed by the mathematician W. R. Hamilton. He asked whether or not a graph has a closed walk which includes each vertex exactly once. Such a walk must be a cycle and is called a *hamiltonian cycle*. A graph with a hamiltonian cycle is called a *hamiltonian graph*. Figure 5-6 gives an example of a graph which is hamiltonian but not eulerian and vice versa. We emphasize that there is no simple criterion to tell whether or not a graph is hamiltonian as there is for eulerian graphs. We note that this problem is closely related to the "traveling salesman problem", i.e. to find a minimum closed walk which includes all vertices where minimum refers to edges which are assigned lengths. (In other words, we view the vertices as cities and the lengths of the edges as distances of roads between the cities.)

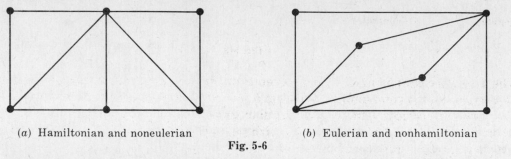

(*a*) Hamiltonian and noneulerian   (*b*) Eulerian and nonhamiltonian

Fig. 5-6

5.6 SPECIAL GRAPHS

There are many different types of graphs. We define four of them here: complete, regular, bipartite and tree graphs.

A graph is *complete* if each vertex is connected to every other vertex. The complete graph with n vertices is denoted by K_n. Figure 5-7 shows the graphs $K_1, K_2, \ldots, K_6$.

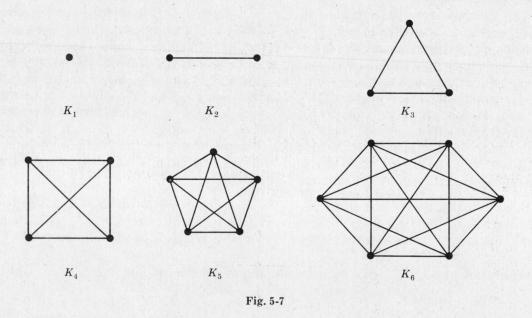

Fig. 5-7

A graph G is *regular of degree k* or *k-regular* if every vertex has degree k. In other words, a graph is regular if every vertex has the same degree.

The connected regular graphs of degrees 0, 1 or 2 are easily described. The connected 0-regular graph is the trivial graph with one vertex and no edges. The connected 1-regular graph is the graph with two vertices and one edge connecting them. The connected 2-regular graph with n vertices is the graph which consists of a single n-cycle. See Fig 5-8.

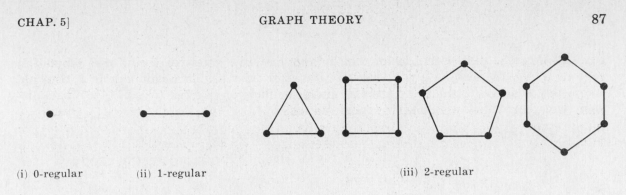

(i) 0-regular   (ii) 1-regular   (iii) 2-regular

Fig. 5-8

The 3-regular graphs must have an even number of vertices since the sum of the degrees of the vertices is an even number (Theorem 5.1).   Figure 5-9 shows two connected 3-regular graphs with six vertices.   In general, regular graphs can be quite complicated.   For example, there are nineteen 3-regular graphs with ten vertices.   We note that the complete graph with n vertices K_n is regular of degree $n - 1$.

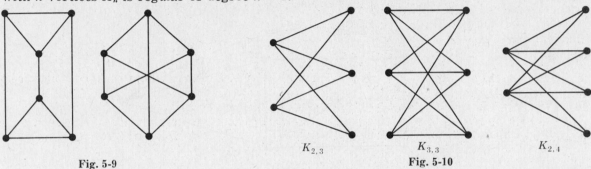

$K_{2,3}$ $K_{3,3}$ $K_{2,4}$

Fig. 5-9   Fig. 5-10

A graph G is said to be *bipartite* if its vertices V can be partitioned into two subsets M and N such that each edge of G connects a vertex of M to a vertex of N.   By a complete bipartite graph, we mean that each vertex of M is connected to each vertex of N; this graph is denoted by $K_{m,n}$ where m is the number of vertices in M and n is the number of vertices in N, and, for standardization, we assume $m \leq n$.   Figure 5-10 shows the graphs $K_{2,3}$, $K_{3,3}$ and $K_{2,4}$.   Clearly the graph $K_{m,n}$ has mn edges.

A graph is said to be *cycle-free* or *acyclic* if it has no cycle.   A connected graph with no cycles is called a *tree*.   Since trees occur in many places in mathematics, we study trees more thoroughly in the next chapter.   Here we simply give its definition and some examples. Figure 5-11 shows the six trees with six vertices.

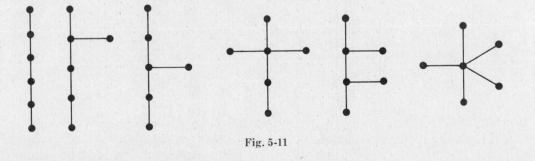

Fig. 5-11

5.7 MATRICES AND GRAPHS

Let G be a graph with vertices $v_1, v_2, \ldots, v_m$ and edges $e_1, e_2, \ldots, e_n$.   It is sometimes practical, especially for computational reasons, to represent G by a matrix.   Note that the edges of G can be represented by an $n \times 2$ integer matrix B where each row of B denotes an edge of G, e.g. the row $(3, 4)$ would denote the edge $\{v_3, v_4\}$. This *edge matrix* B does not completely describe G unless we are also given the number m of vertices of G. We do discuss two other widely used matrix representations of G.

(1) **Adjacency matrix.** Let $A = (a_{ij})$ be the $m \times m$ matrix defined by

$$a_{ij} = \begin{cases} 1 & \text{if } \{v_i, v_j\} \text{ is an edge, i.e. if } v_i \text{ is adjacent to } v_j \\ 0 & \text{otherwise} \end{cases}$$

Then A is called the *adjacency matrix* of G. Observe that $a_{ij} = a_{ji}$; hence A is a symmetric matrix. (We define an adjacency matrix for a multigraph by letting a_{ij} denote the number of edges $\{v_i, v_j\}$.)

(2) **Incidence matrix.** Let $M = (m_{ij})$ be the $m \times n$ matrix defined by

$$m_{ij} = \begin{cases} 1 & \text{if the vertex } v_i \text{ is incident on the edge } e_j \\ 0 & \text{otherwise} \end{cases}$$

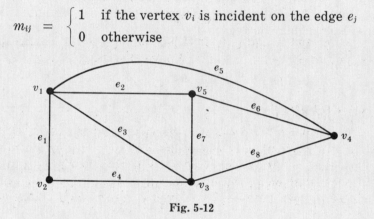

Fig. 5-12

Consider, for example, the graph in Fig. 5-12. Its edge matrix B, adjacency matrix A, and incidence matrix M follow. For easy reading, we have labeled the rows and columns of A and M by the corresponding vertices and edges.

$$B = \begin{pmatrix} 1 & 2 \\ 1 & 5 \\ 1 & 3 \\ 2 & 3 \\ 1 & 4 \\ 4 & 5 \\ 3 & 5 \\ 3 & 4 \end{pmatrix} \qquad A = \begin{matrix} & \begin{matrix} v_1 & v_2 & v_3 & v_4 & v_5 \end{matrix} \\ \begin{matrix} v_1 \\ v_2 \\ v_3 \\ v_4 \\ v_5 \end{matrix} & \begin{pmatrix} 0 & 1 & 1 & 1 & 1 \\ 1 & 0 & 1 & 0 & 0 \\ 1 & 1 & 0 & 1 & 1 \\ 1 & 0 & 1 & 0 & 1 \\ 1 & 0 & 1 & 1 & 0 \end{pmatrix} \end{matrix}$$

$$M = \begin{matrix} & \begin{matrix} e_1 & e_2 & e_3 & e_4 & e_5 & e_6 & e_7 & e_8 \end{matrix} \\ \begin{matrix} v_1 \\ v_2 \\ v_3 \\ v_4 \\ v_5 \end{matrix} & \begin{pmatrix} 1 & 1 & 1 & 0 & 1 & 0 & 0 & 0 \\ 1 & 0 & 0 & 1 & 0 & 0 & 0 & 0 \\ 0 & 0 & 1 & 1 & 0 & 0 & 1 & 1 \\ 0 & 0 & 0 & 0 & 1 & 1 & 0 & 1 \\ 0 & 1 & 0 & 0 & 0 & 1 & 1 & 0 \end{pmatrix} \end{matrix}$$

Although the edge matrix B of a graph G is the most compact representation, it is not always the most useful. In view of the following theorem, the adjacency matrix is very useful in deciding questions of connectivity.

Theorem 5.5:   Let A be the adjacency matrix of a graph G with m vertices where $m > 1$. Then the ij entry of the matrix A^n gives the number of walks of length n from the vertex v_i to the vertex v_j.

(An analogous theorem for directed graphs (Theorem 7.3) is stated and proved in Chapter 7.)

Since G has m vertices, any path from v_i to v_j must have length $m-1$ or less. Hence the matrix

$$A + A^2 + \cdots + A^{m-1}$$

can have a zero ij entry only if there is no path from v_i to v_j.

By the *connection matrix* of a graph G with m vertices, we mean the $m \times m$ matrix $C = (c_{ij})$, where

$$c_{ij} = \begin{cases} 1 & \text{if } i = j \text{ or there is a path from } v_i \text{ to } v_j \\ 0 & \text{otherwise} \end{cases}$$

Note that G is connected if and only if C has no zero entry. The above discussion shows that C and the matrix $A + A^2 + \cdots + A^{m-1}$ have the same zero entries off the main diagonal.

5.8 LABELED GRAPHS

A graph G is called a *labeled graph* if its edges and/or vertices are assigned data of one kind or another. In particular, if each edge e of G is assigned a nonnegative number $\ell(e)$ then $\ell(e)$ is called the *weight* or *length* of e. Figure 5-13 shows a labeled graph where the length of each edge is given in the obvious way. One important problem in graph theory is to find a minimum path between two given points. A minimum path between P and Q in Fig. 5-13 is

$$(P, A_1, A_2, A_5, A_3, A_6, Q)$$

which has length 14. The reader can try to find another minimum path.

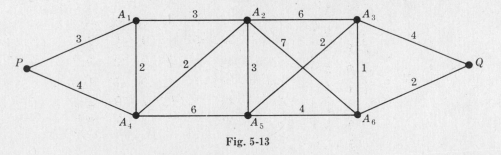

Fig. 5-13

In Chapter 7 we give a "pruning" algorithm which yields a minimum path for the simpler case of a directed graph (which minimizes the possible paths between points). A generalized version of such an algorithm can be used for the nondirected case discussed here.

5.9 ISOMORPHIC GRAPHS

Suppose $G(V, E)$ and $G^*(V^*, E^*)$ are graphs and $f : V \to V^*$ is a one-to-one correspondence between the sets of vertices such that $\{u, v\}$ is an edge of G if and only if $\{f(u), f(v)\}$ is an edge of G^*. Then f is called an *isomorphism* between G and G^*, and G and G^* are said to be *isomorphic* graphs. Normally, we do not distinguish between isomorphic graphs (even though their diagrams may "look different").   Thus we can say that Fig. 5-11 gives all the possible trees with six vertices.

If G and G^* are isomorphic graphs, then corresponding vertices must have the same graphical properties, such as degree, being a cut point and so on.   Figure 5-14 gives ten graphs pictured as letters.   We note that A and R, F and T, K and X, and M, S, V and Z are isomorphic graphs.

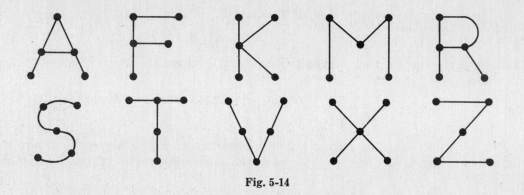

Fig. 5-14

Given any graph G, we can obtain a new graph by dividing an edge of G with additional vertices.   Two graphs G and G^* are said to be *homeomorphic* if they can be obtained from isomorphic graphs by this method.   The graphs (a) and (b) in Fig. 5-15 are not isomorphic; but they are homeomorphic since each can be obtained from (c) by adding appropriate vertices.

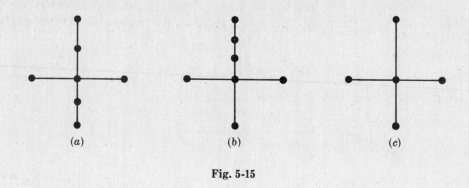

(a) (b) (c)

Fig. 5-15

Solved Problems

GRAPHS, CONNECTIVITY

5.1.   Draw the diagram of each graph $G(V, E)$:

(a) $V = \{A, B, C, D\}$, $E = [\{A, B\}, \{A, C\}, \{B, C\}, \{B, D\}, \{C, D\}]$

(b) $V = \{a, b, c, d, e\}$, $E = [\{a, b\}, \{a, c\}, \{b, c\}, \{d, e\}]$

Which of the graphs, if any, are connected?

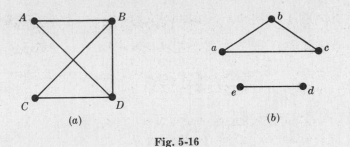

Fig. 5-16

Draw a dot for each vertex v in V, and for each edge $\{x, y\}$ in E draw a curve from the vertex x to the vertex y, as shown in Fig. 5-16. Graph (a) is connected. However, graph (b) is not connected since, for example, there is no path from the vertex a to the vertex d.

5.2.   Consider Fig. 5-17. (a) Describe formally the graph G in the diagram. (b) Find the degree of each vertex and verify Theorem 5.1 for this graph.

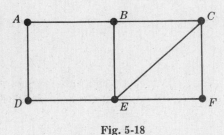

(a)   There are five vertices, so $V = \{a, b, c, d, e\}$. There are seven pairs $\{x, y\}$ of vertices where the vertex x is connected with the vertex y; hence

$$E \;=\; [\{a, b\}, \{a, c\}, \{a, d\}, \{b, c\}, \{b, e\}, \{c, d\}, \{c, e\}]$$

Fig. 5-17

(b)   The degree of a vertex is equal to the number of edges to which it belongs; e.g. $\deg(a) = 3$ since a belongs to the three edges $\{a, b\}$, $\{a, c\}$, $\{a, d\}$. Similarly, $\deg(b) = 3$, $\deg(c) = 4$, $\deg(d) = 2$, $\deg(e) = 2$.

The sum of the degrees of the vertices is

$$3 + 3 + 4 + 2 + 2 \;=\; 14$$

which does equal twice the number of edges.

5.3.   Consider the graph in Fig. 5-18. Find (a) all paths from the vertex A to the vertex F, (b) all trails from A to F, (c) the distance between A and F, (d) the diameter of the graph.

(a)   A path from A to F is a walk such that no vertex and hence no edge is repeated. There are seven such paths:

(A, B, C, F) (A, D, E, F)

(A, B, C, E, F) (A, D, E, B, C, F)

(A, B, E, F) (A, D, E, C, F)

(A, B, E, C, F)

(b)   A trail from A to F is a walk such that no edge is repeated. There are nine such trails, the seven paths from (a) together with

(A, D, E, B, C, E, F) and (A, D, E, C, B, E, F)

(c)   The distance from A to F is 3 since there is a path, e.g. (A, B, C, F), from A to F of length 3 and no shorter path from A to F.

(d)   The distance between any two vertices is not greater than 3, and the distance between A and F is 3; hence the diameter of the graph is 3.

5.4.   Consider the graph in Fig. 5-18. Find the subgraphs obtained when each vertex is deleted. Does the graph have any cut points?

When we delete a vertex from a graph, we also have to delete all edges incident on the vertex. The six graphs obtained by deleting each of the vertices of Fig. 5-18 are shown in Fig. 5-19. All six graphs are connected; hence no vertex is a cut point.

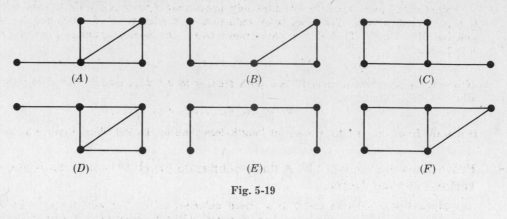

Fig. 5-19

5.5. Which of the multigraphs in Fig. 5-20 are (a) connected, (b) loop-free (i.e. have no loops), (c) graphs?

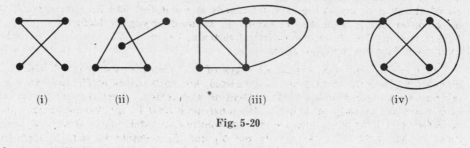

Fig. 5-20

(a) Only (i) and (iii) are connected.

(b) Only (iv) has a loop, i.e. an edge with the same endpoints.

(c) Only (i) and (ii) are graphs. The multigraph (iii) has multiple edges and (iv) has multiple edges and a loop.

5.6. Which of the multigraphs in Fig. 5-21 are traversable?

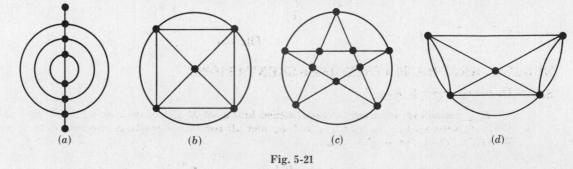

Fig. 5-21

(a) Traversable since six vertices are even and two are odd.

(b) Not traversable since there are four odd vertices.

(c) Traversable since all ten vertices are even.

(d) Traversable since there are two odd vertices. The traversable path must begin at one of the odd vertices. See Fig. 5-22 for example.

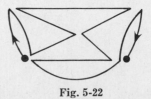

Fig. 5-22

5.7. Prove Theorem 5.2: There is a walk from a vertex u to a vertex v if and only if there is a path from u to v.

Since every path is a walk, we need only prove that if there is a walk W from u to v then there is a path from u to v. The proof is by induction on the length of W. Suppose the length of W is one, i.e. $W = (u, v)$. Then W is a path from u to v. On the other hand, suppose the length of W is $n > 1$, say

$$W = (u = v_0, v_1, v_2, \ldots, v_{n-1}, v = v_n)$$

If no vertex is repeated, then W is a path from u to v. Suppose a vertex is repeated, say $v_i = v_j$ where $i < j$. Then

$$W' = (v_0, v_1, \ldots, v_i, v_{j+1}, \ldots, v_n)$$

is a walk from $u = v_0$ to $v = v_n$ of length less than n. By induction, there is a path from u to v.

5.8. Prove Theorem (Euler) 5.3: A finite connected graph G is eulerian if and only if each vertex has even degree.

Suppose G is eulerian and T is a closed eulerian trail. For any vertex v of G, the trail T enters and leaves v the same number of times without repeating any edge. Hence v has even degree.

Suppose conversely that each vertex of G has even degree. We construct an eulerian trail. We begin a trail T_1 at any edge e. We extend T_1 by adding one edge after the other. If T_1 is not closed at any step, say T_1 begins at u but ends at $v \neq u$, then only an odd number of the edges incident on v appear in T_1; hence we can extend T_1 by another edge incident on v. Thus we can continue to extend T_1 until T_1 returns to its initial vertex u, i.e. until T_1 is closed. If T_1 includes all the edges of G, then T_1 is our eulerian trail.

Suppose T_1 does not include all edges of G. Consider the graph H obtained by deleting all edges of T_1 from G. H may not be connected, but each vertex of H has even degree since T_1 contains an even number of the edges incident on any vertex. Since G is connected, there is an edge e' of H which has an endpoint u' in T_1. We construct a trail T_2 in H beginning at u' and using e'. Since all vertices in H have even degree, we can continue to extend T_2 in H until T_2 returns to u' as pictured in Fig. 5-23. We can clearly put T_1 and T_2 together to form a larger closed trail in G. We continue this process until all the edges of G are used. We finally obtain an eulerian trail, and so G is eulerian.

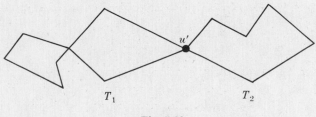

Fig. 5-23

SPECIAL GRAPHS, MATRIX REPRESENTATIONS

5.9. Draw the graph $K_{2,5}$.

$K_{2,5}$ consists of seven vertices partitioned into a set M of two vertices, say u_1 and u_2, and a set N of five vertices, say v_1, v_2, v_3, v_4 and v_5, and all possible edges from a vertex u_i to a vertex v_j. We exhibit the graph in Fig. 5-24.

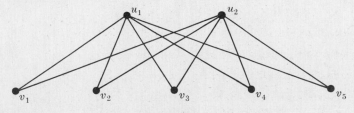

Fig. 5-24

5.10. Which connected graphs can be both regular and bipartite?

The bipartite graph $K_{m,m}$ is regular of degree m since each vertex is connected to m other vertices and hence its degree is m. Subgraphs of $K_{m,m}$ can also be regular if m disjoint edges are deleted. For example, the subgraph of $K_{4,4}$ shown in Fig. 5-25 is 3-regular. We can continue to delete m disjoint edges and each time obtain a regular graph of one less degree. These graphs may be disconnected, but in any case their connected components have the desired properties.

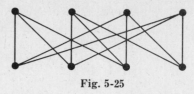

Fig. 5-25

5.11. Draw all trees with five or fewer vertices.

There are eight such trees which are exhibited in Fig. 5-26. The graph with one vertex and no edge is called the *trivial tree*.

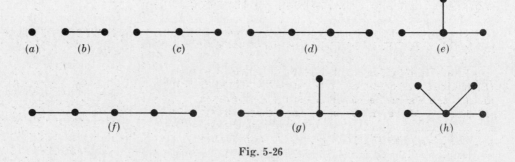

(a) (b) (c) (d) (e)

(f) (g) (h)

Fig. 5-26

5.12. Find the adjacency matrix $A = (a_{ij})$ and the incidence matrix $M = (m_{ij})$ of the graph in Fig. 5-27.

The adjacency matrix $A = (a_{ij})$ is defined by $a_{ij} = 1$ if there is an edge $\{v_i, v_j\}$ and $a_{ij} = 0$ otherwise.   Hence

$$A = \begin{pmatrix} 0 & 1 & 0 & 1 \\ 1 & 0 & 1 & 1 \\ 0 & 1 & 0 & 1 \\ 1 & 1 & 1 & 0 \end{pmatrix}$$

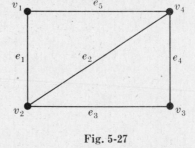

Fig. 5-27

The incidence matrix $M = (m_{ij})$ .is defined by $m_{ij} = 1$ if the vertex v_i is incident on the edge e_j and $m_{ij} = 0$ otherwise.   Thus

$$M = \begin{pmatrix} 1 & 0 & 0 & 0 & 1 \\ 1 & 1 & 1 & 0 & 0 \\ 0 & 0 & 1 & 1 & 0 \\ 0 & 1 & 0 & 1 & 1 \end{pmatrix}$$

5.13. Draw the graph G whose adjacency matrix $A = (a_{ij})$ follows:

$$A = \begin{pmatrix} 0 & 1 & 0 & 1 & 0 \\ 1 & 0 & 0 & 1 & 1 \\ 0 & 0 & 0 & 1 & 1 \\ 1 & 1 & 1 & 0 & 1 \\ 0 & 1 & 1 & 1 & 0 \end{pmatrix}$$

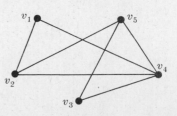

Since A is a 5-square matrix, G has five vertices, say $v_1, \ldots, v_5$. Draw an edge from v_i to v_j if $a_{ij} = 1$. The graph is shown in Fig. 5-28.

Fig. 5-28

5.14. Draw the multigraph G whose adjacency matrix $A = (a_{ij})$ follows:

$$A = \begin{pmatrix} 1 & 3 & 0 & 0 \\ 3 & 0 & 1 & 1 \\ 0 & 1 & 2 & 2 \\ 0 & 1 & 2 & 0 \end{pmatrix}$$

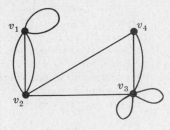

Fig. 5-29

Since A is a 4-square matrix, G has four vertices, say $v_1, \ldots, v_4$. Draw n edges from v_i to v_j if $a_{ij} = n$. Note that v_i has n loops if $a_{ii} = n$. The multigraph is shown in Fig. 5-29.

5.15. Show that the six graphs obtained in Problem 5.4 are distinct, i.e. no two are isomorphic. Also show that (B) and (C) are homeomorphic.

The degrees of the five vertices of any graph cannot be paired off with the degrees of any other graph, except for (B) and (C). Hence none of the graphs are isomorphic except possibly (B) and (C).

However, if we delete the vertex of degree 3 in (B) and in (C), we obtain distinct subgraphs. Thus (B) and (C) are also nonisomorphic; hence all six graphs are distinct. However, (B) and (C) are homeomorphic since they can be obtained, respectively, from the isomorphic graphs in Fig. 5-30 by adding appropriate vertices.

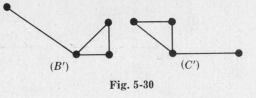

Fig. 5-30

Supplementary Problems

5.16. Let $V = \{u, v, w, x, y\}$. Draw the diagram of each graph $G(V, E)$ where:

(a) $E = [\{u, v\}, \{u, x\}, \{v, w\}, \{v, x\}, \{v, y\}, \{x, y\}]$

(b) $E = [\{u, v\}, \{v, w\}, \{w, x\}, \{w, y\}, \{x, y\}]$

Find the degree of each vertex and the diameter of each graph.

5.17. Consider the graph in Fig. 5-31. Find (a) all paths from the vertex A to the vertex H, (b) the diameter of the graph, and (c) the degree of each vertex. (d) Which vertices, if any, are cut points? (e) An edge e in a connected graph G is called a *bridge* if $G - e$, i.e. the subgraph obtained from G by deleting the edge e, is disconnected. Which edges, if any, are bridges?

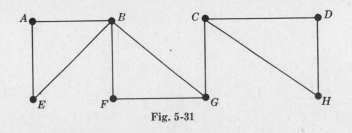

Fig. 5-31

5.18. Which of the multigraphs in Fig. 5-32 are (a) connected, (b) loop-free (i.e. have no loops), (c) graphs?

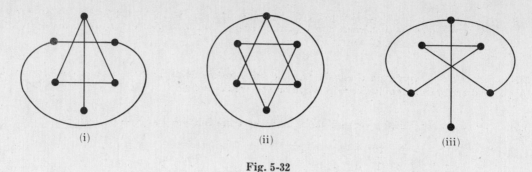

Fig. 5-32

5.19. Which of the multigraphs in Fig. 5-33 are traversable?   Find a traversable trail if the multigraph is traversable?

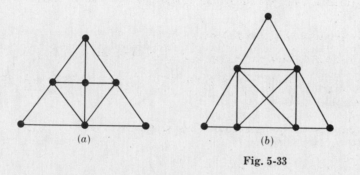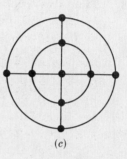

Fig. 5-33

5.20. Draw the following graphs: (a) K_7,   (b) $K_{2,6}$,   (c) $K_{3,4}$.

5.21. Draw all trees with seven vertices.

5.22. Determine the diameter of any complete bipartite graph.

5.23. Show that any tree is a bipartite graph.

5.24. Draw two 3-regular graphs with eight vertices.

5.25. Find the adjacency matrix A and the incidence matrix M for the graph in Fig. 5-34.

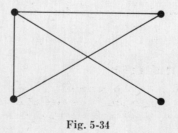

Fig. 5-34

5.26. Suppose a graph G is bipartite. Show that one can order the vertices of G so that its adjacency matrix A is of the form

$$A \;=\; \begin{pmatrix} 0 & B \\ C & 0 \end{pmatrix}$$

5.27. Find all connected graphs with four vertices.

5.28. Consider the following two steps on a graph G: (1) Delete an edge. (2) Delete a vertex and all edges incident on that vertex. Show that every subgraph of a finite graph G can be obtained by a sequence consisting of these two steps.

5.29. Prove that any graph G can be partitioned into maximal disjoint connected subgraphs by choosing the appropriate equivalence relation on the vertices of G.

5.30. Draw the multigraph corresponding to each of the following adjacency matrices:

(a) $\quad A = \begin{pmatrix} 0 & 2 & 0 & 1 \\ 2 & 1 & 1 & 1 \\ 0 & 1 & 0 & 1 \\ 1 & 1 & 1 & 0 \end{pmatrix}$ (b) $\quad A = \begin{pmatrix} 1 & 1 & 1 & 2 \\ 1 & 0 & 0 & 0 \\ 1 & 0 & 0 & 2 \\ 2 & 0 & 2 & 2 \end{pmatrix}$

5.31. Prove that a finite tree (with at least one edge) has at least two vertices of degree 1.

5.32. Suppose G is a connected graph. Prove the following:

(a) If G contains a cycle C which contains an edge e, then $G - e$ is still connected.

(b) If $e = \{u, v\}$ is an edge such that $G - e$ is disconnected, then u and v belong to different components of $G - e$.

5.33. Prove that a connected graph with n vertices must have at least $n - 1$ edges.

5.34. Which two trees in Fig. 5-11 are homeomorphic?

5.35. Suppose G and G^* are homeomorphic graphs. Show that G is traversable if and only if G^* is traversable.

5.36. Suppose G and G^* are nonisomorphic graphs with the same edge matrix B. Show that one of the graphs can be obtained from the other by simply adding isolated vertices.

5.37. Suppose G is a graph with no edge $e = \{v_r, v_s\}$. Suppose we add the edge e to G to obtain the graph $H = G + e$. Let $C = (c_{ij})$ and $D = (d_{ij})$ be the connection matrices of G and H respectively.

(a) Suppose $c_{rs} = 1$, i.e. G contains a path from v_r to v_s. Prove that $D = C$ and H contains a cycle including e.

(b) Suppose $c_{rs} = 0$, i.e. v_r and v_s are not connected in G. Prove:

    (i) $d_{ij} = 1$ if and only if $c_{ij} = 1$ or $c_{ir} = c_{sj} = 1$.

    (ii) If G is cycle-free, then H is cycle-free.

COMPUTER PROGRAMMING PROBLEMS

In Problems 5.38 through 5.40 let G be a graph with six vertices, $v_1, v_2, \ldots, v_6$, and seven edges. Let B be the 7×2 edge matrix of G. (Recall that each row of B denotes an edge of G, e.g. the row $(3, 4)$ would denote the edge $\{v_3, v_4\}$.) Suppose B is punched into a deck of cards.

5.38. Write a program which prints the degree of each vertex of G.

5.39. Write a program which prints the 6×6 adjacency matrix A of G.

5.40. Write a program which decides whether or not G is connected.
(Hint: Use Problem 5.39 and Theorem 5.5.)

Test the above three programs with the following data:

$$
\text{(i)} \quad B = \begin{pmatrix} 1 & 2 \\ 1 & 4 \\ 1 & 6 \\ 5 & 3 \\ 2 & 4 \\ 2 & 6 \\ 4 & 6 \end{pmatrix} \qquad \text{(ii)} \quad B = \begin{pmatrix} 1 & 2 \\ 6 & 3 \\ 4 & 2 \\ 2 & 5 \\ 4 & 1 \\ 3 & 2 \\ 1 & 5 \end{pmatrix} \qquad \text{(iii)} \quad B = \begin{pmatrix} 5 & 2 \\ 1 & 3 \\ 2 & 6 \\ 6 & 5 \\ 1 & 6 \\ 2 & 1 \\ 3 & 5 \end{pmatrix}
$$

In Problems 5.41 and 5.42 suppose A is the $m \times m$ adjacency matrix of a graph H, and suppose A and m are input parameters.

5.41.   Write a subprogram called DEG such that DEG computes the degrees of the vertices of H.

5.42.   Write a subprogram called CON such that CON decides whether or not H is connected.

5.43.   Suppose C is the $m \times m$ connection matrix of a graph G and suppose the edge $\{v_r, v_s\}$ is added to G. Write a subprogram CONMAT such that CONMAT computes the new $m \times m$ connection matrix. (*Hint:* Use Problem 5.37.)

In Problems 5.44 through 5.47 suppose B is the $n \times 2$ edge matrix of a graph G with m vertices, and suppose B, n and m are the input parameters.

5.44.   Write a subprogram called DEGREE such that DEGREE computes the degree of each of the m vertices.

5.45.   Write a subprogram called ADJ such that ADJ computes the $m \times m$ adjacency matrix.

546.   Write a subprogram called CNMTX such that CNMTX computes the connection matrix of G. (*Hint:* Add one edge after the other using Problem 5.43 at each step.)

5.47.   Write a subprogram called CONCT such that CONCT decides whether or not G is connected. (*Hint:* Use Problem 5.46, or use Problems 5.45 and 5.42.)

Answers to Supplementary Problems

5.16.   (*a*) diam $(G) = 2$,   (*b*) diam $(G) = 3$

5.17.   (*a*) There are eight paths:

   (A, B, G, C, H) (A, B, F, G, C, H) (A, B, G, C, D, H) (A, B, F, G, C, D, H)

   (A, E, B, G, C, H) (A, E, B, F, G, C, H) (A, E, B, G, C, D, H) (A, E, B, F, G, C, D, H)

   (*b*) 4   (*d*) B, C and G

   (*c*) deg $(B) = 4$, deg $(C) = \deg (G) = 3$, others have degree 2   (*e*) $\{C, G\}$

5.18. (a) (iii), (b) (i) and (iii), (c) (iii)

5.19. (a) and (b). A traversable trail must begin at an odd vertex since (a) and (b) have two odd vertices.

5.21. There are ten such trees.

5.22. Diam $(K_{1,1}) = 1$; all others have diameter 2.

5.24. The two 3-regular graphs shown in Fig. 5-35 are not isomorphic since (b) has a 5-cycle but (a) does not.

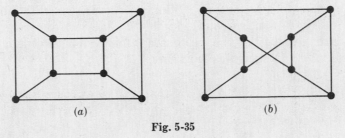

Fig. 5-35

5.25. $A = \begin{pmatrix} 0 & 1 & 1 & 1 \\ 1 & 0 & 0 & 1 \\ 1 & 0 & 0 & 0 \\ 1 & 1 & 0 & 0 \end{pmatrix}$, $M = \begin{pmatrix} 1 & 1 & 1 & 0 \\ 1 & 0 & 0 & 1 \\ 0 & 1 & 0 & 0 \\ 0 & 0 & 1 & 1 \end{pmatrix}$

5.26. Let M and N be the two disjoint sets of vertices determining the bipartite graph. Order the vertices in M first and then those in N.

5.27. There are five of them, as shown in Fig. 5-36.

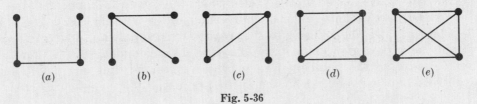

Fig. 5-36

5.29. Let $u \sim v$ if $u = v$ or if there is a path from u to v. Show that $\sim$ is an equivalence relation.

5.30.

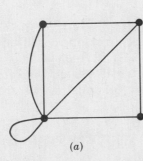 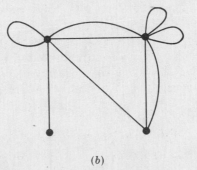

Fig. 5-37

5.34. Second and third

Chapter 6

Planar Graphs, Colorations, Trees

6.1 INTRODUCTION

A graph or multigraph which can be drawn in the plane so that its edges do not cross is said to be *planar*. Although the complete graph with four vertices K_4 is usually pictured with crossing edges as in Fig. 6-1(a), it can also be drawn with noncrossing edges as in Fig. 6-1(b). Thus K_4 is a planar graph. In this chapter we investigate planar graphs and show how they are related to the celebrated four color problem. We also study the important class of planar graphs called trees.

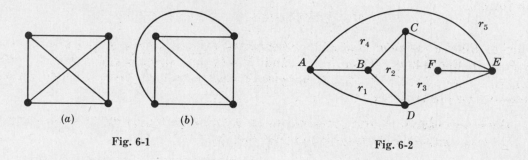

(a)   (b)

Fig. 6-1   Fig. 6-2

6.2 MAPS, REGIONS

A particular planar representation of a finite planar multigraph is called a *map*. We say that the map is *connected* if the underlying multigraph is connected. A given map divides the plane into various regions. For example, the map in Fig. 6-2 with six vertices and nine edges divides the plane into five regions. Observe that four of the regions are bounded, but the fifth region, outside the diagram, is unbounded. Thus there is no loss in generality in counting the number of regions if we assume that our map is contained in some large rectangle rather than in the entire plane.

Observe that the border of each region of a map consists of edges. Sometimes the edges will form a cycle, but sometimes not. For example, in Fig. 6-2 the borders of all the regions are cycles except for r_3. However, if we do move counterclockwise around r_3 starting, say, at the vertex C, then we obtain the closed walk

$$(C, D, E, F, E, C)$$

where the edge $\{E, F\}$ occurs twice. By the *degree* of a region r, written deg (r), we mean the length of the cycle or closed walk which borders r. We note that each edge either borders two regions or is contained in a region and will occur twice in any walk along the border of the region. Thus we have a theorem for regions which is analogous to Theorem 5.1 for vertices.

Theorem 6.1: The sum of the degrees of the regions of a map is equal to twice the number of edges.

The degrees of the regions of Fig. 6-2 are:

$$\deg(r_1) = 3, \quad \deg(r_2) = 3, \quad \deg(r_3) = 5, \quad \deg(r_4) = 4, \quad \deg(r_5) = 3$$

The sum of the degrees is 18, which, as expected, is twice the number of edges.

For notational convenience we shall picture the vertices of a map with dots or small circles, or we shall assume that any intersections of lines or curves in the plane are vertices.

6.3 EULER'S FORMULA

Euler gave a formula which connects the number V of vertices, the number E of edges and the number R of regions of any connected map.

Theorem (Euler) 6.2: $V - E + R = 2$.

We emphasize that the underlying graph of the map must be connected or else the formula does not hold. In Fig. 6-2, we have $V = 6$, $E = 9$ and $R = 5$; and as expected by Euler's formula

$$V - E + R = 6 - 9 + 5 = 2$$

Proof of Euler's formula. Let M be a connected map. Suppose M consists of a single vertex P as in Fig. 6-3(a). Then $V = 1$ and $E = 0$, and there is one region, i.e. $R = 1$. Thus in this case $V - E + R = 2$. Otherwise M can be built up from a single vertex by the following two constructions:

(1) Add a new vertex Q_2 and connect it to an existing vertex Q_1 by an edge which does not cross any existing edge, as in Fig. 6-3(b).

(2) Connect two existing vertices Q_1 and Q_2 by an edge e which does not cross any existing edge, as in Fig. 6-3(c).

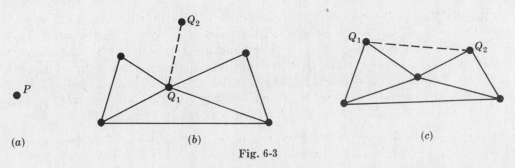

(a)   (b)   (c)

Fig. 6-3

The first operation does not change the value of $V - E + R$ since both V and E are increased by 1, but the number R of regions is not changed. The second operation also does not change the value of $V - E + R$ since V does not change, E is increased by 1, and it can be shown that the number R of regions is also increased by 1. Accordingly, M must have the same value of $V - E + R$ as the map consisting of a single vertex; that is, $V - E + R = 2$, and the theorem is proved.

Let G be a connected planar multigraph with three or more vertices, so G is neither K_1 nor K_2. Let M be a planar representation of G. It is not difficult to see that (1) a region of M can have degree 1 only if its border is a loop, and (2) a region of M can have degree 2 only if its border consists of two multiple edges. Accordingly, if G is a graph, not a multigraph, then every region of M must have degree 3 or more. This comment together with Euler's formula is used to prove the following result on planar graphs.

Theorem 6.3:   Let G be a connected planar graph with p vertices and q edges, where $p \geqq 3$. Then $q \leqq 3p - 6$.

Note that the theorem is not true for K_1 where $p = 1$ and $q = 0$, and is not true for K_2 where $p = 2$ and $q = 1$.

Proof.   Let r be the number of regions in a planar representation of G. By Euler's formula,

$$p - q + r = 2$$

Now the sum of the degrees of the regions equals $2q$ by Theorem 6.1. But each region has degree 3 or more; hence

$$2q \geqq 3r$$

Thus $r \leqq 2q/3$. Substituting this in Euler's formula gives

$$2 = p - q + r \leqq p - q + 2q/3 \quad \text{or} \quad 2 \leqq p - q/3$$

Multiplying the inequality by 3 gives $6 \leqq 3p - q$ which gives us our result.

6.4   NONPLANAR GRAPHS, KURATOWSKI'S THEOREM

We give two examples of nonplanar graphs. Consider first the *utility graph*; that is, three houses A_1, A_2, A_3 are to be connected to outlets for water, gas and electricity, B_1, B_2, B_3, as in Fig. 6-4(a). Observe that this is the graph $K_{3,3}$ and it has $p = 6$ vertices and $q = 9$ edges. Suppose the graph is planar. By Euler's formula a planar representation has $r = 5$ regions. Observe that no three vertices are connected to each other; hence the degree of each region must be 4 or more and so the sum of the degrees of the regions must be 20 or more. By Theorem 6.1, the graph must have 10 or more edges. This contradicts the fact that the graph has $q = 9$ edges. Thus the utility graph $K_{3,3}$ is nonplanar.

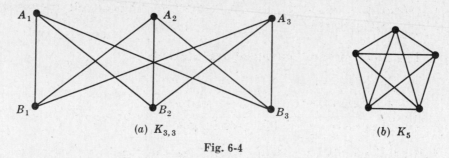

(a) $K_{3,3}$ (b) K_5

Fig. 6-4

Consider next the *star graph* in Fig. 6-4(b). This is the complete graph K_5 on $p = 5$ vertices and has $q = 10$ edges. If the graph is planar, then by Theorem 6.3,

$$10 = q \leqq 3p - 6 = 15 - 6 = 9$$

which is impossible. Thus K_5 is nonplanar.

For many years mathematicians tried to characterize planar and nonplanar graphs. This problem was finally solved in 1930 by the Polish mathematician K. Kuratowski. The proof of this result, stated below, lies beyond the scope of this text.

Theorem (Kuratowski) 6.4:   A graph is nonplanar if and only if it contains a subgraph homeomorphic to $K_{3,3}$ or K_5.

6.5 COLORED GRAPHS

A *vertex coloring*, or simply *coloring*, of a graph G is an assignment of colors to the vertices of G such that adjacent vertices have different colors. We say that G is n-colorable if there exists a coloring of G which uses n colors. (Since the word "color" is used as a noun, we will try to avoid its use as a verb by saying, for example, "paint" G rather than "color" G when we are assigning colors to the vertices of G.) The minimum number of colors needed to paint G is called the *chromatic number* of G and is denoted by $\chi(G)$.

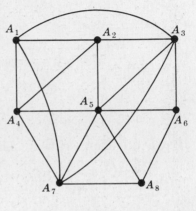

We give an algorithm by Welch and Powell to color a graph G. First order the vertices of G according to decreasing degrees. (Such an ordering need not be unique since some vertices may have the same degree.) Then use the first color to paint the first vertex and to paint, in sequential order, each vertex which is not adjacent to a previously painted vertex (of the same color). Repeat the process using the second color and the subsequence of unpainted vertices. Continue the process with the third color, and so on until all vertices are colored.

We use the Welch-Powell algorithm to color the graph G in Fig. 6-5. Ordering the vertices according to decreasing degrees we obtain the sequence

$$A_5, A_3, A_7, A_1, A_2, A_4, A_6, A_8$$

Fig. 6-5

The first color is used to paint the vertices A_5 and A_1. The second color is used to paint the vertices A_3, A_4 and A_8. The third color is used to paint the vertices A_7, A_2, and A_6. Thus G is 3-colorable. Note that G is not 2-colorable since A_1, A_2 and A_3 must be painted different colors. Accordingly, $\chi(G) = 3$.

There is no simple way to actually determine whether or not an arbitrary graph is n-colorable. However, the following theorem (proved in Problem 6.6) gives a simple characterization of 2-colorable graphs.

Theorem 6.5:   The following are equivalent for a graph G:

    (i) G is 2-colorable.

    (ii) G is bipartite.

    (iii) Every cycle of G has even length.

It is clear that the complete graph K_n with n vertices requires n colors in any coloring since each vertex is adjacent to every other vertex. On the other hand, planar graphs with any number of vertices have the following property. (See Problem 6.8 for the proof.)

Theorem 6.6:   A planar graph G is 5-colorable.

Actually mathematicians conjecture that planar graphs are 4-colorable since every known planar graph is 4-colorable. This conjecture is equivalent to the celebrated four color theorem discussed in the next section.

6.6 FOUR COLOR THEOREM

Consider a map M (i.e. a planar representation of a finite planar multigraph). Two regions of M are said to be adjacent if they have an edge in common. For example, in Fig. 6-6(a) the regions r_2 and r_5 are adjacent but the regions r_3 and r_5 are not. By a coloring of M we mean an assignment of a color to each region of M such that adjacent regions have different colors.   A map M is n-colorable if there exists a coloring of M which uses n colors.   The map in Fig. 6-6(a) is 3-colorable since the regions could be painted as follows:

r_1 red, r_2 white, r_3 red, r_4 white, r_5 red, r_6 blue

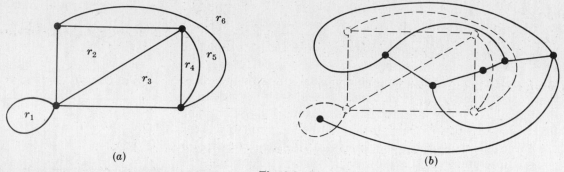

(a)   (b)

Fig. 6-6

Figure 6-7 shows a very simple map which requires four colors for any coloring.

Observe the similarity between the above discussion on coloring maps and the previous discussion on coloring graphs.   In fact, coloring maps is equivalent to vertex coloring of planar graphs in view of the concept of the dual map defined below.

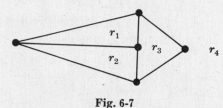

Fig. 6-7

Consider a map M.   In each region of M we choose a point, and if two regions have an edge in common then we connect the corresponding points with a curve through the common edge.   These curves can be drawn so that they are noncrossing.   Thus we obtain a new map M^*, called the *dual* of M, such that each vertex of M^* corresponds to exactly one region of M.   Figure 6-6(b) shows the dual of the map of Fig. 6-6(a).   One can prove that each region of M^* will contain exactly one vertex of M and that each edge of M^* will intersect exactly one edge of M and vice versa.   Thus M will be the dual of the map M^*.

Observe that any coloring of the regions of the map M will correspond to a coloring of the vertices of the dual map M^*.   In other words, a map M is n-colorable if and only if the planar graph of the dual map M^* is vertex n-colorable.   Thus Theorem 6.6 can be restated as follows.

Theorem 6.7:   Every map M is 5-colorable.

On the other hand, no map has been found that requires five colors in every coloring. This has led mathematicians to try to prove the celebrated

Four Color Theorem: If the regions of a map M are colored so that adjacent regions have different colors, then no more than four colors are required.

As noted above, this theorem can be restated as follows:

Four Color Theorem: Every planar graph is (vertex) 4-colorable.

In 1976, Appel and Haken announced a proof of this theorem which uses a computer to analyze almost 2000 graphs involving millions of cases.   Since the computer programs are long and involved and hence error-prone, they are being rechecked by various people throughout the world.

6.7 TREES

Recall that a graph G is said to be *acyclic* or *cycle-free* if it contains no cycles. A *tree* is a connected graph with no cycles. A *forest* is a graph with no cycles; hence the connected components of a forest are trees. Figure 5-11 in the preceding chapter shows all trees with six vertices. The tree consisting of a single vertex with no edges is called the *degenerate tree.*

There are a number of equivalent ways of defining a tree, as shown by the following theorem, which is proved in Problem 6.15.

Theorem 6.8: Let G be a graph with more than one vertex. Then the following are equivalent:

   (i) G is a tree.

   (ii) Each pair of vertices is connected by exactly one path.

   (iii) G is connected, but if any edge is deleted then the resulting graph is not connected.

   (iv) G is cycle-free, but if any edge is added to the graph then the resulting graph has exactly one cycle.

In the case that our graphs are finite, then we have additional ways of defining a tree.

Theorem 6.9: Let G be a finite graph with $n > 1$ vertices. Then the following are equivalent:

   (i) G is a tree.

   (ii) G is cycle-free and has $n-1$ edges.

   (iii) G is connected and has $n-1$ edges.

In particular, this theorem tells us that a finite tree with n vertices has $n-1$ edges. Another property of trees follows.

Theorem 6.10: Trees (and hence forests) are 2-colorable.

Thus trees are bipartite graphs. The converse is not true since, for example, the nonplanar complete bipartite graph $K_{3,3}$ is not a tree.

A subgraph T of a graph G is called a *spanning tree* of G if T is a tree and T includes all the vertices of G. Figure 6-8 shows a graph G and spanning trees T_1, T_2 and T_3 of G. If G is a graph whose edges have lengths, then a minimal spanning tree of G is a spanning tree of G such that the sum of the lengths of the edges is minimal among all spanning trees of G.

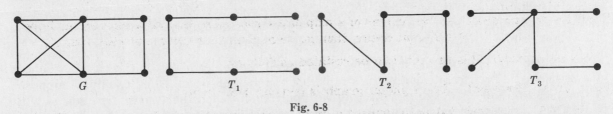

Fig. 6-8

We give two algorithms to find a minimal spanning tree of a finite connected labeled graph G with m vertices. First, order the edges of G according to decreasing lengths. Proceeding sequentially, delete each edge which does not disconnect the graph until $m-1$

edges remain.   These edges will then form a minimal spanning tree of G.   This algorithm depends upon knowing whether or not a graph is connected, which, in general, is not easily programmable.

The second algorithm begins with ordering the edges according to increasing lengths. Then, beginning with only the vertices of G, we add one edge after another where each edge has minimal length and does not form any cycle.   After adding $m - 1$ edges we obtain a minimal spanning tree.   By Problem 5.37, we can add an edge $\{v_r, v_s\}$ whenever the rs entry in the connection matrix is zero.   This algorithm can be efficiently programmed by having the computer keep track of the connection matrix after each edge is added.

We emphasize that since some edges have the same lengths, we can obtain different minimal spanning trees.   Figure 6-9 gives a labeled connected graph G and a minimal spanning tree M.

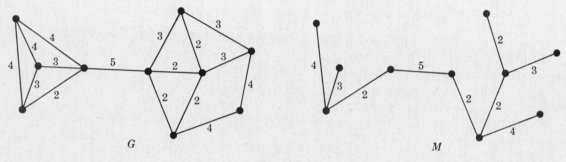

Fig. 6-9

6.8 ROOTED TREES

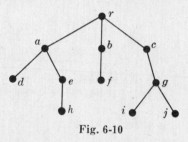

A rooted tree R consists of a tree graph together with a designated vertex r called the *root* of the tree.   Since there is a unique path from r to any other vertex v, this gives a direction to the edges of R.   The length of the path from the root r to v is called the *level* or *depth* of v.   Those vertices with degree one, other than r, are called the *leaves* of the rooted tree.   A directed path from a vertex to a leaf is called

Fig. 6-10

a *branch*.   Figure 6-10 shows a rooted tree; the root r is at the top of the tree.   The tree has five leaves, d, f, h, i and j.   The level of a is 1, the level of f is 2 and the level of j is 3. We emphasize that any tree may be made into a rooted tree by simply picking one of the vertices as the root.

Since a rooted tree gives a direction to the edges, we will say that a vertex u *precedes* a vertex v or that v *follows* u if the path from the root r to v includes u.   In particular, we say that v *immediately follows* u if v follows u and is adjacent to u.   In Fig. 6-10, the vertex j follows c but immediately follows g.   Observe that every vertex other than the root r immediately follows a unique vertex, but can be immediately followed by more than one vertex, e.g. vertices i and j both immediately follow g.

A rooted tree is a useful device to enumerate all the logical possibilities of a sequence of events where each event can occur in a finite number of ways.   For example, suppose two men, Marc and Erik, are playing a tennis tournament such that the first person to win two games in a row or who wins a total of three games wins the tournament.   The rooted tree in Fig. 6-11 shows the various ways the tournament can proceed.   Observe that there are ten leaves which correspond to the ten ways that the tournament can occur:

MM, MEMM, MEMEM, MEMEE, MEE, EMM, EMEMM, EMEME, EMEE, EE

Specifically, the path from the root to the leaf describes who won which game in the particular tournament.

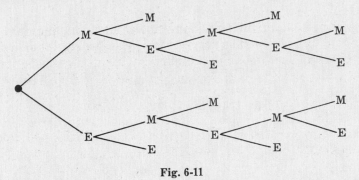

Fig. 6-11

6.9 ORDERED ROOTED TREES

Consider a rooted tree R in which the edges leaving each vertex are ordered. Then we have the concept of an *ordered rooted tree*. One can systematically label (or *address*) the vertices of such a tree as follows: We first assign 0 to the root r. We next assign 1, 2, 3, ... to the vertices immediately following r according as the edges were ordered, We then label the remaining vertices in the following way. If a is the label of a vertex v, then $a.1, a.2, ...$ are assigned to the vertices immediately following v according as the edges were ordered. We illustrate this address system in Fig. 6-12, where edges are pictured from left to right according to their order. Observe that the number of decimal points in any label is one less than the level of the vertex. We will refer to this labeling system as the *universal address system* for an ordered rooted tree.

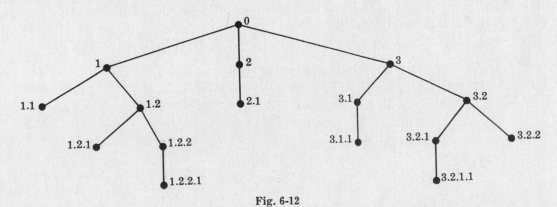

Fig. 6-12

The universal address system gives us an important way of linearly describing (or storing) an ordered rooted tree. Specifically, given addresses a and b, we let $a < b$ if a is an *initial segment* of b, i.e. if $b = a.c$, or if there exist positive integers m and n with $m < n$ such that

$$a = r.m.s \qquad \text{and} \qquad b = r.n.t$$

This order is called *lexicographic order* since it is similar to the way words are arranged in a dictionary. For example, the addresses in Fig. 6-12 are linearly ordered as follows:

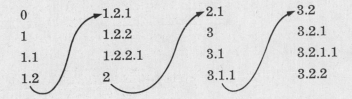

This lexicographic order is identical to the order obtained by moving down the leftmost branch of the tree, then the next branch to the right, then the second branch to the right, and so on.

Any algebraic expression involving binary operations, for example, addition, subtraction, multiplication and division, can be represented by an ordered rooted tree. For example, Fig. 6-13(a) represents the arithmetic expression

$$(a - b)/((c \times d) + e) \tag{6.1}$$

Observe that the variables in the expression, a, b, c, d and e, appear as leaves, and the operations appear as the other vertices. The tree must be ordered since $a - b$ and $b - a$ yield the same tree but not the same ordered tree.

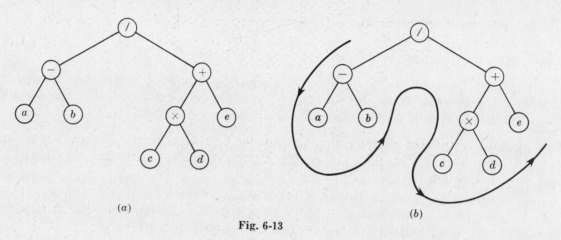

(a)   (b)

Fig. 6-13

The Polish mathematician Lukasiewicz observed that by placing the binary operational symbol before its arguments, e.g.

$$+ a\, b \quad \text{instead of} \quad a + b \quad \text{and} \quad /\, c\, d \quad \text{instead of} \quad c/d$$

one does not need to use any parentheses. This notation is called *Polish notation* in *prefix form*. (Analogously, one can place the symbol after its arguments, called *Polish notation* in *postfix form*.) Rewriting (6.1) in prefix form we obtain

$$/ - a\, b + \times c\, d\, e$$

Observe that this is precisely the lexicographic order of the vertices in its tree which can be obtained by scanning the tree as in Fig. 6-13(b).

Solved Problems

PLANAR GRAPHS, MAPS

6.1. Draw a planar representation of each graph in Fig. 6-14, if possible.

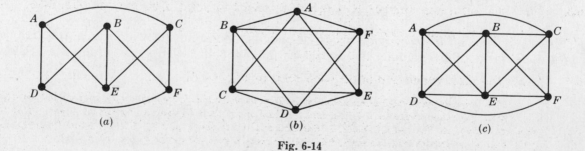

(a)   (b)   (c)

Fig. 6-14

(a) Redrawing the positions of the vertices B and E we get a planar representation of the graph, as in Fig. 6-15(i).

(b) This is not the star graph K_5. This has a planar representation such as in Fig. 6-15(ii).

(c) This graph is nonplanar. The utility graph $K_{3,3}$ is a subgraph, as shown in Fig.6-15(iii), where we have redrawn the positions of C and F.

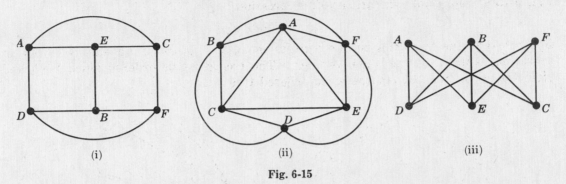

Fig. 6-15

6.2. Count the number V of vertices, the number E of edges and the number R of regions of each map in Fig. 6-16, and verify Euler's formula. Also find the degree of the outside region.

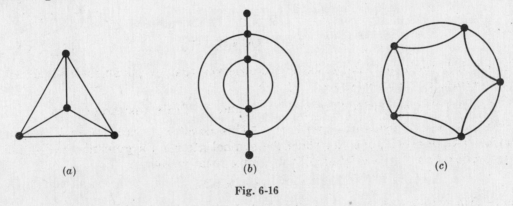

Fig. 6-16

(a) $V = 4$, $E = 6$, $R = 4$. Hence $V - E + R = 4 - 6 + 4 = 2$. Also $d = 3$.

(b) $V = 6$, $E = 9$, $R = 5$; so $V - E + R = 6 - 9 + 5 = 2$. Here $d = 6$ since two of the edges are counted twice.

(c) $V = 5$, $E = 10$, $R = 7$. Hence $V - E + R = 5 - 10 + 7 = 2$. Here $d = 5$.

6.3. Find the minimum number n of colors required to paint each map in Fig. 6-16.

(a) $n = 4$, (b) $n = 3$, (c) Only two colors are needed, i.e. $n = 2$.

6.4. Draw the map which is dual to each map in Fig. 6-16.

Pick a point in each region, and connect two vertices if the corresponding regions have an edge in common. We emphasize that a map and its dual must have the same number of edges. The results are shown in Fig. 6-17. Observe that there are two loops in Fig. 6-17(b), which corresponds to the two edges in the original map that are entirely contained in the "outside" region.

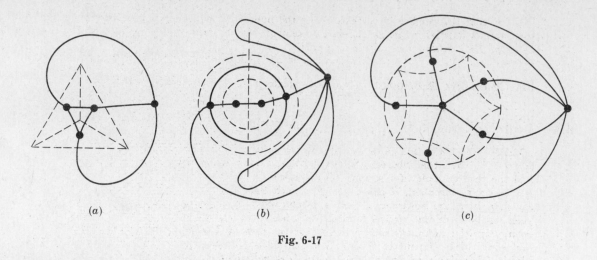

(a)   (b)   (c)

Fig. 6-17

6.5. Use the Welch-Powell algorithm to paint the graph in Fig. 6-18, and find the chromatic number n of the graph.

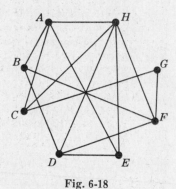

Fig. 6-18

First order the vertices according to decreasing degrees to obtain the sequence

$$H, A, D, F, B, C, E, G$$

Proceeding sequentially, we use the first color to paint the vertices H, B and then G. (We cannot paint A, D or F the first color since each is connected to H, and we cannot paint C or E the first color since each is connected to either H or B.) Proceeding sequentially with the unpainted vertices, we use the second color to paint the vertices A and D. The remaining vertices F, C and E can be painted with the third color. Thus the chromatic number n cannot be greater than 3. However, in any coloring, H, D and E must be painted different colors since they are connected to each other. Hence $n = 3$.

6.6. Prove Theorem 6.5: The following are equivalent for a graph G: (i) G is 2-colorable. (ii) G is bipartite. (iii) Every cycle of G has even length.

(i) *implies* (ii). Suppose G is 2-colorable. Let M be the set of vertices painted the first color, and let N be the set of vertices painted the second color. Then M and N form a bipartite partition of the vertices of G since neither the vertices of M nor the vertices of N can be adjacent to each other since they are of the same color.

(ii) *implies* (iii). Suppose G is bipartite and M and N form a bipartite partition of the vertices of G. If a cycle begins at a vertex u of, say, M, then it will go to a vertex of N, and then to a vertex of M, and then to N and so on. Hence when the cycle returns to u it must be of even length. That is, every cycle of G will have even length.

(iii) *implies* (i). Lastly, suppose every cycle of G has even length. We pick a vertex in each connected component and paint it the first color, say red. We then successively paint all the vertices as follows: If a vertex is painted red, then any vertex adjacent to it will be painted the second color, say blue. If a vertex is painted blue, then any vertex adjacent to it will be painted red. Since every cycle has even length, no adjacent vertices will be painted the same color. Hence G is 2-colorable, and the theorem is proved.

6.7. Let G be a finite connected planar graph with at least three vertices. Show that G has at least one vertex of degree 5 or less.

Let p be the number of vertices and q the number of edges of G, and suppose $\deg(u) \geq 6$ for each vertex u of G. But $2q$ equals the sum of the degrees of the vertices of G (Theorem 5.1); $2q \geq 6p$.　Therefore

$$q \geq 3p > 3p - 6$$

This contradicts Theorem 6.3. Thus some vertex of G has degree 5 or less.

6.8. Prove Theorem 6.6: A planar graph G is 5-colorable.

The proof is by induction on the number p of vertices of G. If $p \leq 5$, then the theorem obviously holds. Suppose $p > 5$, and the theorem holds for graphs with less than p vertices. By the preceding problem, G has a vertex v such that $\deg(v) \leq 5$. By induction, the subgraph $G - v$ is 5-colorable. Assume one such coloring. If the vertices adjacent to v use less than the five colors, then we simply paint v with one of the remaining colors and obtain a 5-coloring of G. We are still left with the case that v is adjacent to five vertices which are painted different colors. Say the vertices, moving counterclockwise about v, are $v_1, \ldots, v_5$ and are painted respectively by the colors $c_1, \ldots, c_5$. (See Fig. 6-19.)

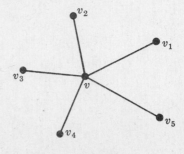

Fig. 6-19

Consider now the subgraph H of G generated by the vertices painted c_1 and c_3. Note H includes v_1 and v_3. If v_1 and v_3 belong to different components of H, then we can interchange the colors c_1 and c_3 in the component containing v_1 without destroying the coloring of $G - v$. Then v_1 and v_3 are painted by c_3, c_1 can be chosen to paint v, and we have a 5-coloring of G. On the other hand, suppose v_1 and v_3 are in the same component of H. Then there is a path P from v_1 to v_3 whose vertices are painted either c_1 or c_3. The path P together with the edges $\{v, v_1\}$ and $\{v, v_3\}$ form a cycle C which encloses either v_2 or v_4. Consider now the subgraph K generated by the vertices painted c_2 or c_4. Since C encloses v_2 or v_4, but not both, the vertices v_2 and v_4 belong to different components of K. Thus we can interchange the colors c_2 and c_4 in the component containing v_2 without destroying the coloring of $G - v$. Then v_2 and v_4 are painted by c_4, and we can choose c_2 to paint v and obtain a 5-coloring of G. Thus G is 5-colorable and the theorem is proved.

TREES

6.9.　Find all spanning trees of the graph G shown in Fig. 6-20.

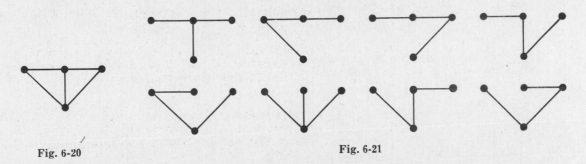

Fig. 6-20　　　　　　　　　Fig. 6-21

There are eight such spanning trees as shown in Fig 6-21. Each spanning tree must have $4 - 1 = 3$ edges since G has four vertices. Thus each tree can be obtained by deleting two of the five edges of G. This can be done in ten ways, except that two of the ways lead to disconnected graphs. Hence the above eight spanning trees are all the spanning trees of G.

6.10.　Find a minimum spanning tree for the graph with labeled edges in Fig. 6-22.

Keep deleting edges with maximum length without disconnecting the graph. Alternately, begin with the nine vertices and keep adding edges with minimum length without forming any cycle. Both methods give a minimum spanning tree such as that shown in Fig. 6-23.

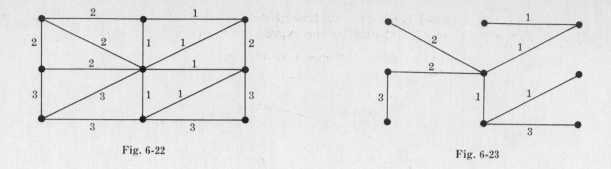

Fig. 6-22   Fig. 6-23

6.11. Consider the tree shown in Fig. 6-24. (a) Which vertices, if any, are cut points? Which edges, if any, are bridges? (b) Find all vertices at level 3 if the vertex picked as a root is: (i) u, (ii) w.

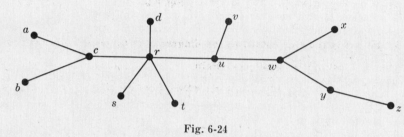

Fig. 6-24

(a) Each vertex of degree more than 1 is a cut point in a tree; hence c, r, u, w and y. Every edge in a tree is a bridge since the removal of any edge disconnects the graph (Theorem 6.8).

(b) Find all paths of length 3 from the root to obtain vertices at level (or depth) 3. Thus: (i) a, b and z, (ii) c, d, s and t.

6.12. Suppose a universal address system contains the address $x = 4.5.2.3$. What other addresses must be in the system?

We need to find all addresses y such that $y = a.m$ and $x = a.n.b$ where m and n are integers with $m \leq n$. These can be found systematically as follows. We first list the initial segments of x which are

$$a = 4.5.2.3, \quad b = 4.5.2, \quad c = 4.5 \quad \text{and} \quad d = 4$$

For each initial segment, we form all preceding addresses with the same length and with the same spelling except for the last number. Corresponding to $a = 4.5.2.3$ we have the three addresses

$$4.5.2.1, \quad 4.5.2.2 \quad \text{and} \quad 4.5.2.3$$

Corresponding to $b = 4.5.2$, we have the two addresses

$$4.5.1 \quad \text{and} \quad 4.5.2$$

Corresponding to $c = 4.5$, we have the five addresses,

$$4.1, \quad 4.2, \quad 4.3, \quad 4.4 \quad \text{and} \quad 4.5$$

Corresponding to the initial segment $d = 4$, we have the addresses

$$0, \quad 1, \quad 2, \quad 3 \quad \text{and} \quad 4$$

The above fifteen addresses must be in the system if x is in the system.

6.13. Consider the algebraic expression $(2x + y)(5a - b)^3$. (a) Draw the corresponding ordered rooted tree. (b) Find the *scope* of the exponentiation operation. (The scope of a vertex v in a rooted tree is the subtree generated by v and the vertices which follow v with v as the root.) (c) Rewrite the expression in prefix Polish notation.

(a) Use an arrow ($\uparrow$) for exponentiation and an asterisk ($*$) for multiplication to obtain the tree shown in Fig. 6-25.

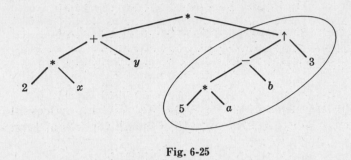

Fig. 6-25

(b) The scope of $\uparrow$ is the tree circled in the diagram. It corresponds to the expression $(5a - b)^3$.

(c) Scan the tree as in Fig. 6-13(b) to obtain

$$* + * 2 \, x \, y \uparrow - * 5 \, a \, b \, 3$$

6.14. Suppose there are two distinct paths, say P_1 and P_2, from a vertex u to a vertex v in a graph G. Prove that G contains a cycle.

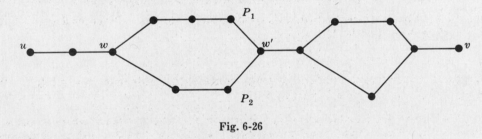

Fig. 6-26

Let w be a vertex on P_1 and P_2 such that the next vertices on P_1 and P_2 are distinct. Let w' be the first vertex following w which lies on both P_1 and P_2. (See Fig. 6-26.) Then the subpaths of P_1 and P_2 between w and w' have no vertices in common except w and w'; hence these two subpaths form a cycle.

6.15. Prove Theorem 6.8: Let G be a graph with more than one vertex. Then the following are equivalent: (i) G is a tree. (ii) Each pair of vertices is connected by exactly one path. (iii) G is connected; but if any edge is deleted then the resulting graph is disconnected. (iv) G is cycle-free; but if any edge is added to the graph then the resulting graph has exactly one cycle.

(i) *implies* (ii). Let u and v be two vertices in G. Since G is a tree, G is connected so there is at least one path between u and v. By Problem 6.14, there can only be one path between u and v, otherwise G will contain a cycle.

(ii) *implies* (iii). Suppose we delete an edge $e = \{u, v\}$ from G. Note e is a path from u to v. Suppose the resulting graph $G - e$ has a path P from u to v. Then P and e are two distinct paths from u to v, which contradicts the hypothesis. Thus there is no path between u and v in $G - e$, so $G - e$ is disconnected.

(iii) *implies* (iv). Suppose G contains a cycle C which contains an edge $e = \{u, v\}$. By hypothesis, G is connected but $G' = G - e$ is disconnected, with u and v belonging to different components of G' (Problem 5.32). This contradicts the fact that u and v are connected by the path $P = C - e$ which lies in G'. Hence G is cycle-free. Now let x and y be vertices of G and let H be the graph obtained by adjoining the edge $e = \{x, y\}$ to G. Since G is connected, there is a path P from x to y in G; hence $C = Pe$ forms a cycle in H. Suppose H contains another cycle C'. Since G is cycle-free, C' must contain the edge e, say $C' = P'e$. Then P and P' are two paths in G from x to y. (See Fig. 6-27.) By Problem 6.14, G contains a cycle, which contradicts the fact that G is cycle-free. Hence H contains only one cycle.

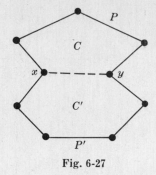

Fig. 6-27

(iv) *implies* (i). Since adding any edge $e = \{x, y\}$ to G produces a cycle, the vertices x and y must already be connected in G. Hence G is connected and by hypothesis G is cycle-free; that is, G is a tree.

6.16. Prove Theorem 6.9: Let G be a finite graph with n vertices. Then the following are equivalent: (i) G is a tree. (ii) G is cycle-free and has $n-1$ edges. (iii) G is connected and has $n-1$ edges.

The proof is by induction on the number n of vertices of G. Suppose $n = 1$, i.e. G has only one vertex. Then G has $0 = 1 - 1$ edges and so G is connected and cycle-free. Thus the theorem holds for $n = 1$.

Suppose $n > 1$, i.e. G has more than one vertex. We show that (i), (ii) and (iii) are equivalent for G where we assume they are equivalent for all graphs with less than n vertices.

(i) *implies* (ii). Suppose G is a tree. Then G is cycle-free, so we only need to show that G has $n-1$ edges. By Problem 5.31, G has a vertex of degree 1. Deleting this vertex and its edge, we obtain a tree T which has $n-1$ vertices. The theorem holds for T, so T has $n-2$ edges. Hence G has $n-1$ edges.

(ii) *implies* (iii). Suppose G is cycle-free and has $n-1$ edges. We only need show that G is connected. Suppose G is disconnected and has k components, $T_1, \ldots, T_k$, which are trees since each is connected and cycle-free. Say T_i has n_i vertices. Note $n_i < n$. Hence the theorem holds for T_i, so T_i has $n_i - 1$ edges. Thus

$$n = n_1 + n_2 + \cdots + n_k$$

and

$$n - 1 = (n_1 - 1) + (n_2 - 1) + \cdots + (n_k - 1) = n_1 + n_2 + \cdots + n_k - k = n - k$$

Hence $k = 1$. But this contradicts the assumption that G is disconnected and has $k > 1$ components. Hence G is connected.

(iii) *implies* (i). Suppose G is connected and has $n-1$ edges. We only need to show that G is cycle-free. Suppose G has a cycle containing an edge e. Deleting e we obtain the graph $H = G - e$ which is also connected. But H has n vertices and $n-2$ edges, and this contradicts Problem 5.33. Thus G is cycle-free and hence is a tree.

Supplementary Problems

PLANAR GRAPHS, MAPS, COLORING

6.17. Draw a planar representation of each graph in Fig. 6-28, if possible; otherwise show that it has a subgraph homeomorphic to K_5 or $K_{3,3}$.

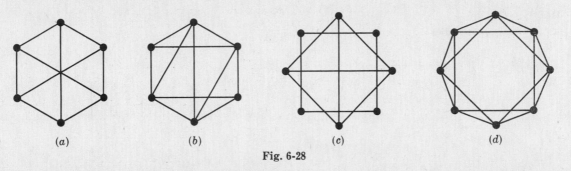

(a)   (b)   (c)   (d)

Fig. 6-28

6.18. For the map in Fig. 6-29, find the degree of each region and verify that the sum of the degrees of the regions is equal to twice the number of edges.

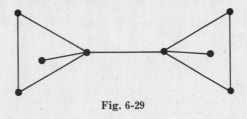

Fig. 6-29

6.19. Count the number V of vertices, the number E of edges, and the number R of regions of each map in Fig. 6-30, and verify Euler's formula.

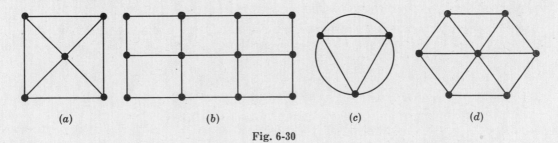

(a)   (b)   (c)   (d)

Fig. 6-30

6.20. Find the minimum number of colors needed to paint the regions of each map in Fig. 6-30.

6.21. Draw the map which is dual to each map in Fig. 6-30.

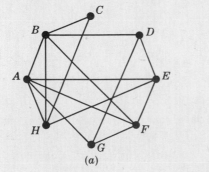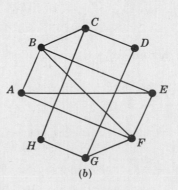

(a)   (b)

Fig. 6-31

6.22.   Use the Welch-Powell algorithm to paint each graph in Fig. 6-31. Find the chromatic number n of the graph.

6.23.   Suppose n is the maximum degree of any of the vertices of a graph G. Prove that $\chi(G) \leqq n + 1$, where $\chi(G)$ is the chromatic number of G.

TREES

6.24.   Find the number of spanning trees in Fig. 6-32.

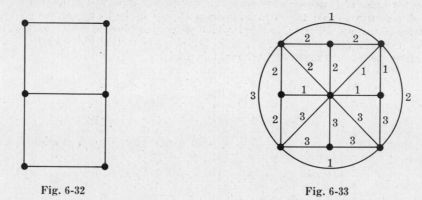

Fig. 6-32   Fig. 6-33

6.25.   Find a minimal spanning tree in Fig. 6-33.

6.26.   Suppose a universal address system contains the address $x = 4.3.6.2$. What other addresses must be in the system?

6.27.   An address x *tree-precedes* an address y if x precedes y as vertices in the underlying rooted tree (not in the lexicographical ordering).

(a) Show that an address x tree-precedes an address y if x and y are of the form $x = a.m$ and $y = a.n.b$ where m and n are integers with $m \leqq n$ and a and b are partial addresses (possibly empty).

(b) Determine whether or not x tree-precedes y where:

   (i) $x = 3.5.4$, $y = 3.2.6.7$ (iii) $x = 3.4$, $y = 3.5.1$

   (ii) $x = 4.2$, $y = 5.3.6$ (iv) $x = 4.1.3$, $y = 4.1.4$

6.28.   Addresses $x_1, x_2, \ldots, x_k$ are said to be *independent* if no address tree-precedes any other address. Determine whether or not the following sets of addresses are independent:

(a) $x_1 = 3.2.5$, $x_2 = 3.1$, $x_3 = 2.3$

(b) $x_1 = 2.5.4$, $x_2 = 3.1$, $x_3 = 1.2.3$

6.29.   Prove that any vertex in a tree of degree 2 or more is a cut point.

6.30.   Consider the algebraic expression

$$\frac{(3x - 5z)^4}{a(2b + c^2)}$$

(a) Draw the corresponding ordered rooted tree, using an arrow ($\uparrow$) for exponentiation, an asterisk ($*$) for multiplication and a slash ($/$) for division.

(b) Rewrite the expression in (i) prefix Polish notation, (ii) postfix Polish notation.

(c) Find the scope of each multiplication operation.

COMPUTER PROGRAMMING PROBLEMS

Let G be a graph with six vertices, $v_1, v_2, \ldots, v_6$ and ten edges. Let B be the 10×2 edge matrix of G. Suppose B is punched onto a deck of 10 cards so that each card contains a row of B.

6.31.   Write a program which colors the vertices of G according to the Welch-Powell algorithm.   Test the program with matrices B_1 and B_2 below.

6.32.   Suppose the edges of G are assigned lengths (positive integers), and suppose a third column is added to the edge matrix so that the third entry in each row is the length of the edge given by the first two entries of the row.   Write a program which prints the edges of a minimal spanning tree T of G. Test the program with matrices B_3 and B_4 below.   (*Hint*: Use the second minimal spanning tree algorithm in Section 6.8 together with Problem 5.43.)

$$B_1 = \begin{pmatrix} 1 & 2 \\ 2 & 3 \\ 5 & 3 \\ 3 & 6 \\ 4 & 1 \\ 5 & 6 \\ 1 & 3 \\ 4 & 5 \\ 5 & 2 \\ 6 & 4 \end{pmatrix}, \quad B_2 = \begin{pmatrix} 1 & 3 \\ 2 & 1 \\ 4 & 6 \\ 4 & 3 \\ 5 & 4 \\ 3 & 6 \\ 2 & 4 \\ 3 & 2 \\ 3 & 5 \\ 1 & 4 \end{pmatrix}, \quad B_3 = \begin{pmatrix} 1 & 2 & 3 \\ 2 & 3 & 2 \\ 5 & 3 & 3 \\ 3 & 6 & 1 \\ 4 & 1 & 2 \\ 5 & 6 & 2 \\ 1 & 3 & 1 \\ 4 & 5 & 2 \\ 5 & 2 & 1 \\ 6 & 4 & 2 \end{pmatrix}, \quad B_4 = \begin{pmatrix} 1 & 3 & 2 \\ 2 & 1 & 4 \\ 4 & 6 & 3 \\ 4 & 3 & 2 \\ 5 & 4 & 3 \\ 1 & 6 & 2 \\ 2 & 4 & 4 \\ 3 & 2 & 4 \\ 5 & 6 & 2 \\ 1 & 4 & 2 \end{pmatrix}$$

Answers to Supplementary Problems

6.17.   Only (*a*) is nonplanar.

6.18.   The outside region has degree 8, and the other two regions have degree 5.

6.19.   (*a*) $V = 5, E = 8, R = 5$ (*c*) $V = 3, E = 6, R = 5$

         (*b*) $V = 12, E = 17, R = 7$ (*d*) $V = 7, E = 12, R = 7$

6.20.   (*a*) 3, (*b*) 3, (*c*) 2, (*d*) 3

6.21.   See Fig. 6-34.

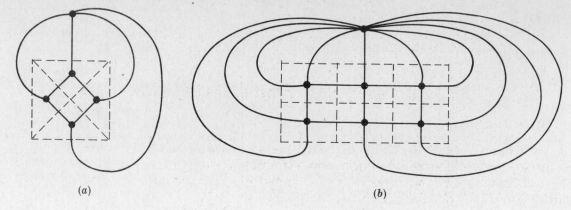

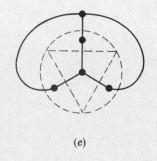

(a)

(b)

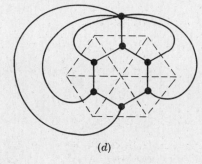

(c)

(d)

Fig. 6-34

6.22.   (a) $n = 3$,   (b) $n = 4$

6.24.   15

6.26.   0, 1, 2, 3, 4, 4.1, 4.2, 4.3, 4.3.1, 4.3.2, 4.3.3,   4.3.4, 4.3.5, 4.3.6, 4.3.6.1

6.27.   (b) (i) no, (ii) no, (iii) yes, (iv) yes

6.28.   (a) no, since x_2 precedes x_1,   (b) yes

6.30.   (a)   See Fig. 6-35.

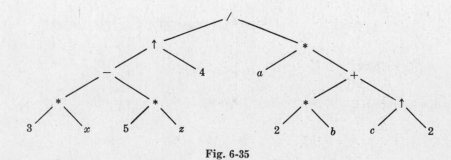

Fig. 6-35

(b)   (i) $/ \uparrow - * 3\,x * 5\,z\,4 * a + * 2\,b \uparrow c\,2$

(ii) $3\,x * 5\,z * - 4 \uparrow a\,2\,b * c\,2 \uparrow + */$

(c) $3x,\ 5z,\ a(2b + c^2),\ 2b$

Chapter 7

Directed Graphs, Finite State Machines

7.1 INTRODUCTION

In dynamic situations, e.g. digital computers or flow systems, the concept of a directed graph is frequently more useful than that of a nondirected graph. In this chapter we give the basic definitions and properties of directed graphs, and an important example of a labeled directed graph, a finite state machine.

7.2 DIRECTED GRAPHS

A *directed graph* D or *digraph* consists of two things:

(i) A set V whose elements are called *vertices, points* or *nodes*.

(ii) A set A of ordered pairs of vertices called *arcs*.

We will denote the digraph by $D(V, A)$ when we want to emphasize two parts of D.

We also picture digraphs by diagrams in the plane. Again each vertex v in V is represented by a dot (or small circle); but each arc $a = \langle u, v \rangle$ is represented by an arrow, i.e. directed curve, from the *initial point* u of a to its *terminal point* v. For example Fig. 7-1 represents the digraph $D(V, A)$ where

(i) V consists of four vertices A, B, C, D; and

(ii) A consists of seven arcs:

$$a_1 = \langle A, D \rangle, \quad a_2 = \langle B, A \rangle, \quad a_3 = \langle B, A \rangle, \quad a_4 = \langle D, B \rangle,$$
$$a_5 = \langle B, C \rangle, \quad a_6 = \langle D, C \rangle, \quad a_7 = \langle B, B \rangle$$

The arc a_7 is called a *loop* since its initial point B is the same vertex as its terminal point. The arcs a_2 and a_3 are called *parallel* arcs since they have the same initial point B and the same terminal point A.

We will usually denote a digraph by drawing its diagram rather than explicitly listing its vertices and arcs.

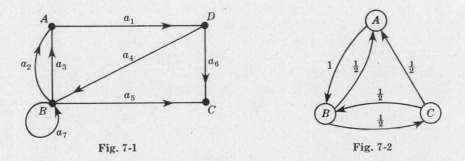

Fig. 7-1   Fig. 7-2

If the arcs and/or vertices of a directed graph are labeled with some type of data, then we say we have a *labeled directed graph*. Such graphs are frequently used to picture dynamic situations. For example, suppose three boys, A, B, C are throwing a ball to each

119

other such that A always throws the ball to B, but B and C are just as likely to throw the ball to A as they are to each other. Figure 7-2 illustrates this dynamic situation where arcs are labeled with the respective probabilities, i.e. A throws the ball to B with probability 1, B throws the ball to A and C each with probability $\frac{1}{2}$, and C throws the ball to A and B each with probability $\frac{1}{2}$.

Let $D(V, A)$ be a digraph. We say $D(V, A)$ is *finite* if its set V of vertices is finite and its set A of arcs is finite. Let V' be a subset of V and let A' be a subset of A whose endpoints belong to V'. Then $D(V', A')$ is a directed graph and is called a *subgraph* of $D(V, A)$. If A' contains all the arcs of A whose endpoints belong to V', then $D(V', A')$ is called the subgraph *generated* by V'.

7.3 BASIC DEFINITIONS

Let D be a directed graph. An arc $a = \langle u, v \rangle$ is said to *begin* at its initial point u and *end* at its terminal point v. The *outdegree* and *indegree* of a vertex v are equal respectively to the number of arcs beginning and ending at v. Since each arc begins and ends at a vertex, we see that the sum of the outdegrees of the vertices equals the sum of the indegrees of the vertices, which equals the number of arcs. A vertex with zero indegree is called a *source*, and a vertex with zero outdegree is called a *sink*. In Fig. 7.1, for example, the vertex C is a sink, but the digraph has no sources. Also, the outdegree of D is 2 and its indegree is 1.

The concepts of walk, trail, path and cycle carry over from undirected graphs except that the directions of the arcs must agree with the direction of the walk. Specifically, a (directed) *walk* W in a digraph D is an alternating sequence of vertices and arcs,

$$W = (v_0, a_1, v_1, a_2, v_2, \ldots, a_n, v_n)$$

such that each arc a_i begins at v_{i-1} and ends at v_i. If there is no ambiguity, we denote W by its sequence of vertices or its sequence of arcs. The *length* of W is n, its number of arcs. A *closed walk* has the same first and last vertices, and a *spanning walk* contains all the vertices of the digraph. A *trail* is a walk with distinct arcs, and a *path* is a walk with distinct vertices. A *cycle* is a closed walk with distinct vertices (except the first and last). A vertex u is *reachable* from a vertex v if there is a path from v to u. A *semiwalk* is the same as a walk except that the arc a_i may begin at either v_{i-1} or v_i and end at the other vertex. Semitrails and semipaths have analogous definitions.

There are three types of connectivity in a digraph D. We say that D is *weakly connected* or *weak* if there is a semipath between any two vertices u and v of D. We say that D is *unilaterally connected* or *unilateral* if, for any vertices u and v of D, either u is reachable from v or v is reachable from u, i.e. there is a path from one of them to the other. We say that D is *strongly connected* or *strong* if, for any vertices u and v of D, both u is reachable from v and v is reachable from u, i.e. there is a path from u to v and one from v to u. Observe that strongly connected implies unilaterally connected, and that unilaterally connected implies weakly connected. We say that D is *strictly unilateral* if it is unilateral but not strong, and we say that D is *strictly weak* if it is weak but not unilateral. In Fig 7-3, (a) is strictly weak, (b) is strictly unilateral and (c) is strong.

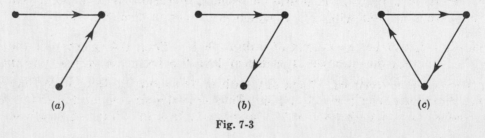

(a)   (b)   (c)

Fig. 7-3

Connectivity can be characterized in terms of spanning walks as follows:

Theorem 7.1: Let D be a finite directed graph. Then

(a) D is weak if and only if D has a spanning semiwalk.

(b) D is unilateral if and only if D has a spanning walk.

(c) D is strong if and only if D has a closed spanning walk.

Digraphs with sources and sinks appear in many applications (e.g. flow diagrams). A sufficient condition for such vertices to exist follows:

Theorem 7.2: Suppose a finite digraph D contains no (directed) cycles. Then D contains a source and a sink.

Proof. Let $P = (v_0, v_1, \ldots, v_n)$ be a path of maximum length that exists since D is finite. Then the last vertex v_n must be a sink; otherwise an arc $\langle v_n, u \rangle$ will either extend P or form a cycle if $u = v_i$ for some i. Similarly, the first vertex v_0 must be a source.

7.4 DIGRAPHS, RELATIONS, NONNEGATIVE INTEGER SQUARE MATRICES

Let $D(V, A)$ be a directed graph without parallel arcs. Then A is simply a subset of $V \times V$ and hence A is a relation on V. Conversely, if R is a relation on a set V, then $D(V, R)$ is a directed graph without parallel edges. Thus the concepts of relations on a set and digraphs without parallel edges are one and the same. In fact, in Chapter 2 we have already introduced the directed graph corresponding to a relation on a set.

Now let D be a directed graph with vertices $v_1, v_2, \ldots, v_m$. The matrix of D is the $m \times m$ matrix $M_D = (m_{ij})$ where

$$m_{ij} = \text{the number of arcs beginning at } v_i \text{ and ending at } v_j$$

If D has no parallel arcs, then the entries of M_D will only be zeros and ones; otherwise the entries will be nonnegative integers. Conversely, every $m \times m$ matrix M with nonnegative integer entries uniquely defines a digraph with m vertices. Figure 7-4 shows a digraph D and its matrix M.

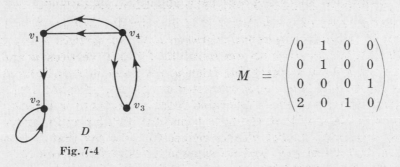

$$M = \begin{pmatrix} 0 & 1 & 0 & 0 \\ 0 & 1 & 0 & 0 \\ 0 & 0 & 0 & 1 \\ 2 & 0 & 1 & 0 \end{pmatrix}$$

Fig. 7-4

The matrix of a digraph is useful in deciding questions of reachability in the graph. This comes from the following result which is proved in Problem 7.5.

Theorem 7.3: Let M be the matrix of a digraph D. Then the ij entry of the matrix M^n gives the number of walks of length n from the vertex v_i to the vertex v_j.

Suppose D has m vertices. Then any path of D cannot contain more than m vertices and hence must have length $m - 1$ or less. Thus D is strongly connected if for any vertices u and v there exists a path from u to v and one from v to u, each of length $m - 1$ or less.

Corollary 7.4: Let M be the matrix of a digraph D with m vertices. Let

$$C = M + M^2 + M^3 + \cdots + M^{m-1}$$

Then D is strongly connected if and only if C has no zero entry off the main diagonal.

Proof. The ij entry of C is zero if and only if the ij entries of $M, M^2, \ldots, M^{m-1}$ are all zero. This means there is no walk from v_i to v_j of length $m-1$ or less, and so there is no path from v_i to v_j. Thus D is strongly connected if and only if the ij entries of C, for $i \neq j$, are nonzero.

7.5 PRUNING ALGORITHM FOR MINIMAL PATH

Consider a directed graph D with edges labeled with lengths (i.e. positive numbers). We seek the shortest path $P(D)$ between two given vertices, say u and w. We assume that D is finite so at each step there are a finite number of possible moves, and we assume that D has no cycles. Hence all the possible paths between u and w can be given by a rooted tree. For example, Fig. 7-5(b) enumerates all the paths between u and w in the *digraph* in Fig. 7-5(a).

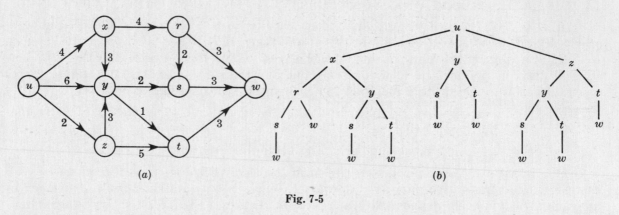

(a)   (b)

Fig. 7-5

One way of finding the shortest path is to compute all the paths of the tree. Observe that if two partial paths lead to an intermediate vertex v then we need only consider from then on the shorter partial path, i.e. we prune the tree at the vertex corresponding to the longer partial path. This pruning algorithm is formally described below.

During the algorithm, each vertex v' is assigned a number $\ell(v')$ denoting the current minimal length from the initial point u to v' and a path $p(v')$ from u to v' which has length $\ell(v')$. At each step of the algorithm, we examine an arc $\langle v', v \rangle$ from v' to v which, say, has length k. If this is the first time that we enter v then we assign

$$\ell(v) = \ell(v') + k \quad \text{and} \quad p(v) = p(v')v \tag{7.1}$$

If we have been at v before, then we check to see whether

$$\ell(v') + k < \ell(v)$$

If it is, then we have found a shorter path to v, and we change $\ell(v)$ and $p(v)$ as in (7.1); otherwise we leave $\ell(v)$ and $p(v)$ alone. Also, if there are no other unexamined arcs entering v, then we say that $p(v)$ *has been determined*. We begin by setting $\ell(u) = 0$ and $p(u) = u$ where u is the initial point. It is usually best to examine an arc which begins at a vertex whose path has been determined.

We apply the above algorithm to the graph in Fig. 7-5(a):

From u: The successive vertices, entered for the first time, are x, y and z.   Hence

$$\ell(x) = 4,\ \ p(x) = ux;\ \ \ \ell(y) = 6,\ \ p(y) = uy;\ \ \ \ell(z) = 2,\ \ p(z) = uz$$

Note that $p(x)$ and $p(z)$ have been determined.

From x: The successive vertices are r, entered for the first time, and y.   Hence

$$\ell(r) = 4 + 4 = 8 \quad \text{and} \quad p(r) = p(x)r = uxr$$

Also, $p(r)$ has been determined.   We next check that

$$\ell(x) + k = 4 + 3 \quad \text{is not less than} \quad \ell(y) = 6$$

Thus we leave $\ell(y)$ and $p(y)$ alone.

From z: The successive vertices are t, entered for the first time, and y.   Hence

$$\ell(t) = 2 + 5 = 7 \quad \text{and} \quad p(t) = p(z)t = uzt$$

We next check that

$$\ell(z) + k = 2 + 3 = 5 \quad \text{is less than} \quad \ell(y) = 6$$

Thus we have found a shorter path to y and we change $\ell(y)$ and $p(y)$ accordingly:

$$\ell(y) = \ell(z) + k = 5 \quad \text{and} \quad p(y) = p(z)y = uzy$$

Now $p(y)$ has been determined.

From y: The successive vertices are s, entered for the first time, and t.   Hence

$$\ell(s) = \ell(y) + k = 5 + 2 = 7 \quad \text{and} \quad p(s) = p(y)s = uzys$$

We next check that

$$\ell(y) + k = 5 + 1 = 6 \quad \text{is less than} \quad \ell(t) = 7$$

Hence we change $\ell(t)$ and $p(t)$ to read:

$$\ell(t) = \ell(y) + 1 = 6 \quad \text{and} \quad p(t) = p(y)t = uzyt$$

Now $p(t)$ has been determined.

From r: The successive vertices are w, entered for the first time, and s.   Hence

$$\ell(w) = \ell(r) + 3 = 11 \quad \text{and} \quad p(w) = uxrw$$

We next check that

$$\ell(r) + k = 8 + 2 = 10 \quad \text{is not less than} \quad \ell(s) = 7$$

Hence we change neither $\ell(s)$ nor $p(s)$.   Note $p(s)$ has now been determined.

From s: The successive vertex is w.   We note that

$$\ell(s) + k = 7 + 3 \quad \text{is less than} \quad \ell(w) = 11$$

Hence we change $\ell(w)$ and $p(w)$ to read:

$$\ell(w) = \ell(s) + 3 = 10 \quad \text{and} \quad p(w) = p(s)w = uzysw$$

From t: The successive vertex is w.   We note that

$$\ell(t) + k = 6 + 3 = 9 \quad \text{is less than} \quad \ell(w) = 10$$

Hence we change $\ell(w)$ and $p(w)$ to read:

$$\ell(w) = \ell(t) + 3 = 9 \quad \text{and} \quad p(w) = p(t)w = uzytw$$

Thus $p(w)$ has been determined and is the required shortest path; it has length $\ell(w) = 9$.

Figure 7-6 gives the tree of examined arcs. This is the tree in Fig. 7-5(b) which has been pruned at vertices belonging to longer partial paths. Observe that only thirteen of the original twenty-three arcs of the tree had to be examined.

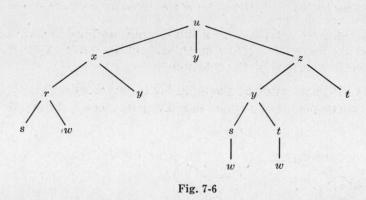

Fig. 7-6

When the above algorithm is used, it is convenient to initially assign $\ell(v) = \infty$ for every vertex v other than the initial vertex. Then it is always true that

$$\ell(v') + k \ < \ \ell(v)$$

when we first enter v.

7.6 FINITE STATE MACHINES

We may view a digital computer as a machine which is in a certain "internal state" at any given moment. The computer "reads" an input symbol, and then "prints" an output symbol and changes its "state". The output symbol depends solely upon the input symbol and the internal state of the machine, and the internal state of the machine depends solely upon the preceding state of the machine and the preceding input symbol. The number of states, input symbols and output symbols are assumed to be finite. These ideas are formalized in the following definition.

A *finite state machine* (or *complete sequential machine*) M consists of five things:

(1) A finite set A of input symbols.

(2) A finite set S of internal states.

(3) A finite set Z of output symbols.

(4) A next-state function f from $S \times A$ into S.

(5) An output function g from $S \times A$ into Z.

This machine M is denoted by $M = \langle A, S, Z, f, g \rangle$ when we want to designate its five parts. Sometimes we are also given an *initial state* q_0 in S, and then the machine M is designated by the six-tuple $M = \langle A, S, Z, q_0, f, g \rangle$.

EXAMPLE 7.1. The following defines a finite state machine with two input symbols, three internal states and three output symbols:

(1) $A \ = \ \{a, b\}$

(2) $S \ = \ \{q_0, q_1, q_2\}$

(3) $Z \ = \ \{x, y, z\}$

(4)   Next-state function $f : S \times A \to S$ defined by

$$f(q_0, a) \ = \ q_1 \qquad f(q_1, a) \ = \ q_2 \qquad f(q_2, a) \ = \ q_0$$
$$f(q_0, b) \ = \ q_2 \qquad f(q_1, b) \ = \ q_1 \qquad f(q_2, b) \ = \ q_1$$

(5)   Output function $g : S \times A \to Z$ defined by

$$g(q_0, a) = x \qquad g(q_1, a) = x \qquad g(q_2, a) = z$$
$$g(q_0, b) = y \qquad g(q_1, b) = z \qquad g(q_2, b) = y$$

It is traditional to use the letter q for the states of a machine and to use the symbol q_0 for the initial state.

There are two ways of representing a finite state machine in compact form. One way is by a table called the *state table* of the machine, and the other way is by a labeled directed graph called the *state diagram* of the machine.

The state table of the machine in Example 7.1 is shown with Fig. 7-7. Observe that the entry in the table corresponding to q_i and a_j is the pair $f(q_i, a_j)$, $g(q_i, a_j)$.

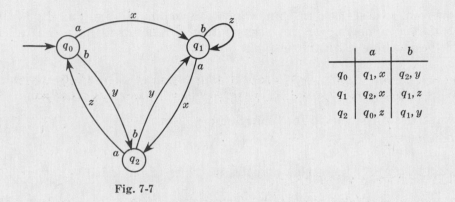

	a	b
q_0	q_1, x	q_2, y
q_1	q_2, x	q_1, z
q_2	q_0, z	q_1, y

Fig. 7-7

The state diagram D of a finite machine $M = \langle A, S, Z, f, g \rangle$ is a labeled directed graph. The vertices of D are the states of M, and if

$$f(q_i, a_j) = q_k \qquad \text{and} \qquad g(q_i, a_j) = z_r$$

then there is an arc from q_i to q_k which is labeled with the pair a_j, z_r. We usually put the input symbol a_j near the base of the arrow (near q_i) and the output symbol z_r near the center of the arrow. In case an initial state q_0 is given, then we label the vertex q_0 by drawing an extra arrow into q_0. For example, Fig. 7-7 is the state diagram of the machine in Example 7.1 with q_0 being the initial state.

7.7 STRINGS. INPUT AND OUTPUT TAPES

The previous section did not show the dynamic quality of a machine. Suppose a finite state machine M is given a string of input symbols:

$$U = a_1 a_2 \cdots a_n$$

We visualize these symbols on an "input tape"; the machine M "reads" these input symbols one by one and, simultaneously, changes through a string of states

$$V = s_0 s_1 s_2 \cdots s_n$$

where s_0 is the initial state, while printing a string of output symbols

$$W = z_1 z_2 \cdots z_n$$

on an "output tape". Formally, the initial state s_0 and the input string U determine the strings V and W by

$$s_i = f(s_{i-1}, a_i) \qquad \text{and} \qquad z_i = g(s_{i-1}, a_i)$$

where $i = 1, 2, \ldots, n$. (Observe that the word "string" is used for a finite sequence instead of "n-tuple" or "list.")

EXAMPLE 7.2. Suppose q_0 is the initial state of the machine in Example 7.1, and suppose the machine is given the input string

abaab

We calculate the string of states and the string of output symbols from the state diagram by beginning at the vertex q_0 and following the arrows which are labeled with the input symbols:

$$q_0 \xrightarrow{a,\,x} q_1 \xrightarrow{b,\,z} q_1 \xrightarrow{a,\,x} q_2 \xrightarrow{a,\,z} q_0 \xrightarrow{b,\,y} q_2$$

This yields the following strings of states and output symbols:

$q_0 q_1 q_1 q_2 q_0 q_2$ and $xzxzy$

We now want to describe a machine which can do binary addition. By adding 0's at the beginning of our numbers, we can assume that our numbers have the same number of digits. If the machine is given the input

1101011
+ 0111011

then we want the output to be the binary sum

10100110

Specifically, the input is the string of pairs of digits to be added:

11, 11, 00, 11, 01, 11, 10, b

where b denotes blank spaces, and the output should be the string

0, 1, 1, 0, 0, 1, 0, 1

We also want the machine to enter a state called "stop" when the machine finishes the addition.

The input symbols are
$$A = \{00, 01, 10, 11, b\}$$
and the output symbols are
$$Z = \{0, 1, b\}$$

The machine that we "construct" will have three states:

$$S = \{\text{carry } (c), \text{ no carry } (n), \text{ stop } (s)\}$$

Here n is the initial state. The machine is shown in Fig. 7-8.

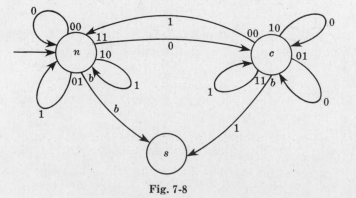

Fig. 7-8

In order to show the limitations of our machines, we state the following theorem.

Theorem 7.5:   There is no finite machine which can do binary multiplication.

If we limit the size of the numbers that we multiply, then such machines do exist. Computers are important examples of finite machines which multiply numbers, but the numbers are limited as to their size.

7.8   FINITE AUTOMATA

A finite automaton is similar to a finite state machine except that an automaton has "accepting" and "rejecting" states rather than an output.   Specifically, a finite automaton M consists of five things:

(1) A finite set A of *input symbols*.

(2) A finite set S of *internal states*.

(3) A subset T of S (whose elements are called *accepting states*).

(4) An *initial state* q_0 in S.

(5) A *next-state function* f from $S \times A$ into S.

The automaton M is denoted by $M = \langle A, S, T, q_0, f \rangle$ when we want to designate its five parts.

EXAMPLE 7.3.   The following defines a finite automaton with two input symbols and three states:

f	a	b
q_0	q_0	q_1
q_1	q_0	q_2
q_2	q_2	q_2

(1) $A = \{a, b\}$, input symbols

(2) $S = \{q_0, q_1, q_2\}$, states

(3) $T = \{q_0, q_1\}$, accepting states

(4) q_0, the initial state

(5)   Next-state function $f : S \times A \to S$ defined by the table at right.

We can concisely describe a finite automaton M by its state diagram as was done with finite state machines, except that here we use double circles for accepting states and each edge is labeled only by the input symbol.   Specifically, the *state diagram* D of M is a labeled directed graph whose vertices are the states of S where accepting states are labeled by having a double circle, and if $f(q_j, a_i) = q_k$ then there is an arc from q_j to q_k which is labeled with a_i.   Also the initial state q_0 is denoted by having an arrow entering the vertex q_0. For example, the state diagram of the automaton M of Example 7.3 is given in Fig. 7-9.

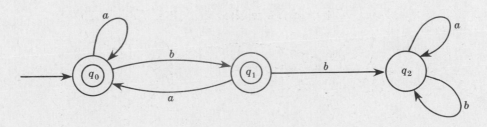

Fig. 7-9

Given a finite string $W = a_1 a_2 a_3 \cdots a_n$ of input symbols of an automaton M, we obtain a sequence of states $s_0 s_1 s_2 \cdots s_n$ where s_0 is the initial state and $s_i = f(s_{i-1}, a_i)$ for $i > 0$. We say that M *recognizes* or *accepts* the string W if the final state s_n is an accepting state, i.e. if $s_n \in T$.   We will let $L(M)$ denote the set of all strings which are recognized by M. For example, one can show that the automaton M in Example 7.3 will recognize those strings which do not have two successive b's. Thus M will accept

$$aababaaba, \quad aaa, \quad baab, \quad abaaababab, \quad b, \quad aabaaab$$

but will reject

$$aabaabba, \quad bbaaa, \quad ababbaab, \quad bb, \quad abbbbaa$$

We will show the relationship between finite automata and languages in Chapter 9.

Remark: We may also view a finite automaton M as a finite state machine with two output symbols, say, YES and NO, where the output is YES if M goes into an accepting state and the output is NO if M goes into a nonaccepting state. In other words, we make M into a finite state machine by defining an output function g from $S \times A$ into $Z = \{$YES, NO$\}$ as follows:

$$g(q_i, a_j) = \begin{cases} \text{YES}, & \text{if } f(q_i, a_j) \text{ is accepting (belongs to } T) \\ \text{NO}, & \text{if } f(q_i, a_j) \text{ is nonaccepting} \end{cases}$$

Conversely, a finite state machine with two output symbols may be viewed as a finite automaton in an analogous way.

Solved Problems

DIRECTED GRAPHS

7.1. Consider the digraph D pictured in Fig. 7-10. (*a*) Describe D formally. (*b*) Find the number of paths from X to Z. (*c*) Find the number of paths from Y to Z. (*d*) Are there any sources or sinks? (*e*) Find the matrix M_D of the digraph D.

(*f*) Is D weakly connected? Unilaterally connected? Strongly connected?

(*a*) There are four vertices: X, Y, Z, W, and there are seven arcs: $\langle X, Y \rangle$, $\langle X, W \rangle$, $\langle X, Z \rangle$, $\langle Y, W \rangle$, $\langle Z, Y \rangle$, $\langle Z, W \rangle$, $\langle W, Z \rangle$.

(*b*) There are three paths from X to Z: (X, Z), (X, W, Z) and (X, Y, W, Z).

(*c*) There is only one path from Y to Z: (Y, W, Z).

(*d*) X is a source since it is not the terminal point of any arc, i.e. its indegree is zero. There are no sinks since every vertex has a nonzero outdegree, i.e. each vertex is the initial point of some arc.

(*e*) The matrix M_D of D follows:

$$M_D = \begin{pmatrix} 0 & 1 & 1 & 1 \\ 0 & 0 & 0 & 1 \\ 0 & 1 & 0 & 1 \\ 0 & 0 & 1 & 0 \end{pmatrix}$$

(Here the rows and columns of M_D are labeled by X, Y, Z, W respectively.) The entry m_{ij} denotes the number of arcs from the ith vertex to the jth vertex.

(*f*) The digraph is not strongly connected since X is a source and hence there is no path from any other vertex, say Y, to X. However, D is unilaterally connected since the path (X, Y, W, Z) passes through all the vertices, and so there is a subpath connecting any pair of vertices.

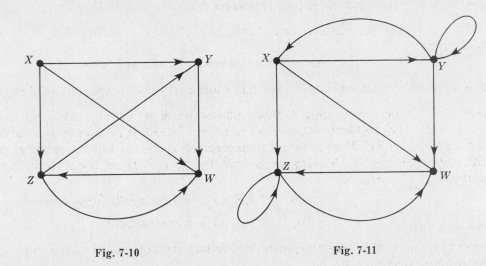

Fig. 7-10   Fig. 7-11

7.2.   Consider the digraph D pictured in Fig. 7-11.

      (a) Are there any sources or sinks?

      (b) Use Corollary 7.4 to show that D is not strongly connected.

      (a) There are no sources or sinks.

      (b) Find the matrix M of D. Evaluate M^2, M^3 and $C = M + M^2 + M^3$.

$$M = \begin{pmatrix} 0 & 1 & 1 & 1 \\ 1 & 1 & 0 & 1 \\ 0 & 0 & 1 & 1 \\ 0 & 0 & 1 & 0 \end{pmatrix} \qquad M^2 = \begin{pmatrix} 1 & 1 & 2 & 2 \\ 1 & 2 & 2 & 2 \\ 0 & 0 & 2 & 1 \\ 0 & 0 & 1 & 1 \end{pmatrix}$$

$$M^3 = \begin{pmatrix} 1 & 2 & 5 & 4 \\ 2 & 3 & 5 & 5 \\ 0 & 0 & 3 & 2 \\ 0 & 0 & 2 & 1 \end{pmatrix} \qquad C = \begin{pmatrix} 2 & 4 & 8 & 7 \\ 4 & 6 & 7 & 8 \\ 0 & 0 & 6 & 4 \\ 0 & 0 & 4 & 2 \end{pmatrix}$$

The rows and columns of the matrices are labeled by the vertices X, Y, Z, W respectively. The zero entries in C (off the main diagonal) show that D is not strongly connected. In particular, C tells us that there are no paths from Z or W to X or Y.

7.3.   Let $V = \{2, 3, 4, 5, 6\}$. Let R be the relation on V defined by xRy if x is less than y and x is relatively prime to y.

      (a) Write R as a set of ordered pairs.

      (b) Draw a diagram of the directed graph which corresponds to R.

      (a) $R = \{(2,3), (2,5), (3,4), (3,5), (4,5), (5,6)\}$

      (b) We draw an arc from x to y if (x,y) belongs to R, as shown in Fig. 7-12.

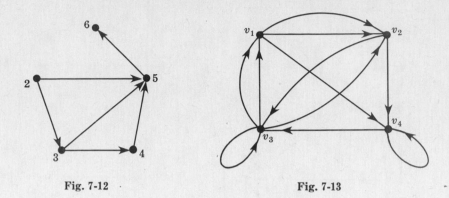

Fig. 7-12   Fig. 7-13

7.4. Draw the digraph D corresponding to the following matrix M which has nonnegative integer entries:

$$M = \begin{pmatrix} 0 & 2 & 0 & 1 \\ 0 & 0 & 1 & 1 \\ 2 & 1 & 1 & 0 \\ 0 & 0 & 1 & 1 \end{pmatrix}$$

Since M is a 4×4 matrix, D has four vertices, say, v_1, v_2, v_3, v_4. For each entry m_{ij} draw m_{ij} arcs from the vertex v_i to the vertex v_j to obtain the digraph in Fig. 7-13.

7.5. Prove Theorem 7.3: Let M be the matrix of a digraph D. Then the ij entry of the matrix M^n gives the number of walks of length n from the vertex v_i to the vertex v_j.

The proof is by induction on n. Note first that a walk of length 1 from v_i to v_j is precisely an arc $\langle v_i, v_j \rangle$. Thus the theorem holds for $n = 1$ since the ij entry of M gives the number of arcs $\langle v_i, v_j \rangle$ which is the number of walks of length 1 from v_i to v_j.

Suppose $n > 1$ and $M^{n-1} = (a_{ik})$, $M = (b_{kj})$ and $M^n = (c_{ij})$. Then $c_{ij} = \sum_{k=1}^{n} a_{ik} b_{kj}$. By induction a_{ik} equals the number of walks of length $n-1$ from v_i to v_k. Also, b_{kj} equals the number of arcs from v_k to v_j. Thus $a_{ik} b_{kj}$ gives the number of walks of length n from v_i to v_j where v_k is the next-to-last vertex in the walk. Thus all walks of length n from v_i to v_j can be obtained by summing up the $a_{ik} b_{kj}$ for all k. That is, c_{ij} is the number of walks of length n from v_i to v_j. Thus the theorem is proved.

FINITE MACHINES

7.6. Consider a finite state machine M with input symbols a, b and output symbols x, y, z and the state diagram given in Fig. 7-14. (a) Find the state table for M. (b) Determine the output if the input is the string of symbols $W = aababaabbab$.

(a) We label the rows of the table by the four states q_0, q_1, q_2, q_3 and the columns by the input symbols a, b. Using the state diagram given in Fig. 7-14, we find the entries in the table as follows. From state q_0, the arrow labeled a goes to state q_1 and is labeled with the output symbol x. Hence q_1, x is put in the table in the position corresponding to the row q_0 and column a. The other entries in the state table below are obtained similarly.

	a	b
q_0	q_1, x	q_2, y
q_1	q_3, y	q_1, z
q_2	q_1, z	q_0, x
q_3	q_0, z	q_2, x

(b) Beginning at q_0, the initial state, we go from state to state by the arrows which are labeled respectively by the given input symbols:

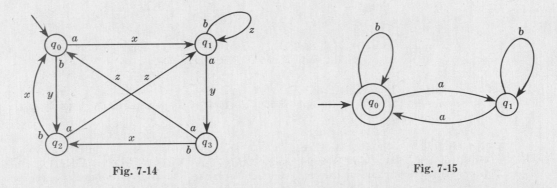

$$q_0 \xrightarrow{a} q_1 \xrightarrow{a} q_3 \xrightarrow{b} q_2 \xrightarrow{a} q_1 \xrightarrow{b} q_1 \xrightarrow{a} q_3 \xrightarrow{a} q_0 \xrightarrow{b} q_2 \xrightarrow{b} q_0 \xrightarrow{a} q_1 \xrightarrow{b} q_1$$

The output symbols on the above arrows are respectively *xyxzzyzyxxz*.

Fig. 7-14   Fig. 7-15

7.7. Let a and b be the input symbols. Construct a finite automaton M which will accept precisely those strings in a and b which have an even number of a's. (For example, *aababbab, aa, bbb, ababaa* will be accepted but *ababa, aaa, bbabb* will not be accepted.)

  We need only two states, q_0 and q_1. We assume that M is in state q_0 or q_1 according as the number of a's up to the given step is even or odd. (Thus q_0 is an accepting state but q_1 is non-accepting.) Then only a will change the state. Also, q_0 is the initial state. The state diagram of M is Fig. 7-15.

7.8. Let $M = \langle A, S, T, q_0, f \rangle$ and $M' = \langle A, S', T', q'_0, f' \rangle$ be two finite automata with the same set A of input symbols. Let $L(M)$ and $L(M')$ be the sets of strings accepted by

are accepting states of M and M' respectively, and we let (q_0, q'_0) be the initial state of N. The next-state function

$$g : (S \times S') \times A \to (S \times S')$$

of N is defined as follows:

$$g((q, q'), a) = (f(q, a), f'(q', a))$$

Then N will accept precisely those strings in $L(M) \cap L(M')$.

Supplementary Problems

DIRECTED GRAPHS

7.9. Consider the digraph D pictured in Fig. 7-16.

  (a) Find the indegree and outdegree of each vertex.

  (b) Find the number of paths from v_1 to v_4.

  (c) Are there any sources or sinks?

  (d) Find the matrix M of D.

  (e) Find the number of walks of length 3 or less from v_1 to v_3.

  (f) Is D unilaterally connected? Strongly connected?

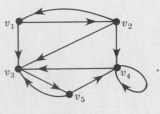

Fig. 7-16

7.10. Let D be the digraph with vertices v_1, v_2, v_3, v_4 corresponding to the matrix

$$M = \begin{pmatrix} 0 & 1 & 1 & 1 \\ 0 & 1 & 1 & 0 \\ 1 & 1 & 0 & 2 \\ 1 & 0 & 0 & 0 \end{pmatrix}$$

(a) Draw the diagram of D. (b) Find the number of walks of length 3 from v_1 to (i) v_1, (ii) v_2, (iii) v_3, (iv) v_4. (c) Is D unilaterally connected? Strongly connected?

7.11. Let R be the relation on $V = \{2, 3, 4, 9, 15\}$ defined by "x is less than and relatively prime to y". (a) Draw the diagram of the digraph of R. (b) Is R weakly connected? Unilaterally connected? Strongly connected?

7.12. A digraph D is *complete* if for each pair of distinct vertices v_i and v_j, either $\langle v_i, v_j \rangle$ is an arc or $\langle v_j, v_i \rangle$ is an arc. Show that a finite, complete digraph D has a path which includes all vertices. (This obviously holds for nondirected, complete graphs.) Thus D is unilaterally connected.

7.13. Use the pruning algorithm (Section 7.5) to find the shortest path from s to t in Fig. 7-17.

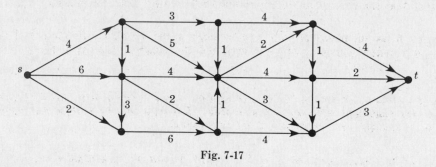

Fig. 7-17

FINITE MACHINES

7.14. Consider the finite state machine M with input symbols a, b, c and output symbols x, y, z and the state diagram shown in Fig. 7-18. (a) Construct the state table of M. (b) Find the output if the input is the string $W = caabbaccab$.

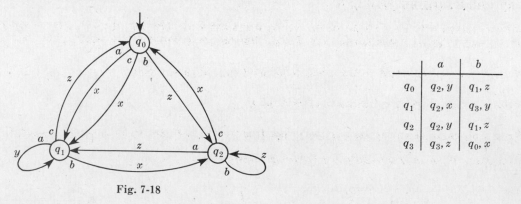

Fig. 7-18

	a	b
q_0	q_2, y	q_1, z
q_1	q_2, x	q_3, y
q_2	q_2, y	q_1, z
q_3	q_3, z	q_0, x

7.15. Let M be the finite state machine with the state table shown next to Fig. 7-18. (a) Draw the state diagram of M given that q_0 is the initial state. (b) Find the output if the input is the string $W = aabbabbaab$.

7.16. For each of the machines in Fig. 7-19 with input symbols a, b and output symbols x, y, z, find the output if the input is the string $W = abaabbabbaabaa$.

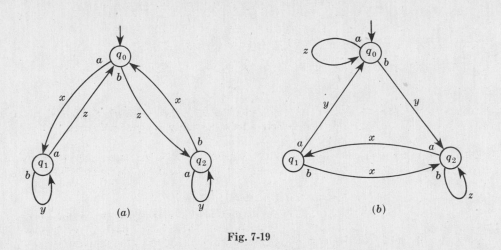

Fig. 7-19

7.17. Construct a finite automaton M with input symbols a, b which will only accept those strings in a and b such that the number of b's is divisible by 3. (*Hint*: Three states are required.)

7.18. Construct a finite automaton M with input symbols a and b which will only accept those strings in a and b such that the number of a's is even and the number of b's is divisible by 3. (*Hint*: Use Problems 7.7, 7.8 and 7.17.)

7.19. Construct a finite automaton M with input symbols a and b which will only accept those strings in a and b such that $aabb$ appears as a substring. For example, $ba(aabb)ba$ and $bab(aabb)a$ will be accepted, but $babbaa$ and $aababaa$ will not be accepted.

7.20. Let $M = \langle A, S, T, q_0, f \rangle$ and $M' = \langle A, S', T', q_0', f' \rangle$ be two finite automata with the ~~same~~ input symbols. Let $L(M)$ and $L(M')$ ~~be~~ Construct a finite automaton N ~~set of strings that do not belong to $L(M)$~~, (b) $L(M) \cup L(M')$ i.e. the set of strings that belong to $L(M)$ or $L(M')$.

COMPUTER PROGRAMMING PROBLEMS

Let D be a digraph with six vertices, $v_1, v_2, \ldots, v_6$, and eight arcs. Let B be the 8×2 arc matrix of D. That is, a row, say $(4, 2)$, of B denotes the arc $\langle v_4, v_2 \rangle$. Suppose B is punched into a deck of cards.

7.21. Write a program which prints the indegree and outdegree of each vertex.

7.22. Write a program which prints the matrix M of D.

7.23. Write a program which decides whether or not D is weakly, unilaterally or strongly connected.

Test the above three programs with the following matrices:

$$
\text{(i)} \quad B = \begin{pmatrix} 1 & 4 \\ 2 & 2 \\ 6 & 5 \\ 1 & 5 \\ 3 & 2 \\ 4 & 1 \\ 4 & 3 \\ 4 & 6 \end{pmatrix}
\qquad
\text{(ii)} \quad B = \begin{pmatrix} 1 & 4 \\ 5 & 4 \\ 4 & 2 \\ 6 & 1 \\ 2 & 5 \\ 3 & 5 \\ 2 & 6 \\ 4 & 3 \end{pmatrix}
\qquad
\text{(iii)} \quad B = \begin{pmatrix} 1 & 6 \\ 5 & 6 \\ 2 & 3 \\ 4 & 1 \\ 5 & 2 \\ 3 & 5 \\ 1 & 5 \\ 6 & 3 \end{pmatrix}
$$

Answers to Supplementary Problems

7.9.   (a) Indegrees: 1, 1, 4, 3, 1. Outdegrees: 2, 3, 1, 2, 2.

    (b) Three: (v_1, v_2, v_4), (v_1, v_3, v_5, v_4), $(v_1, v_2, v_3, v_5, v_4)$.   (c) No.

$$(d) \quad M = \begin{pmatrix} 0 & 1 & 1 & 0 & 0 \\ 1 & 0 & 1 & 1 & 0 \\ 0 & 0 & 0 & 0 & 1 \\ 0 & 0 & 1 & 1 & 0 \\ 0 & 0 & 1 & 1 & 0 \end{pmatrix}, \quad (e)\ 5, \quad (f)\ \text{Unilateral, but not strong}$$

7.10.   (a) See Fig. 7-20.   (b) 3, 5, 4, 4   (c) Unilateral and strong

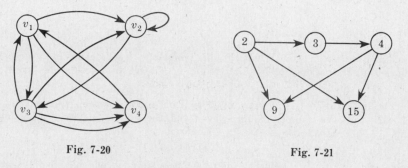

Fig. 7-20   Fig. 7-21

7.11.   (a) See Fig. 7-21.   (b) Only weakly connected

7.12.   *Hint*: Suppose $(v_1, v_2, \ldots, v_m)$ is the longest path in D and does not include the vertex u. Then $\langle u, v_1 \rangle$ and $\langle v_m, u \rangle$ are not arcs since, if they were, the path could be extended. Hence $\langle v_1, u \rangle$ and $\langle u, v_m \rangle$ are arcs. Let k be the smallest integer such that $\langle v_k, u \rangle$ and $\langle u, v_{k+1} \rangle$ are arcs. Then $(v_1, \ldots, v_k, u, v_{k+1}, \ldots, v_m)$ is a longer path.

7.13.

$$s \;\overset{4}{\text{---}}\; \bullet \;\overset{1}{\text{---}}\; \bullet \;\overset{2}{\text{---}}\; \bullet \;\overset{1}{\text{---}}\; \bullet \;\overset{2}{\text{---}}\; \bullet \;\overset{1}{\text{---}}\; \bullet \;\overset{2}{\text{---}}\; t$$

7.14.   (a)

	a	b	c
q_0	q_1, x	q_2, z	q_1, x
q_1	q_1, y	q_2, x	q_0, z
q_2	q_1, z	q_2, z	q_0, x

    (b) $xyyxzzzxyx$

7.15.   (a) See Fig. 7-22.   (b) $yyzyzxzxyz$

7.16.   (a) $xyzxyyzzxxzzyy$,   (b) $zyxyyzxxzxyyxy$

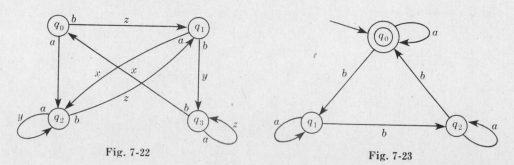

Fig. 7-22   Fig. 7-23

7.17.   See Fig. 7-23.

7.18.   See Fig. 7-24.

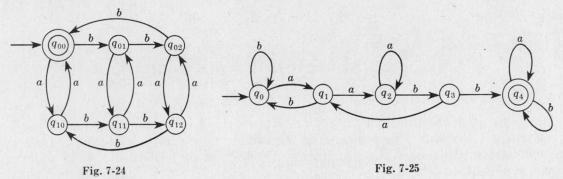

Fig. 7-24   Fig. 7-25

7.19.   See Fig. 7-25.

7.20.   (a) $N = \langle A, S, S \setminus T, q_0, f \rangle$

(b) $N = \langle A, S \times S', (S \times T') \cup (T \times S'), (q_0, q_0'), g \rangle$ where

$$g((q, q'), a) = (f(q, a), f'(q', a))$$

Chapter 8

Combinatorial Analysis

8.1 FUNDAMENTAL PRINCIPLE OF COUNTING

Combinatorial analysis, which includes the study of permutations, combinations and partitions, is concerned with determining the number of logical possibilities of some event without necessarily enumerating each case. The following basic principle of counting is used throughout.

Fundamental Principle of Counting: If some event can occur in n_1 different ways, and if, following this event, a second event can occur in n_2 different ways, and, following this second event, a third event can occur in n_3 different ways, ..., then the number of ways the events can occur in the order indicated is $n_1 \cdot n_2 \cdot n_3 \cdots$.

EXAMPLE 8.1.

(a) Suppose a license plate contains two letters followed by three digits with the first digit not zero. How many different license plates can be printed?

Each letter can be printed in twenty-six different ways, the first digit in nine ways and each of the other two digits in ten ways. Hence

$$26 \cdot 26 \cdot 9 \cdot 10 \cdot 10 \ = \ 608,400$$

different plates can be printed.

(b) In how many ways can an organization containing twenty-six members elect a president, treasurer and secretary (assuming no person is elected to more than one position)?

The president can be elected in twenty-six different ways; following this, the treasurer can be elected in twenty-five different ways (since the person chosen president is not eligible to be treasurer); and, following this, the secretary can be elected in twenty-four different ways. Thus, by the above principle of counting, there are

$$26 \cdot 25 \cdot 24 \ = \ 15,600$$

different ways in which the organization can elect the officers.

8.2 FACTORIAL NOTATION

The product of the positive integers from 1 to n inclusive is denoted by $n!$ (read "n factorial"):

$$n! \ = \ 1 \cdot 2 \cdot 3 \cdots (n-2)(n-1)n$$

In other words, $n!$ is defined by

$$1! = 1 \quad \text{and} \quad n! = n \cdot (n-1)!$$

It is also convenient to define $0! = 1$.

EXAMPLE 8.2.

(a) $2! = 1 \cdot 2 = 2 \quad 3! = 1 \cdot 2 \cdot 3 = 6 \quad 4! = 1 \cdot 2 \cdot 3 \cdot 4 = 24$

$5! = 5 \cdot 4! = 5 \cdot 24 = 120 \quad 6! = 6 \cdot 5! = 6 \cdot 120 = 720$

(b) $\dfrac{8!}{6!} = \dfrac{8 \cdot 7 \cdot 6!}{6!} = 8 \cdot 7 = 56 \qquad 12 \cdot 11 \cdot 10 = \dfrac{12 \cdot 11 \cdot 10 \cdot 9!}{9!} = \dfrac{12!}{9!}$

$\dfrac{12 \cdot 11 \cdot 10}{1 \cdot 2 \cdot 3} = 12 \cdot 11 \cdot 10 \cdot \dfrac{1}{3!} = \dfrac{12!}{3! \, 9!}$

(c) $n(n-1)\cdots(n-r+1) = \dfrac{n(n-1)\cdots(n-r+1)(n-r)(n-r-1)\cdots 3 \cdot 2 \cdot 1}{(n-r)(n-r-1)\cdots 3 \cdot 2 \cdot 1} = \dfrac{n!}{(n-r)!}$

$\dfrac{n(n-1)\cdots(n-r+1)}{1 \cdot 2 \cdot 3 \cdots (r-1)r} = n(n-1)\cdots(n-r+1) \cdot \dfrac{1}{r!} = \dfrac{n!}{(n-r)!} \cdot \dfrac{1}{r!} = \dfrac{n!}{r!\,(n-r)!}$

8.3 BINOMIAL COEFFICIENTS

The symbol $\dbinom{n}{r}$ (read "nCr"), where r and n are positive integers with $r \leqq n$, is defined as follows

$$\binom{n}{r} = \frac{n(n-1)(n-2)\cdots(n-r+1)}{1 \cdot 2 \cdot 3 \cdots (r-1)r}$$

By Example 8.2(c), we see that

$$\binom{n}{r} = \frac{n(n-1)\cdots(n-r+1)}{1 \cdot 2 \cdot 3 \cdots (r-1)r} = \frac{n!}{r!\,(n-r)!}$$

But $n - (n-r) = r$; hence we have the following important relation:

$$\binom{n}{n-r} = \binom{n}{r} \quad \text{or, in other words, if } a + b = n \text{ then } \binom{n}{a} = \binom{n}{b}$$

EXAMPLE 8.3.

(a) $\dbinom{8}{2} = \dfrac{8 \cdot 7}{1 \cdot 2} = 28 \qquad \dbinom{9}{4} = \dfrac{9 \cdot 8 \cdot 7 \cdot 6}{1 \cdot 2 \cdot 3 \cdot 4} = 126 \qquad \dbinom{12}{5} = \dfrac{12 \cdot 11 \cdot 10 \cdot 9 \cdot 8}{1 \cdot 2 \cdot 3 \cdot 4 \cdot 5} = 792$

$\dbinom{10}{3} = \dfrac{10 \cdot 9 \cdot 8}{1 \cdot 2 \cdot 3} = 120 \qquad \dbinom{13}{1} = \dfrac{13}{1} = 13$

Note that $\dbinom{n}{r}$ has exactly r factors in both the numerator and the denominator.

(b) Compute $\dbinom{10}{7}$. By definition,

$$\binom{10}{7} = \frac{10 \cdot 9 \cdot 8 \cdot 7 \cdot 6 \cdot 5 \cdot 4}{1 \cdot 2 \cdot 3 \cdot 4 \cdot 5 \cdot 6 \cdot 7} = 120$$

On the other hand, $10 - 7 = 3$ and so we can also compute $\dbinom{10}{7}$ as follows:

$$\binom{10}{7} = \binom{10}{3} = \frac{10 \cdot 9 \cdot 8}{1 \cdot 2 \cdot 3} = 120$$

Observe that the second method saves space and time.

The numbers $\dbinom{n}{r}$ are called the *binomial coefficients* since they appear as the coefficients in the expansion of $(a+b)^n$. Specifically, one can prove that

$$(a+b)^n = \sum_{k=0}^{n} \binom{n}{k} a^{n-k} b^k$$

The coefficients of the successive powers of $a + b$ can be arranged in a triangular array of numbers, called Pascal's triangle, as follows:

$$(a + b)^0 = 1$$
$$(a + b)^1 = a + b$$
$$(a + b)^2 = a^2 + 2ab + b^2$$
$$(a + b)^3 = a^3 + 3a^2b + 3ab^2 + b^3$$
$$(a + b)^4 = a^4 + 4a^3b + 6a^2b^2 + 4ab^3 + b^4$$
$$(a + b)^5 = a^5 + 5a^4b + 10a^3b^2 + 10a^2b^3 + 5ab^4 + b^5$$
$$(a + b)^6 = a^6 + 6a^5b + 15a^4b^2 + 20a^3b^3 + 15a^2b^4 + 6ab^5 + b^6$$

```
                    1
                  1   1
                1   2   1
              1   3   3   1
            1   4   6   4   1
          1   5  (10) 10   5   1
        1   6  (15)(20) 15   6   1
```

Pascal's triangle has the following interesting properties.

(i)   The first number and the last number in each row is 1.

(ii)   Every other number in the array can be obtained by adding the two numbers appearing directly above it. For example $10 = 4 + 6$, $15 = 5 + 10$, $20 = 10 + 10$.

Since the numbers appearing in Pascal's triangle are the binomial coefficients, property (ii) of Pascal's triangle comes from the following theorem (proved in Problem 8.8).

Theorem 8.1: $\quad \dbinom{n+1}{r} = \dbinom{n}{r-1} + \dbinom{n}{r}$

8.4 PERMUTATIONS

Any arrangement of a set of n objects in a given order is called a *permutation* of the objects (taken all at a time). Any arrangement of any $r \leq n$ of these objects in a given order is called an *r-permutation* or a *permutation of the n objects taken r at a time*. Consider, for example, the set of letters a, b, c and d. Then:

(i)   *bdca, dcba* and *acdb* are permutations of the four letters (taken all at a time);

(ii)   *bad, adb, cbd* and *bca* are permutations of the four letters taken three at a time;

(iii)   *ad, cb, da* and *bd* are permutations of the four letters taken two at a time.

The number of permutations of n objects taken r at a time is denoted by

$$P(n, r), \quad {}_nP_r, \quad P_{n,r}, \quad P_r^n, \quad \text{or} \quad (n)_r$$

We shall use $P(n, r)$. Before we derive the general formula for $P(n, r)$ we consider a particular case.

EXAMPLE 8.4. How many permutations are there of six objects, say a, b, c, d, e and f, taken three at a time? In other words, we want to find the number of "three-letter words" using the above six letters without repetitions.

Let the general three-letter word be represented by the following three boxes:

Now the first letter can be chosen in six different ways; following this, the second letter can be chosen in five different ways; and, following this, the last letter can be chosen in four different ways. Write each number in its appropriate box as follows:

Thus by the fundamental principle of counting there are $6 \cdot 5 \cdot 4 = 120$ possible three-letter words without repetitions from the six letters, or there are 120 permutations of six objects taken three at a time:

$$P(6, 3) = 120$$

The derivation of the formula for the number of permutations of n objects taken r at a time, or the number of r-permutations of n objects, $P(n, r)$, follows the procedure in the preceding example. The first element in an r-permutation of n objects can be chosen in n different ways; following this, the second element in the permutation can be chosen in $n - 1$ ways; and, following this, the third element in the permutation can be chosen in $n - 2$ ways. Continuing in this manner, we have that the rth (last) element in the r-permutation can be chosen in $n - (r - 1) = n - r + 1$ ways. Thus, by the fundamental principle of counting, we have

$$P(n, r) = n(n - 1)(n - 2) \cdots (n - r + 1)$$

By Example 8.2(c), we see that

$$n(n - 1)(n - 2) \cdots (n - r + 1) = \frac{n(n - 1)(n - 2) \cdots (n - r + 1) \cdot (n - r)!}{(n - r)!} = \frac{n!}{(n - r)!}$$

Thus we have proven

Theorem 8.2: $P(n, r) = \dfrac{n!}{(n - r)!}$

In the special case in which $r = n$, we have

$$P(n, n) = n(n - 1)(n - 2) \cdots 3 \cdot 2 \cdot 1 = n!$$

Accordingly,

Corollary 8.3:   There are $n!$ permutations of n objects (taken all at a time).

For example, there are $3! = 1 \cdot 2 \cdot 3 = 6$ permutations of the three letters a, b and c. These are abc, acb, bac, bca, cab, cba.

8.5 PERMUTATIONS AND REPETITIONS

Frequently we want to find the number of permutations of objects some of which are alike, as illustrated in the examples below. The general formula is as follows:

Theorem 8.4:   The number of permutations of n objects of which n_1 are alike, n_2 are alike, $\ldots$, n_r are alike is

$$\frac{n!}{n_1! \, n_2! \cdots n_r!}$$

We indicate the proof of the above theorem by a particular example. Suppose we want to form all possible five-letter "words" using the letters from the word "DADDY". Now there are $5! = 120$ permutations of the objects D_1, A, D_2, D_3, Y, where the three D's are distinguished. Observe that the following six permutations

$$D_1D_2D_3AY, \quad D_2D_1D_3AY, \quad D_3D_1D_2AY, \quad D_1D_3D_2AY, \quad D_2D_3D_1AY, \quad \text{and} \quad D_3D_2D_1AY$$

produce the same word when the subscripts are removed. The 6 comes from the fact that there are $3! = 3 \cdot 2 \cdot 1 = 6$ different ways of placing the three D's in the first three positions in the permutation. This is true for each set of three positions in which the D's can appear. Accordingly there are

$$\frac{5!}{3!} = \frac{120}{6} = 20$$

different five-letter words that can be formed using the letters from the word "DADDY".

EXAMPLE 8.5.

(a) How many seven-letter words can be formed using the letters of the word "BENZENE"? We seek the number of permutations of seven objects of which three are alike (the three E's), and two are alike (the two N's). By Theorem 8.4, there are

$$\frac{7!}{3!\,2!} = \frac{7\cdot6\cdot5\cdot4\cdot3\cdot2\cdot1}{3\cdot2\cdot1\cdot2\cdot1} = 420$$

such words.

(b) How many different signals, each consisting of eight flags hung in a vertical line, can be formed from a set of four indistinguishable red flags, three indistinguishable white flags, and a blue flag? We seek the number of permutations of eight objects of which four are alike and three are alike. There are

$$\frac{8!}{4!\,3!} = \frac{8\cdot7\cdot6\cdot5\cdot4\cdot3\cdot2\cdot1}{4\cdot3\cdot2\cdot1\cdot3\cdot2\cdot1} = 280$$

different signals.

8.6 COMBINATIONS

Suppose we have a collection of n objects. A *combination* of these n objects taken r at a time is any selection of r of the objects where order doesn't count. In other words, an *r-combination* of a set of n objects is any subset of r elements. For example, the combinations of the letters a, b, c, d taken three at a time are

$$\{a,b,c\},\ \{a,b,d\},\ \{a,c,d\},\ \{b,c,d\} \quad \text{or simply} \quad abc,\ abd,\ acd,\ bcd$$

Observe that the following combinations are equal:

$$abc,\ acb,\ bac,\ bca,\ cab \text{ and } cba$$

That is, each denotes the same set $\{a,b,c\}$.

The number of combinations of n objects taken r at a time is denoted by $C(n,r)$. The symbols $_nC_r$, $C_{n,r}$ and C_r^n also appear in various texts. Before we give the general formula for $C(n,r)$, we consider a special case.

EXAMPLE 8.6.   How many combinations are there of the four objects, a, b, c and d, taken three at a time?

Each combination consisting of three objects determines $3! = 6$ permutations of the objects in the combination

Combinations	Permutations
abc	abc, acb, bac, bca, cab, cba
abd	abd, adb, bad, bda, dab, dba
acd	acd, adc, cad, cda, dac, dca
bcd	bcd, bdc, cbd, cdb, dbc, dcb

Thus the number of combinations multiplied by 3! equals the number of permutations:

$$C(4,3)\cdot3! = P(4,3) \quad \text{or} \quad C(4,3) = \frac{P(4,3)}{3!}$$

But $P(4,3) = 4\cdot3\cdot2 = 24$ and $3! = 6$; hence $C(4,3) = 4$ as noted above.

Since any combination of n objects taken r at a time determines $r!$ permutations of the objects in the combination, we can conclude that

$$P(n,r) = r!\,C(n,r)$$

Thus we obtain

Theorem 8.5: $C(n, r) = \dfrac{P(n, r)}{r!} = \dfrac{n!}{r!\,(n-r)!}$

Recall that the binomial coefficient $\begin{pmatrix} n \\ r \end{pmatrix}$ was defined to be $\dfrac{n!}{r!\,(n-r)!}$; hence

$$C(n, r) = \begin{pmatrix} n \\ r \end{pmatrix}$$

We shall use $C(n, r)$ and $\begin{pmatrix} n \\ r \end{pmatrix}$ interchangeably.

EXAMPLE 8.7.

(a) How many committees of three can be formed from eight people? Each committee is, essentially, a combination of the eight people taken three at a time.   Thus

$$C(8, 3) = \begin{pmatrix} 8 \\ 3 \end{pmatrix} = \frac{8 \cdot 7 \cdot 6}{1 \cdot 2 \cdot 3} = 56$$

different committees can be formed.

(b) A farmer buys three cows, two pigs and four hens from a man who has six cows, five pigs and eight hens. How many choices does the farmer have?

The farmer can choose the cows in $\begin{pmatrix} 6 \\ 3 \end{pmatrix}$ ways, the pigs in $\begin{pmatrix} 5 \\ 2 \end{pmatrix}$ ways, and the hens in $\begin{pmatrix} 8 \\ 4 \end{pmatrix}$ ways.

Hence altogether he can choose the animals in

$$\begin{pmatrix} 6 \\ 3 \end{pmatrix}\begin{pmatrix} 5 \\ 2 \end{pmatrix}\begin{pmatrix} 8 \\ 4 \end{pmatrix} = \frac{6 \cdot 5 \cdot 4}{1 \cdot 2 \cdot 3} \cdot \frac{5 \cdot 4}{1 \cdot 2} \cdot \frac{8 \cdot 7 \cdot 6 \cdot 5}{1 \cdot 2 \cdot 3 \cdot 4} = 20 \cdot 10 \cdot 70 = 14{,}000 \text{ ways}$$

8.7 ORDERED PARTITIONS

Suppose an urn A contains seven marbles numbered 1 through 7. We compute the number of ways we can draw, first, two marbles from the urn, then three marbles from the urn, and lastly two marbles from the urn. In other words, we want to compute the number of *ordered partitions*

$$(A_1, A_2, A_3)$$

of the set of seven marbles into cells A_1 containing two marbles, A_2 containing three marbles and A_3 containing two marbles. We call these ordered partitions since we distinguish between

$$(\{1, 2\}, \{3, 4, 5\}, \{6, 7\}) \quad \text{and} \quad (\{6, 7\}, \{3, 4, 5\}, \{1, 2\})$$

each of which determines the same partition of A.

Now we begin with seven marbles in the urn, so there are $\begin{pmatrix} 7 \\ 2 \end{pmatrix}$ ways of drawing the first two marbles, i.e. of determining the first cell A_1; following this, there are five marbles left in the urn and so there are $\begin{pmatrix} 5 \\ 3 \end{pmatrix}$ ways of drawing the three marbles, i.e. of determining the second cell A_2; finally, there are two marbles left in the urn and so there are $\begin{pmatrix} 2 \\ 2 \end{pmatrix}$ ways of determining the last cell A_3.   Hence there are

$$\begin{pmatrix} 7 \\ 2 \end{pmatrix}\begin{pmatrix} 5 \\ 3 \end{pmatrix}\begin{pmatrix} 2 \\ 2 \end{pmatrix} = \frac{7 \cdot 6}{1 \cdot 2} \cdot \frac{5 \cdot 4 \cdot 3}{1 \cdot 2 \cdot 3} \cdot \frac{2 \cdot 1}{1 \cdot 2} = 210$$

different ordered partitions of A into cells A_1 containing two marbles, A_2 containing three marbles, and A_3 containing two marbles.

Now observe that

$$\binom{7}{2}\binom{5}{3}\binom{2}{2} = \frac{7!}{2!5!} \cdot \frac{5!}{3!2!} \cdot \frac{2!}{2!0!} = \frac{7!}{2!3!2!}$$

since each numerator after the first is canceled by the second term in the denominator of the previous factor.

The above discussion can be shown to hold in general.   Namely,

Theorem 8.6:   Let A contain n elements and let $n_1, n_2, \ldots, n_r$ be positive integers with $n_1 + n_2 + \cdots + n_r = n$.   Then there exist

$$\frac{n!}{n_1!\, n_2!\, n_3! \cdots n_r!}$$

different ordered partitions of A of the form $(A_1, A_2, \ldots, A_r)$ where A_1 contains n_1 elements, A_2 contains n_2 elements, $\ldots$, and A_r contains n_r elements.

We apply this theorem in the next example.

EXAMPLE 8.8.   In how many ways can nine toys be divided between four children if the youngest child is to receive three toys and each of the others two toys?

We wish to find the number of ordered partitions of the nine toys into four cells containing 3, 2, 2 and 2 toys respectively.   By Theorem 8.6, there are

$$\frac{9!}{3!\,2!\,2!\,2!} = 7560$$

such ordered partitions.

8.8   TREE DIAGRAMS

A (rooted) tree diagram is a useful device to enumerate all the logical possibilities of a sequence of events where each event can occur in a finite number of ways.   We illustrate this method with an example.

EXAMPLE 8.9.   A man has time to play roulette at most five times.   At each play he wins or loses a dollar. The man begins with one dollar and will stop playing before the five times if he loses all his money or if he wins three dollars, i.e. if he has four dollars.   The tree diagram describes the way the betting can occur. Each number in the diagram denotes the number of dollars he has at that point.

The betting can occur in eleven different ways.   Note that he will stop betting before the five times are up in only three of the cases.

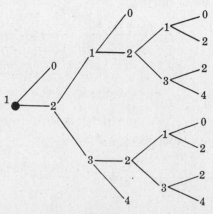

Fig. 8-1

Solved Problems

FACTORIAL NOTATION AND BINOMIAL COEFFICIENTS

8.1. Compute $4!$, $5!$, $6!$, $7!$ and $8!$

$$4! = 1 \cdot 2 \cdot 3 \cdot 4 = 24 \qquad\qquad 7! = 7 \cdot 6! = 7 \cdot 720 = 5040$$

$$5! = 1 \cdot 2 \cdot 3 \cdot 4 \cdot 5 = 5 \cdot 4! = 5 \cdot 24 = 120 \qquad 8! = 8 \cdot 7! = 8 \cdot 5040 = 40{,}320$$

$$6! = 1 \cdot 2 \cdot 3 \cdot 4 \cdot 5 \cdot 6 = 6 \cdot 5! = 6 \cdot 120 = 720$$

8.2. Compute: $(a)\ \dfrac{13!}{11!}$, $(b)\ \dfrac{7!}{10!}$

$(a)\quad \dfrac{13!}{11!} = \dfrac{13 \cdot 12 \cdot 11 \cdot 10 \cdot 9 \cdot 8 \cdot 7 \cdot 6 \cdot 5 \cdot 4 \cdot 3 \cdot 2 \cdot 1}{11 \cdot 10 \cdot 9 \cdot 8 \cdot 7 \cdot 6 \cdot 5 \cdot 4 \cdot 3 \cdot 2 \cdot 1} = 13 \cdot 12 = 156$

$\quad$ or $\quad \dfrac{13!}{11!} = \dfrac{13 \cdot 12 \cdot 11!}{11!} = 13 \cdot 12 = 156$

$(b)\quad \dfrac{7!}{10!} = \dfrac{7!}{10 \cdot 9 \cdot 8 \cdot 7!} = \dfrac{1}{10 \cdot 9 \cdot 8} = \dfrac{1}{720}$

8.3. Write in terms of factorials: $(a)\ 27 \cdot 26$, $(b)\ \dfrac{1}{14 \cdot 13 \cdot 12}$

$(a)\quad 27 \cdot 26 = \dfrac{27 \cdot 26 \cdot 25!}{25!} = \dfrac{27!}{25!}$

$(b)\quad \dfrac{1}{14 \cdot 13 \cdot 12} = \dfrac{11!}{14 \cdot 13 \cdot 12 \cdot 11!} = \dfrac{11!}{14!}$

8.4. Simplify: $(a)\ \dfrac{n!}{(n-1)!}$, $(b)\ \dfrac{(n+2)!}{n!}$

$(a)\quad \dfrac{n!}{(n-1)!} = \dfrac{n(n-1)(n-2)\cdots 3 \cdot 2 \cdot 1}{(n-1)(n-2)\cdots 3 \cdot 2 \cdot 1} = n \quad$ or simply $\quad \dfrac{n!}{(n-1)!} = \dfrac{n(n-1)!}{(n-1)!} = n$

$(b)\quad \dfrac{(n+2)!}{n!} = \dfrac{(n+2)(n+1)n(n-1)(n-2)\cdots 3 \cdot 2 \cdot 1}{n(n-1)(n-2)\cdots 3 \cdot 2 \cdot 1} = (n+2)(n+1) = n^2 + 3n + 2$

$\quad$ or simply

$$\dfrac{(n+2)!}{n!} = \dfrac{(n+2)(n+1) \cdot n!}{n!} = (n+2)(n+1) = n^2 + 3n + 2$$

8.5. Compute: $(a)\ \dbinom{16}{3}$, $(b)\ \dbinom{12}{4}$, $(c)\ \dbinom{15}{5}$

Recall that there are as many factors in the numerator as in the denominator.

$(a)\quad \dbinom{16}{3} = \dfrac{16 \cdot 15 \cdot 14}{1 \cdot 2 \cdot 3} = 560 \qquad (c)\quad \dbinom{15}{5} = \dfrac{15 \cdot 14 \cdot 13 \cdot 12 \cdot 11}{1 \cdot 2 \cdot 3 \cdot 4 \cdot 5} = 3003$

$(b)\quad \dbinom{12}{4} = \dfrac{12 \cdot 11 \cdot 10 \cdot 9}{1 \cdot 2 \cdot 3 \cdot 4} = 495$

8.6. Compute: (a) $\binom{8}{5}$, (b) $\binom{9}{7}$

(a) $\binom{8}{5} = \dfrac{8 \cdot 7 \cdot 6 \cdot 5 \cdot 4}{1 \cdot 2 \cdot 3 \cdot 4 \cdot 5} = 56$

Note that $8 - 5 = 3$; hence we could also compute $\binom{8}{5}$ as follows:

$$\binom{8}{5} = \binom{8}{3} = \frac{8 \cdot 7 \cdot 6}{1 \cdot 2 \cdot 3} = 56$$

(b) Now $9 - 7 = 2$; hence $\binom{9}{7} = \binom{9}{2} = \dfrac{9 \cdot 8}{1 \cdot 2} = 36.$

8.7. Prove: $\binom{17}{6} = \binom{16}{5} + \binom{16}{6}$

Now $\binom{16}{5} + \binom{16}{6} = \dfrac{16!}{5!\,11!} + \dfrac{16!}{6!\,10!}$. Multiply the first fraction by $\dfrac{6}{6}$ and the second by $\dfrac{11}{11}$ to obtain the same denominator in both fractions, and then add:

$$\binom{16}{5} + \binom{16}{6} = \frac{6 \cdot 16!}{6 \cdot 5! \cdot 11!} + \frac{11 \cdot 16!}{6! \cdot 11 \cdot 10!} = \frac{6 \cdot 16!}{6! \cdot 11!} + \frac{11 \cdot 16!}{6! \cdot 11!}$$

$$= \frac{6 \cdot 16! + 11 \cdot 16!}{6! \cdot 11!} = \frac{(6+11) \cdot 16!}{6! \cdot 11!} = \frac{17 \cdot 16!}{6! \cdot 11!} = \frac{17!}{6! \cdot 11!} = \binom{17}{6}$$

8.8. Prove Theorem 8.1: $\binom{n+1}{r} = \binom{n}{r-1} + \binom{n}{r}$

(The technique in this proof is similar to that of the preceding problem.)

Now $\binom{n}{r-1} + \binom{n}{r} = \dfrac{n!}{(r-1)! \cdot (n-r+1)!} + \dfrac{n!}{r! \cdot (n-r)!}$. To obtain the same denominator in both fractions, multiply the first fraction by $\dfrac{r}{r}$ and the second fraction by $\dfrac{n-r+1}{n-r+1}$. Hence

$$\binom{n}{r-1} + \binom{n}{r} = \frac{r \cdot n!}{r \cdot (r-1)! \cdot (n-r+1)!} + \frac{(n-r+1) \cdot n!}{r! \cdot (n-r+1) \cdot (n-r)!}$$

$$= \frac{r \cdot n!}{r! \, (n-r+1)!} + \frac{(n-r+1) \cdot n!}{r! \, (n-r+1)!}$$

$$= \frac{r \cdot n! + (n-r+1) \cdot n!}{r! \, (n-r+1)!} = \frac{[r + (n-r+1)] \cdot n!}{r! \, (n-r+1)!}$$

$$= \frac{(n+1)n!}{r! \, (n-r+1)!} = \frac{(n+1)!}{r! \, (n-r+1)!} = \binom{n+1}{r}$$

PERMUTATIONS

8.9. There are four bus lines between A and B; and three bus lines between B and C. In how many ways can a man travel (a) by bus from A to C by way of B? (b) roundtrip by bus from A to C by way of B? (c) roundtrip by bus from A to C by way of B, if he doesn't want to use a bus line more than once?

(a) There are four ways to go from A to B and three ways to go from B to C; hence there are $4 \cdot 3 = 12$ ways to go from A to C by way of B.

(b) There are twelve ways to go from A to C by way of B, and 12 ways to return. Hence there are $12 \cdot 12 = 144$ ways to travel roundtrip.

(c) The man will travel from A to B to C to B to A. Enter these letters with connecting arrows as follows:

$$A \longrightarrow B \longrightarrow C \longrightarrow B \longrightarrow A$$

The man can travel four ways from A to B and three ways from B to C; but he can only travel two ways from C to B and three ways from B to A since he doesn't want to use a bus line more than once. Enter these numbers above the corresponding arrows as follows:

$$A \overset{4}{\longrightarrow} B \overset{3}{\longrightarrow} C \overset{2}{\longrightarrow} B \overset{3}{\longrightarrow} A$$

Thus there are $4 \cdot 3 \cdot 2 \cdot 3 = 72$ ways to travel roundtrip without using the same bus line more than once.

8.10. Suppose repetitions are not permitted. (a) How many three-digit numbers can be formed from the six digits 2, 3, 5, 6, 7 and 9? (b) How many of these numbers are less than 400? (c) How many are even? (d) How many are odd? (e) How many are multiples of 5?

In each case draw three boxes ☐ ☐ ☐ to represent an arbitrary number, and then write in each box the number of digits that can be placed there.

(a) The box on the left can be filled in six ways; following this, the middle box can be filled in five ways; and, lastly, the box on the right can be filled in four ways: $\boxed{6}\ \boxed{5}\ \boxed{4}$. Thus there are $6 \cdot 5 \cdot 4 = 120$ numbers.

(b) The box on the left can be filled in only two ways, by 2 or 3, since each number must be less than 400; the middle box can be filled in five ways; and, lastly, the box on the right can be filled in four ways: $\boxed{2}\ \boxed{5}\ \boxed{4}$. Thus there are $2 \cdot 5 \cdot 4 = 40$ numbers.

(c) The box on the right can be filled in only two ways, by 2 or 6, since the numbers must be even; the box on the left can then be filled in five ways; and, lastly, the middle box can be filled in four ways: $\boxed{5}\ \boxed{4}\ \boxed{2}$. Thus there are $5 \cdot 4 \cdot 2 = 40$ numbers.

(d) The box on the right can be filled in only four ways, by 3, 5, 7 or 9, since the numbers must be odd; the box on the left can then be filled in five ways; and, lastly, the box in the middle can be filled in four ways: $\boxed{5}\ \boxed{4}\ \boxed{4}$. Thus there are $5 \cdot 4 \cdot 4 = 80$ numbers.

(e) The box on the right can be filled in only one way, by 5, since the numbers must be multiples of 5; the box on the left can then be filled in five ways; and, lastly, the box in the middle can be filled in four ways: $\boxed{5}\ \boxed{4}\ \boxed{1}$. Thus there are $5 \cdot 4 \cdot 1 = 20$ numbers.

8.11. In how many ways can a party of seven persons arrange themselves (a) in a row of seven chairs? (b) around a circular table?

(a) The seven persons can arrange themselves in a row in $7 \cdot 6 \cdot 5 \cdot 4 \cdot 3 \cdot 2 \cdot 1 = 7!$ ways.

(b) One person can sit at any place in the circular table. The other six persons can then arrange themselves in $6 \cdot 5 \cdot 4 \cdot 3 \cdot 2 \cdot 1 = 6!$ ways around the table.

This is an example of a *circular permutation*. In general, n objects can be arranged in a circle in $(n-1)(n-2)\cdots 3 \cdot 2 \cdot 1 = (n-1)!$ ways.

8.12. (a) In how many ways can three boys and two girls sit in a row? (b) In how many ways can they sit in a row if the boys and girls are each to sit together? (c) In how many ways can they sit in a row if just the girls are to sit together?

(a) The five persons can sit in a row in $5 \cdot 4 \cdot 3 \cdot 2 \cdot 1 = 5! = 120$ ways.

(b) There are two ways to distribute them according to sex: BBBGG or GGBBB. In each case the boys can sit in $3 \cdot 2 \cdot 1 = 3! = 6$ ways, and the girls can sit in $2 \cdot 1 = 2! = 2$ ways. Thus, altogether, there are $2 \cdot 3! \cdot 2! = 2 \cdot 6 \cdot 2 = 24$ ways.

(c) There are four ways to distribute them according to sex: GGBBB, BGGBB, BBGGB, BBBGG. Note that each way corresponds to the number, 0, 1, 2 or 3, of boys sitting to the left of the girls. In each case, the boys can sit in 3! ways, and the girls in 2! ways. Thus, altogether, there are $4 \cdot 3! \cdot 2! = 4 \cdot 6 \cdot 2 = 48$ ways.

8.13. Solve the preceding problem in the case of r boys and s girls. (Answers are to be left in factorials.)

(a) The $r + s$ persons can sit in a row in $(r + s)!$ ways.

(b) There are still two ways to distribute them according to sex, the boys on the left or the girls on the left. In each case the boys can sit in $r!$ ways and the girls in $s!$ ways. Thus, altogether, there are $2 \cdot r! \cdot s!$ ways.

(c) There are $r + 1$ ways to distribute them according to sex, each way corresponding to the number, $0, 1, \ldots, r$, of boys sitting to the left of the girls. In each case the boys can sit in $r!$ ways and the girls in $s!$ ways. Thus, altogether, there are $(r + 1) \cdot r! \cdot s!$ ways.

8.14. Find the number of distinct permutations that can be formed from all the letters of each word: (a) THEM, (b) THAT, (c) RADAR, (d) UNUSUAL, (e) SOCIOLOGICAL.

(a) $4! = 24$, since there are four letters and no repetition.

(b) $\dfrac{4!}{2!} = 12$, since there are four letters of which two are T.

(c) $\dfrac{5!}{2!\,2!} = 30$, since there are five letters of which two are R and two are A.

(d) $\dfrac{7!}{3!} = 840$, since there are seven letters of which three are U.

(e) $\dfrac{12!}{3!\,2!\,2!\,2!}$, since there are twelve letters of which three are O, two are C, two are I, and two are L.

8.15. In how many ways can four mathematics books, three history books, three chemistry books and two sociology books be arranged on a shelf so that all books of the same subject are together?

First the books must be arranged on the shelf in four units according to subject matter: ☐ ☐ ☐ ☐ . The box on the left can be filled by any of the four subjects; the next by any three subjects remaining; the next by any two subjects remaining; and the box on the right by the last subject: [4] [3] [2] [1] . Thus there are $4 \cdot 3 \cdot 2 \cdot 1 = 4!$ ways to arrange the books on the shelf according to subject matter.

In each of the above cases, the mathematics books can be arranged in 4! ways, the history books in 3! ways, the chemistry books in 3! ways, and the sociology books in 2! ways. Thus, altogether, there are $4!\,4!\,3!\,3!\,2! = 41{,}472$ arrangements.

8.16. Find the total number of positive integers that can be formed from the digits 1, 2, 3 and 4 if no digit is repeated in any one integer.

Note that no integer can contain more than four digits. Let s_1, s_2, s_3 and s_4 denote the number of integers containing 1, 2, 3 and 4 digits respectively. We compute each s_i separately.

Since there are four digits, there are four integers containing exactly one digit, i.e. $s_1 = 4$. Also, since there are four digits, there are $4 \cdot 3 = 12$ integers containing two digits, i.e. $s_2 = 12$. Similarly, there are $4 \cdot 3 \cdot 2 = 24$ integers containing three digits and $4 \cdot 3 \cdot 2 \cdot 1 = 24$ integers containing four digits, i.e. $s_3 = 24$ and $s_4 = 24$. Thus, altogether, there are $s_1 + s_2 + s_3 + s_4 = 4 + 12 + 24 + 24 = 64$ integers.

8.17. Find n if (a) $P(n, 2) = 72$, (b) $P(n, 4) = 42 P(n, 2)$, (c) $2P(n, 2) + 50 = P(2n, 2)$.

(a) $P(n, 2) = n(n-1) = n^2 - n$; hence $n^2 - n = 72$ or $n^2 - n - 72 = 0$ or $(n-9)(n+8) = 0$.

Since n must be positive, the only answer is $n = 9$.

(b) $P(n, 4) = n(n-1)(n-2)(n-3)$ and $P(n, 2) = n(n-1)$. Hence

$$n(n-1)(n-2)(n-3) = 42n(n-1) \text{ or, if } n \neq 0, n \neq 1, \ (n-2)(n-3) = 42$$

$$\text{or } n^2 - 5n + 6 = 42 \text{ or } n^2 - 5n - 36 = 0 \text{ or } (n-9)(n+4) = 0$$

Since n must be positive, the only answer is $n = 9$.

(c) $P(n, 2) = n(n-1) = n^2 - n$ and $P(2n, 2) = 2n(2n-1) = 4n^2 - 2n$. Hence

$$2(n^2 - n) + 50 = 4n^2 - 2n \text{ or } 2n^2 - 2n + 50 = 4n^2 - 2n \text{ or } 50 = 2n^2 \text{ or } n^2 = 25$$

Since n must be positive, the only answer is $n = 5$.

COMBINATIONS

8.18. In how many ways can a committee consisting of three men and two women be chosen from seven men and five women?

The three men can be chosen from the seven men in $\binom{7}{3}$ ways, and the two women can be chosen from the five women in $\binom{5}{2}$ ways. Hence the committee can be chosen in

$$\binom{7}{3}\binom{5}{2} = \frac{7 \cdot 6 \cdot 5}{1 \cdot 2 \cdot 3} \cdot \frac{5 \cdot 4}{1 \cdot 2} = 350 \text{ ways}$$

8.19. A bag contains six white marbles and five black marbles. Find the number of ways four marbles can be drawn from the bag if (a) they can be any color, (b) two must be white and two black, (c) they must all be of the same color.

(a) The four marbles (of any color) can be chosen from the eleven marbles in

$$\binom{11}{4} = \frac{11 \cdot 10 \cdot 9 \cdot 8}{1 \cdot 2 \cdot 3 \cdot 4} = 330 \text{ ways}$$

(b) The two white marbles can be chosen in $\binom{6}{2}$ ways, and the two black marbles can be chosen in $\binom{5}{2}$ ways. Thus there are $\binom{6}{2}\binom{5}{2} = \frac{6 \cdot 5}{1 \cdot 2} \cdot \frac{5 \cdot 4}{1 \cdot 2} = 150$ ways of drawing two white marbles and two black marbles.

(c) There are $\binom{6}{4} = 15$ ways of drawing four white marbles, and $\binom{5}{4} = 5$ ways of drawing four black marbles. Thus there are $15 + 5 = 20$ ways of drawing four marbles of the same color.

8.20. A delegation of four students is selected each year from a college to attend the National Student Association annual meeting. (a) In how many ways can the delegation be chosen if there are twelve eligible students? (b) In how many ways if two of the eligible students will not attend the meeting together? (c) In how many ways if two of the eligible students are married and will only attend the meeting together?

(a) The four students can be chosen from the twelve students in $\binom{12}{4} = \dfrac{12 \cdot 11 \cdot 10 \cdot 9}{1 \cdot 2 \cdot 3 \cdot 4} = 495$ ways.

(b) Let A and B denote the students who will not attend the meeting together.

Method 1.

If neither A nor B is included, then the delegation can be chosen in $\binom{10}{4} = \dfrac{10 \cdot 9 \cdot 8 \cdot 7}{1 \cdot 2 \cdot 3 \cdot 4} =$ 210 ways. If either A or B, but not both, is included, then the delegation can be chosen in $2 \cdot \binom{10}{3} = 2 \cdot \dfrac{10 \cdot 9 \cdot 8}{1 \cdot 2 \cdot 3} = 240$ ways. Thus, altogether, the delegation can be chosen in $210 + 240 = 450$ ways.

Method 2.

If A and B are both included, then the other two members of the delegation can be chosen in $\binom{10}{2} = 45$ ways. Thus there are $495 - 45 = 450$ ways the delegation can be chosen if A and B are not both included.

(c) Let C and D denote the married students. If C and D do not go, then the delegation can be chosen in $\binom{10}{4} = 210$ ways. If both C and D go, then the delegation can be chosen in $\binom{10}{2} = 45$ ways. Altogether, the delegation can be chosen in $210 + 45 = 255$ ways.

8.21. A student is to answer eight out of ten questions on an exam. (a) How many choices has he? (b) How many if he must answer the first three questions? (c) How many if he must answer at least four of the first five questions?

(a) The eight questions can be selected in $\binom{10}{8} = \binom{10}{2} = \dfrac{10 \cdot 9}{1 \cdot 2} = 45$ ways.

(b) If he answers the first three questions, then he can choose the other five questions from the last seven questions in $\binom{7}{5} = \binom{7}{2} = \dfrac{7 \cdot 6}{1 \cdot 2} = 21$ ways.

(c) If he answers all the first five questions, then he can choose the other three questions from the last five in $\binom{5}{3} = 10$ ways. On the other hand, if he answers only four of the first five questions, then he can choose these four in $\binom{5}{4} = \binom{5}{1} = 5$ ways, and he can choose the other four questions from the last five in $\binom{5}{4} = \binom{5}{1} = 5$ ways; hence he can choose the eight questions in $5 \cdot 5 = 25$ ways. Thus he has a total of thirty-five choices.

8.22. There are twelve points $A, B, \ldots$ in a given plane, no three on the same line. (a) How many lines are determined by the points? (b) How many lines pass through the point A? (c) How many triangles are determined by the points? (d) How many of these triangles contain the point A as a vertex?

(a) Since two points determine a line, there are $\binom{12}{2} = \dfrac{12 \cdot 11}{1 \cdot 2} = 66$ lines.

(b) To determine a line through A, one other point must be chosen; hence there are eleven lines through A.

(c) Since three points determine a triangle, there are $\binom{12}{3} = \dfrac{12 \cdot 11 \cdot 10}{1 \cdot 2 \cdot 3} = 220$ triangles.

(d) **Method 1.**

   To determine a triangle with vertex A, two other points must be chosen; hence there are $\binom{11}{2} = \frac{11 \cdot 10}{1 \cdot 2} = 55$ triangles with A as a vertex.

Method 2.

   There are $\binom{11}{3} = \frac{11 \cdot 10 \cdot 9}{1 \cdot 2 \cdot 3} = 165$ triangles without A as a vertex. Thus $220 - 165 = 55$ of the triangles do have A as a vertex.

8.23. How many committees of five with a given chairman can be selected from twelve persons?

   The chairman can be chosen in twelve ways and, following this, the other four on the committee can be chosen from the eleven remaining in $\binom{11}{4}$ ways. Thus there are $12 \cdot \binom{11}{4} = 12 \cdot 330 = 3960$ such committees.

8.24. Find the number of subsets of a set X containing n elements.

Method 1.

   The number of subsets of X with $r \leq n$ elements is given by $\binom{n}{r}$. Hence, altogether. there are

$$\binom{n}{0} + \binom{n}{1} + \binom{n}{2} + \cdots + \binom{n}{n-1} + \binom{n}{n}$$

subsets of X. The above sum is equal to 2^n (Problem 8.41), and so there are 2^n subsets of X.

Method 2.

   There are two possibilities for each element of X: either it belongs to the subset or it doesn't; hence there are

$$\overbrace{2 \cdot 2 \cdot \cdots \cdot 2}^{n \text{ times}} = 2^n$$

ways to form a subset of X, i.e. there are 2^n different subsets of X.

8.25. In how many ways can a teacher choose one or more students from six eligible students?

Method 1.

   By the preceding problem, there are $2^6 = 64$ subsets of the set consisting of the six students. However, the empty set must be deleted since one or more students are chosen. Accordingly, there are $2^6 - 1 = 64 - 1 = 63$ ways to choose the students.

Method 2.

   Either 1, 2, 3, 4, 5 or 6 students are chosen. Hence the number of choices is

$$\binom{6}{1} + \binom{6}{2} + \binom{6}{3} + \binom{6}{4} + \binom{6}{5} + \binom{6}{6} = 6 + 15 + 20 + 15 + 6 + 1 = 63$$

8.26. In how many ways can three or more persons be selected from twelve persons?

   There are $2^{12} - 1 = 4096 - 1 = 4095$ ways of choosing one or more of the twelve persons. Now there are $\binom{12}{1} + \binom{12}{2} = 12 + 66 = 78$ ways of choosing one or two of the twelve persons. Hence there are $4095 - 78 = 4017$ ways of choosing three or more.

ORDERED AND UNORDERED PARTITIONS

8.27. In how many ways can seven toys be divided among three children if the youngest gets three toys and each of the others gets two?

We seek the number of ordered partitions of seven objects into cells containing 3, 2 and 2 objects, respectively. By Theorem 8.6, there are $\dfrac{7!}{3!\,2!\,2!} = 210$ such partitions.

8.28. There are twelve students in a class. In how many ways can the twelve students take three different tests if four students are to take each test?

Method 1.

We seek the number of ordered partitions of the twelve students into cells containing four students each. By Theorem 8.6, there are $\dfrac{12!}{4!\,4!\,4!} = 34{,}650$ such partitions.

Method 2.

There are $\binom{12}{4}$ ways to choose four students to take the first test; following this, there are $\binom{8}{4}$ ways to choose four students to take the second test. The remaining students take the third test. Thus, altogether, there are $\binom{12}{4} \cdot \binom{8}{4} = 495 \cdot 70 = 34{,}650$ ways for the students to take the tests.

8.29. In how many ways can twelve students be partitioned into three teams, A_1, A_2 and A_3, so that each team contains four students?

Method 1.

Observe that each partition $\{A_1, A_2, A_3\}$ of the students can be arranged in $3! = 6$ ways as an ordered partition. Since (see the preceding problem) there are $\dfrac{12!}{4!\,4!\,4!} = 34{,}650$ such ordered partitions, there are $34{,}650/6 = 5775$ (unordered) partitions.

Method 2.

Let A denote one of the students. Then there are $\binom{11}{3}$ ways to choose three other students to be on the same team as A. Now let B denote a student who is not on the same team as A; then there are $\binom{7}{3}$ ways to choose three students of the remaining students to be on the same team as B. The remaining four students constitute the third team. Thus, altogether, there are $\binom{11}{3} \cdot \binom{7}{3} = 165 \cdot 35 = 5775$ ways to partition the students.

8.30. In how many ways can six students be partitioned into (a) two teams containing three students each, (b) three teams containing two students each?

(a) **Method 1.**

There are $\dfrac{6!}{3!\,3!} = 20$ ordered partitions into two cells containing three students each. Since each unordered partition determines $2! = 2$ ordered partitions, there are $20/2 = 10$ unordered partitions.

Method 2.

Let A denote one of the students; then there are $\binom{5}{2} = 10$ ways to choose two other students to be on the same team as A. The other three students constitute the other team. In other words, there are ten ways to partition the students.

(b) **Method 1.**

There are $\dfrac{6!}{2!\,2!\,2!} = 90$ ordered partitions into three cells containing two students each. Since each unordered partition determines $3! = 6$ ordered partitions, there are $90/6 = 15$ unordered partitions.

Method 2.

Let A denote one of the students. Then there are five ways to choose the other student to be on the same team as A. Let B denote a student who isn't on the same team as A; then there are three ways to choose another student to be on the same team as B. The remaining two students constitute the other team. Thus, altogether, there are $5 \cdot 3 = 15$ ways to partition the students.

8.31. In how many ways can a class X with ten students be partitioned into four teams A_1, A_2, B_1, B_2 where A_1 and A_2 contain two students each and B_1 and B_2 contain three students each?

Method 1.

There are $\dfrac{10!}{2!\,2!\,3!\,3!} = 25{,}200$ ordered partitions of X into four cells containing 2, 2, 3 and 3 students, respectively. However, each unordered partition $\{A_1, A_2, B_1, B_2\}$ of X determines $2! \cdot 2! = 4$ ordered partitions of X. Thus, altogether, there are $25{,}200/4 = 6300$ unordered partitions.

Method 2.

There are $\dbinom{10}{4}$ ways to choose four students who will be on the teams A_1 and A_2 and there are three ways in which each four students can be partitioned into two teams of two students each. On the other hand, there are ten ways (see Problem 8.30) to partition the remaining six students into two teams containing three students each. Thus, altogether, there are

$$\binom{10}{4} \cdot 3 \cdot 10 \;=\; 210 \cdot 3 \cdot 10 \;=\; 6300$$

ways to partition the students.

8.32. In how many ways can nine students be partitioned into three teams containing 4, 3 and 2 students, respectively?

Since all the cells contain different numbers of students, the number of unordered partitions equals the number of ordered partitions, $\dfrac{9!}{4!\,3!\,2!} = 1260$.

TREE DIAGRAMS

8.33. Teams A and B play in a basketball tournament. The team that first wins three games wins the tournament. Find the number of possible ways in which the tournament can occur.

Construct the appropriate tree diagram, as shown in Fig. 8-2.

The tournament can occur in twenty ways:

AAA, AABA, AABBA, AABBB, ABAA, ABABA, ABABB, ABBAA, ABBAB, ABBB

BAAA, BAABA, BAABB, BABAA, BABAB, BABB, BBAAA, BBAAB, BBAB, BBB

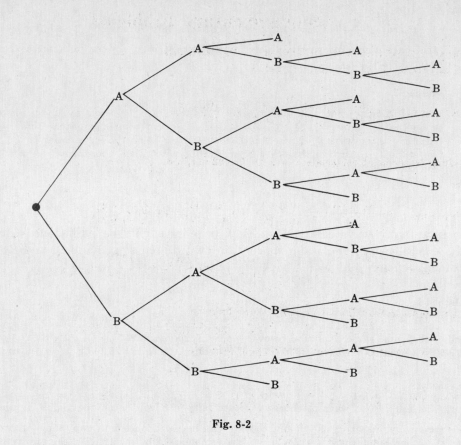

Fig. 8-2

8.34. Find the permutations of $\{a, b, c\}$.

Corollary 8.3 tells us that there are $3! = 3 \cdot 2 \cdot 1 = 6$ such permutations. However, a tree diagram can be used to find them as illustrated below.

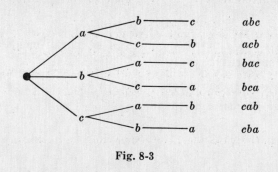

Fig. 8-3

The six permutations are listed on the right of the diagram.

Supplementary Problems

FACTORIAL NOTATION AND BINOMIAL COEFFICIENTS

8.35.   Evaluate: (a) 9!, (b) 10!, (c) 11!

8.36.   Evaluate: (a) $\dfrac{16!}{14!}$, (b) $\dfrac{14!}{11!}$, (c) $\dfrac{8!}{10!}$, (d) $\dfrac{10!}{13!}$

8.37.   Write in terms of factorials: (a) $24 \cdot 23 \cdot 22 \cdot 21$, (b) $\dfrac{1}{10 \cdot 11 \cdot 12}$

8.38.   Simplify: (a) $\dfrac{(n+1)!}{n!}$, (b) $\dfrac{n!}{(n-2)!}$, (c) $\dfrac{(n-1)!}{(n+2)!}$, (d) $\dfrac{(n-r+1)!}{(n-r-1)!}$

8.39.   Evaluate: (a) $\dbinom{5}{2}$, (b) $\dbinom{7}{3}$, (c) $\dbinom{14}{2}$, (d) $\dbinom{6}{4}$, (e) $\dbinom{20}{17}$, (f) $\dbinom{18}{15}$

8.40.   The eighth row of Pascal's triangle is as follows:

$$1 \quad 8 \quad 28 \quad 56 \quad 70 \quad 56 \quad 28 \quad 8 \quad 1$$

Compute the ninth and tenth rows of the triangle.

8.41.   Show that $\dbinom{n}{0} + \dbinom{n}{1} + \dbinom{n}{2} + \dbinom{n}{3} + \cdots + \dbinom{n}{n} = 2^n$

8.42.   Show that $\dbinom{n}{0} - \dbinom{n}{1} + \dbinom{n}{2} - \dbinom{n}{3} + \cdots \pm \dbinom{n}{n} = 0$

PERMUTATIONS

8.43.   (a) How many automobile license plates can be made if each plate contains two different letters followed by three different digits?   (b) Solve the problem if the first digit cannot be 0.

8.44.   There are six roads between A and B and four roads between B and C.

(a) In how many ways can one drive from A to C by way of B?

(b) In how many ways can one drive roundtrip from A to C by way of B?

(c) In how many ways can one drive roundtrip from A to C without using the same road more than once?

8.45.   Find the number of ways in which six people can ride a toboggan if one of three must drive.

8.46.   (a) Find the number of ways in which five persons can sit in a row.

(b) How many ways are there if two of the persons insist on sitting next to one another?

8.47.   Solve the preceding problem if they sit around a circular table.

8.48.   Find the number of ways in which a judge can award first, second and third places in a contest with ten contestants.

8.49.   How many different signals, each consisting of eight flags hung in a vertical line, can be formed from four red flags, two blue flags and two green flags?

8.50. Find the number of permutations that can be formed from all the letters of each word:

(a) QUEUE, (b) COMMITTEE, (c) PROPOSITION, (d) BASEBALL.

8.51. (a) Find the number of ways in which four boys and four girls can be seated in a row if the boys and girls are to have alternate sets.

(b) Find the number of ways if they sit alternately and if one boy and one girl are to sit in adjacent seats.

(c) Find the number of ways if they sit alternately and if one boy and one girl must not sit in adjacent seats.

8.52. Find the number of ways in which five large books, four medium-size books and three small books can be placed on a shelf so that all books of the same size are together.

8.53. Consider all positive integers with three different digits. (Note that 0 cannot be the first digit.) (a) How many are greater than 700? (b) How many are odd? (c) How many are even? (d) How many are divisible by 5?

8.54. (a) Find the number of permutations that can be formed from the letters of the word ELEVEN.

(b) How many of them begin and end with E? (c) How many of them have the three E's together?

(d) How many begin with E and end with N?

COMBINATIONS

8.55. A class contains nine boys and three girls. (a) In how many ways can the teacher choose a committee of four? (b) How many of them will contain at least one girl? How many of them will contain exactly one girl?

8.56. A woman has eleven close friends. (a) In how many ways can she invite five of them to dinner? (b) In how many ways if two of the friends are married and will not attend separately? (c) In how many ways if two of them are not on speaking terms and will not attend together?

8.57. A woman has eleven close friends of whom six are also women. (a) In how many ways can she invite three or more to a party? (b) In how many ways can she invite three or more of them if she wants the same number of men as women (including herself)?

8.58. There are ten points $A, B, \ldots$ in a plane, no three on the same line. (a) How many lines are determined by the points? (b) How many of these lines do not pass through A or B? (c) How many triangles are determined by the points? (d) How many of these triangles contain the point A? (e) How many of these triangles contain the side AB?

8.59. A student is to answer ten out of thirteen questions on an exam. (a) How many choices has he? (b) How many if he must answer the first two questions? (c) How many if he must answer the first or second question but not both? (d) How many if he must answer exactly three out of the first five questions? (e) How many if he must answer at least three of the first five questions?

8.60. (a) How many triangles are determined by the vertices of an octagon?

(b) How many if the sides of the octagon are not to be sides of any triangle?

8.61. How many triangles are determined by the vertices of a regular polygon with n sides if the sides of the polygon are not to be sides of any triangle?

8.62. A man is dealt a poker hand (five cards) from an ordinary playing deck. In how many ways can he be dealt (a) a straight flush, (b) four of a kind, (c) a straight, (d) a pair of aces, (e) two of a kind (a pair)?

8.63.   The English alphabet has twenty-six letters of which five are vowels. Consider only the "words" consisting of five letters including three different consonants and two different vowels. How many of such words (a) can be formed? (b) contain the letter B? (c) contain the letters B and C? (d) begin with B and contain the letter C? (e) begin with B and end with C? (f) contain the letters A and B? (g) begin with A and contain B? (h) begin with B and contain A? (i) begin with A and end with B? (j) contain the letters A, B and C?

ORDERED AND UNORDERED PARTITIONS

8.64.   In how many ways can nine toys be divided evenly among three children?

8.65.   In how many ways can nine students be evenly divided into three teams?

8.66.   In how many ways can ten students be divided into three teams, one containing four students and the others three?

8.67.   There are twelve marbles in an urn. In how many ways can three marbles be drawn from the urn four times in succession?

8.68.   In how many ways can a club with twelve members be partitioned into three committees containing 5, 4 and 3 members respectively?

8.69.   In how many ways can n students be partitioned into two teams containing at least one student?

8.70.   In how many ways can fourteen men be partitioned into six committees where two of the committees contain three men and the others two?

8.71.   (a) Assuming that a cell can be empty, in how many ways can a set with three elements be partitioned into (i) three ordered cells, (ii) three unordered cells? (b) In how many ways can a set with four elements be partitioned into (i) three ordered cells, (ii) three unordered cells?

8.72.   In bridge, thirteen cards are dealt to each of four players who are called North, South, East and West. The distribution of the cards is called a bridge hand. (a) How many bridge hands are there? (b) In how many ways will one player be dealt all four aces? (c) In how many of them will each player be dealt an ace? (d) In how many of them will North be dealt eight spades and South the other five spades? (e) In how many of them will North and South have, together, all four aces? (Leave answers in factorial notation.)

TREE DIAGRAMS

8.73.   Find the product set $\{1, 2\} \times \{a, b, c\} \times \{3, 4\}$ by constructing the appropriate tree diagram.

8.74.   Teams A and B play in baseball's world series where the team that first wins four games wins the series. Assuming that A wins the first game and the team that wins the second game also wins the fourth game, how many ways can the series occur?

8.75.   A man is at the origin on the x-axis and takes a one-unit step either to the left or to the right. He stops if he reaches 3 or −3, or if he occupies any position, other than the origin, more than once. Find the number of different paths the man can travel.

8.76.   A man has time to play roulette five times. He wins or loses a dollar at each play. The man begins with two dollars and will stop playing before the five times if he loses all his money or wins three dollars (i.e. has five dollars). Find the number of ways the playing can occur.

8.77. In the following diagram let $A, B, \ldots, F$ denote islands, and the lines connecting them bridges. A man begins at A and walks from island to island. He stops for lunch when he cannot continue to walk without crossing the same bridge twice. Find the number of ways that he can take his walk before eating lunch.

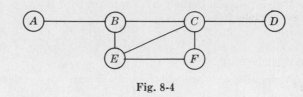

Fig. 8-4

8.78. Teams A and B play in a basketball tournament. The first team that wins two games in a row or a total of four games wins the tournament. Find the number of ways the tournament can occur.

COMPUTER PROGRAMMING PROBLEMS

8.79. Write a subprogram called NFACT such that NFACT(K) computes K factorial.

8.80. Write a subprogram called NPERM such that NPERM(M, K) computes the number of permutations of m objects taken k at a time, i.e. $P(m, k)$.

8.81. Write a subprogram called NCOMB such that NCOMB(M, K) computes the number of combinations of m objects taken k at a time, i.e. $\binom{m}{k}$.

8.82. A data card contains five letters which may or may not be distinct. Write a program which prints all distinct five-letter words using the five letters. Test the program with the following input data: (i) A, B, A, C, D, (ii) A, B, B, C, A, (iii) A, B, B, A, A

Answers to Supplementary Problems

8.35. (a) 362,880   (b) 3,628,800   (c) 39,916,800

8.36. (a) 240   (b) 2184   (c) 1/90   (d) 1/1716

8.37. (a) 24!/20!   (b) 9!/12!

8.38. (a) $n+1$ (b) $n(n-1) = n^2 - n$ (c) $1/[n(n+1)(n+2)]$ (d) $(n-r)(n-r+1)$

8.39. (a) 10   (b) 35   (c) 91   (d) 15   (e) 1140   (f) 816

8.40.

	1	8	28	56	70	56	28	8	1	
1	9	36	84	126	126	84	36	9	1	
1	10	45	120	210	252	210	120	45	10	1

8.41. *Hint:* Expand $(1+1)^n$.

8.42.   *Hint*: Expand $(1-1)^n$.

8.43.   (*a*) $26 \cdot 25 \cdot 10 \cdot 9 \cdot 8 = 468,000$ (*b*) $26 \cdot 25 \cdot 9 \cdot 9 \cdot 8 = 421,200$

8.44.   (*a*) 24   (*b*) 576   (*c*) 360

8.45.   360

8.46.   (*a*) 120   (*b*) 48

8.47.   (*a*) 24   (*b*) 12

8.48.   720

8.49.   420

8.50.   (*a*) 30   (*b*) $\dfrac{9!}{2!\,2!\,2!} = 45,360$ (*c*) $\dfrac{11!}{2!\,3!\,2!} = 1,663,200$ (*d*) $\dfrac{8!}{2!\,2!\,2!} = 5040$

8.51.   (*a*) 1152   (*b*) 504   (*c*) 648

8.52. $3!\,5!\,4!\,3! = 103,680$

8.53.   (*a*) 216   (*b*) 320   (*c*) 328   (*d*) 136

8.54.   (*a*) 120   (*b*) 24   (*c*) 24   (*d*) 12

8.55.   (*a*) 495   (*b*) 369   (*c*) 252

8.56.   (*a*) 462   (*b*) 210   (*c*) 252

8.57.   (*a*) $2^{11} - 1 - \dbinom{11}{1} - \dbinom{11}{2} = 1981$ or $\dbinom{11}{3} + \dbinom{11}{4} + \cdots + \dbinom{11}{11} = 1981$

  (*b*) $\dbinom{5}{5}\dbinom{6}{4} + \dbinom{5}{4}\dbinom{6}{3} + \dbinom{5}{3}\dbinom{6}{2} + \dbinom{5}{2}\dbinom{6}{1} = 325$

8.58.   (*a*) 45   (*b*) 28   (*c*) 120   (*d*) 36   (*e*) 8

8.59.   (*a*) 286   (*b*) 165   (*c*) 110   (*d*) 80   (*e*) 276

8.60.   (*a*) 56   (*b*) 16

8.61. $\dbinom{n}{3} - n\dbinom{n-4}{1} - n = \dfrac{n}{6}(n-5)(n-4)$

8.62.   (*a*) $4 \cdot 10 = 40$, (*b*) $13 \cdot 48 = 624$, (*c*) $10 \cdot 4^5 - 40 = 10,200$. (We subtract the number of straight flushes.)

  (*d*) $\dbinom{4}{2}\dbinom{12}{3} \cdot 4^3 = 84,480$ (*e*) $13 \cdot \dbinom{4}{2}\dbinom{12}{3} \cdot 4^3 = 1,098,240$

8.63. (a) $\binom{21}{3}\binom{5}{2} \cdot 5! = 1,596,000$ (e) $19 \cdot \binom{5}{2} \cdot 3! = 1140$ (i) $4 \cdot \binom{20}{2} \cdot 3! = 4456$

(b) $\binom{20}{2}\binom{5}{2} \cdot 5! = 228,000$ (f) $4 \cdot \binom{20}{2} \cdot 5! = 91,200$ (j) $4 \cdot 19 \cdot 5! = 9120$

(c) $19 \cdot \binom{5}{2} \cdot 5! = 22,800$ (g) $4 \cdot \binom{20}{2} \cdot 4! = 18,240$

(d) $19 \cdot \binom{5}{2} \cdot 4! = 4560$ (h) $18,240$

8.64. $\dfrac{9!}{3!\,3!\,3!} = 1680$

8.65. $\dfrac{1680}{3!} = 280$ or $\binom{8}{2}\binom{5}{2} = 280$

8.66. $\dfrac{10!}{4!\,3!\,3!} \cdot \dfrac{1}{2!} = 2100$ or $\binom{10}{4}\binom{5}{2} = 2100$

8.67. $\dfrac{12!}{3!\,3!\,3!\,3!} = 369,600$

8.68. $\dfrac{12!}{5!\,4!\,3!} = 27,720$

8.69. $2^{n-1} - 1$

8.70. $\dfrac{14!}{3!\,3!\,2!\,2!\,2!\,2!} \cdot \dfrac{1}{2!\,4!} = 3,153,150$

8.71. (a) (i) $3^3 = 27$ (Each element can be placed in any of the three cells.)

(ii) The number of elements in the three cells can be distributed as follows:

(a) $[\{3\}, \{0\}, \{0\}]$,   (b) $[\{2\}, \{1\}, \{0\}]$,   (c) $[\{1\}, \{1\}, \{1\}]$

Thus the number of partitions is $1 + 3 + 1 = 5$.

(b) (i) $3^4 = 81$.

(ii) The number of elements in the three cells can be distributed as follows:

(a) $[\{4\}, \{0\}, \{0\}]$,   (b) $[\{3\}, \{1\}, \{0\}]$,   (c) $[\{2\}, \{2\}, \{0\}]$,   (d) $[\{2\}, \{1\}, \{1\}]$

Thus the number of partitions is $1 + 4 + 3 + 6 = 14$.

8.72. (a) $\dfrac{52!}{13!\,13!\,13!\,13!}$ (b) $4 \cdot \dfrac{48!}{9!\,13!\,13!\,13!}$ (c) $4! \cdot \dfrac{48!}{12!\,12!\,12!\,12!}$ (d) $\binom{13}{8} \dfrac{39!}{5!\,8!\,13!\,13!}$

(e) $2 \cdot \dfrac{48!}{9!\,13!\,13!\,13!} + 2 \cdot 4 \cdot \dfrac{48!}{10!\,12!\,13!\,13!} + 2 \cdot 3 \cdot \dfrac{48!}{11!\,11!\,13!\,13!} = 2300 \dfrac{48!}{11!\,13!\,13!\,13!}$

8.74. 15

8.75. 14

8.76. 20

8.77. 11

8.78. 14

Chapter 9

Algebraic Systems, Formal Languages

9.1 OPERATIONS AND SEMIGROUPS

An *n-ary operation* on a set S is a mapping from S^n into S. In particular, a *binary operation* (i.e. 2-ary operation) is a mapping from $S \times S$ into S. We will mainly study binary operations in this chapter, so we use the word operation for binary operation unless otherwise stated.

Let $*$ be a binary operation on a set S. We usually write

$$a * b \quad \text{or simply} \quad ab$$

instead of $*(a, b)$. If S is a finite set, then the operation can be given by its operation table where the entry in the row labeled a and the column labeled b is $a * b$. If A is a subset of S, then A is said to be *closed under* $*$ if $a * b$ belongs to A for any elements a and b in A. For example, suppose S is the set of integers, and A is the subset of positive integers. Then A is closed under addition, but not under subtraction.

The operation $*$ on S is said to satisfy the *associative law*, or to be *associative*, if

$$(a * b) * c = a * (b * c)$$

for any elements a, b, c in S. An element e in S is called an *identity element* for $*$ if

$$a * e = e * a = a$$

for any element a in S. More generally, e is called a *right identity* if $a * e = a$ for any a in S, and a *left identity* if $e * a = a$ for any a in S.

Theorem 9.1: Let e be a left identity and f be a right identity for an operation. Then $e = f$.

The proof is very simple. Since e is a left identity, $ef = f$; but since f is a right identity, $ef = e$. Hence $e = f$. This theorem tells us, in particular, that an identity element is unique, and that if an operation has more than one left identity then it has no right identity and vice versa.

The operation $*$ on S is said to satisfy the *left cancellation law* if

$$a * b = a * c \quad \text{implies} \quad b = c$$

and the *right cancellation law* if

$$b * a = c * a \quad \text{implies} \quad b = c$$

The operation is said to be *commutative* or satisfy the *commutative law* if

$$a * b = b * a$$

for any a, b in S.

159

Suppose an operation $*$ on a set S has an identity element e. The *inverse* of an element a in S, usually denoted by a^{-1}, is an element with the property that

$$a * a^{-1} = a^{-1} * a = e$$

If the operation is associative, then the inverse of a, if it exists, is unique (Problem 9.3).

A set S together with an associative operation $*$ is called a *semigroup*. We denote the semigroup by $(S, *)$ or simply by S when the operation is understood.

EXAMPLE 9.1.

(a) Consider the set $\mathbf{Z} = \{\ldots, -1, 0, 1, \ldots\}$ of integers. Then $(\mathbf{Z}, +)$ is a semigroup since addition is associative, i.e.

$$(a + b) + c = a + (b + c)$$

for any a, b, c in $\mathbf{Z}$. In fact, $(\mathbf{Z}, +)$ is a commutative semigroup with identity element 0 since

$$a + b = b + a \quad \text{and} \quad a + 0 = 0 + a = a$$

for any integers a, b in $\mathbf{Z}$.

On the other hand, $(\mathbf{Z}, -)$ is not a semigroup since subtraction is not associative. For example,

$$(12 - 6) - 2 \neq 12 - (6 - 2)$$

(b) Consider the set $\mathbf{N} = \{1, 2, 3, \ldots\}$ of positive integers. Let

$$a * b = \gcd(a, b)$$
$$a \circ b = a^b$$

where $\gcd(a, b)$ denotes the *greatest common divisor* of a and b. Then $(\mathbf{N}, *)$ is a semigroup since the operation $*$ is associative. However, $(\mathbf{N}, \circ)$ is not a semigroup since the operation $\circ$ is nonassociative. For example,

$$(2 \circ 2) \circ 3 = (2^2)^3 = 4^3 = 64 \quad \text{but} \quad 2 \circ (2 \circ 3) = 2^{2^3} = 2^8 = 256$$

(c) Let S be any nonempty set with the operation

$$a * b = a$$

Then S is a semigroup since the operation is associative. In fact,

$$(a * b) * c = a * c = a \quad \text{and} \quad a * (b * c) = a * b = a$$

9.2 FREE SEMIGROUPS, LANGUAGES

Consider a set S of symbols. A *word* on S is a finite sequence of its elements. For example,

$$U = ababb \quad \text{and} \quad V = accba$$

are words on $S = \{a, b, c\}$. When discussing words on S, we frequently call S the *alphabet* and its elements *letters*. For convenience, the empty sequence, denoted by ϵ or 1, is also considered a word on S. We shall also abbreviate our notation by writing a^2 for aa, a^3 for aaa, and so on. The set of all words on S is usually denoted by S^*.

Now consider two words U and V on S. We can form the word UV obtained by writing the letters of V after the letters of U. For example, if U and V are the words above, then

$$UV = ababbaccba = abab^2ac^2ba$$

This operation is called *concatenation*. Clearly the operation is associative. Thus the set of words on S is a semigroup under the concatenation operation. This semigroup is called the *free semigroup* on S or generated by S. Clearly the empty word ϵ is an identity element for the semigroup, and the semigroup satisfies both right and left cancellation laws.

Again let S be a nonempty set of symbols. We now interpret the elements of S as words, and finite sequences of elements of S as sentences. We may also think of a language as a set of meaningful sentences. In fact, we formally define a language L on a set S to be a collection of words on S. For example, suppose $S = \{a, b\}$. Then the following are languages on S:

$$L_1 = \{a, ab, ab^2, ab^3, \ldots\}$$

$$L_2 = \{a^n b^m : n \text{ and } m \text{ nonnegative integers}\}$$

$$L_3 = \{b^n ab^m : n \text{ and } m \text{ nonnegative integers}\}$$

Observe that L_1 is a sublanguage of L_2 which consists of all words which begin with a's and end with b's. Also L_1 is a sublanguage of L_3 which consists of all words with exactly one a.

9.3 GRAMMARS AND LANGUAGES

Figure 9-1 shows the grammatical construction of a specific sentence. Observe that there are (1) various variables, e.g. ⟨sentence⟩, ⟨noun phrase⟩, ..., (2) various terminal words, e.g. "The", "boy", ..., (3) a beginning variable ⟨sentence⟩, and (4) various substitutions or productions, e.g.

$$\langle \text{sentence} \rangle \rightarrow \langle \text{noun phrase} \rangle \langle \text{verb phrase} \rangle$$

$$\langle \text{object phrase} \rangle \rightarrow \langle \text{article} \rangle \langle \text{noun} \rangle$$

$$\langle \text{noun} \rangle \rightarrow \text{apple}$$

The final sentence only contains terminals, although both variables and terminals appear in its construction by the productions. This intuitive description is given in order to motivate the following definition of a grammar and the language it generates.

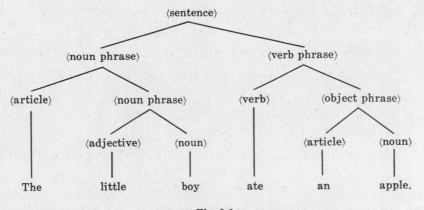

Fig. 9-1

A *grammar* G consists of four things:

(1) A finite set V of elements called *variables* or *nonterminals*.

(2) A finite set T of elements called *terminals*.

(3) An element S in V called the *start symbol*.

(4) A finite set P of productions. A *production* is an ordered pair (α, β) usually written $\alpha \rightarrow \beta$ where α and β are words in $V \cup T$. At least one of the α's must contain a variable.

We denote the above grammar by $G = G(V, T, S, P)$ when we want to emphasize the parts of G. We also assume that no element can be both a terminal and nonterminal, i.e. that T is disjoint from V.

Let G be a grammar and let W and W' be words in $V \cup T$. We write

$$W \Rightarrow W'$$

if W' can be obtained from W using one of the productions, i.e. if

$$W = L\alpha M, \quad W' = L\beta M \quad \text{and} \quad \alpha \to \beta$$

is a production. We write

$$W \Rightarrow\Rightarrow W'$$

if W' can be obtained from W using a finite sequence of productions. The language of G, denoted by $L(G)$, consists of all words in the terminals that can be obtained from the start symbol S by the above process, i.e.

$$L(G) = \{W : S \Rightarrow\Rightarrow W\}$$

EXAMPLE 9.2.   Consider the following grammar G:

$$V = \{A, B, S\}, \quad T = \{a, b\}$$
$$P = \{S \to AB, \ A \to Aa, \ B \to Bb, \ A \to a, \ B \to b\}$$

It is not difficult to see that $L(G)$ consists of all words of the form

$$a^n b^m$$

i.e. words which begin with a string of a's followed by a string of b's. For example $a^2 b^4$ can be obtained from S as follows:

$$S \Rightarrow AB \Rightarrow AaB \Rightarrow aaB \Rightarrow aaBb \Rightarrow aaBbb \Rightarrow aaBbbb \Rightarrow aabbbb = a^2 b^4$$

Here we used the productions 1, 2, 4, 3, 3, 3 and 5 respectively. Thus we can write $S \Rightarrow\Rightarrow a^2 b^4$.

The grammar in the above definition is said to be of type 0. By placing restrictions on the type of productions permissible, we obtain grammars of types 1, 2 and 3. These are defined below where A and B denote variables, a denotes a terminal and α, α' and β denote words (possibly empty) on $V \cup T$:

(1) A grammar G is said to be *context sensitive* or of *type 1* if the productions are of the form

$$\alpha A \alpha' \to \alpha \beta \alpha'$$

The name "context sensitive" comes from the fact that we can replace the variable A by β in a word only when A lies between α and α'.

(2) A grammar G is said to be *context free* or of *type 2* if the productions are of the form

$$A \to \beta$$

The name "context free" comes from the fact that we can now replace the variable A by β regardless of where A appears.

(3) A grammar G is said to be *regular* or of *type 3* if the productions are of the form

$$A \to aB \quad \text{or} \quad A \to a$$

Observe that each grammar belongs to a previous type, i.e. any type 3 grammar is of type 2, any type 2 grammar is of type 1, and any type 1 grammar is of type 0.

A language L is said to be *context sensitive, context free* or *regular* according as it can be generated by a context-sensitive, context-free or regular grammar respectively. For example, the language $L = \{a^n b^m\}$, where n and m are positive integers, is a context-free language since it is generated by the grammar G in Example 9.2 which is context free. In fact, Problem 9.10 shows that L can also be generated by a type 3 grammar, i.e. that L is regular.

A fundamental relationship between regular grammars and finite automata (see Chapter 7) follows:

Theorem 9.2:   A language L can be generated by a type 3 grammar, i.e. L is regular, if and only if there exists a finite automaton M which accepts L.

In discussing languages and grammars, we use the following notation unless otherwise stated or implied.   Terminals will be denoted by italic lowercase Latin letters, $a, b, c, \ldots$, and variables will still be denoted by italic capital Latin letters, $A, B, C, \ldots$, with S denoting the start variable.   Also Greek letters $\alpha, \beta, \ldots$, will denote words in both variables and terminals.   Furthermore, we will write

$$\alpha \rightarrow (\beta_1, \ldots, \beta_k)$$

for the productions $\alpha \rightarrow \beta_1, \ldots, \alpha \rightarrow \beta_k$.

9.4 GROUPS

Let G be a nonempty set with a binary operation (denoted by juxtaposition).   Then G is called a *group* if the folowing axioms hold:

[G$_1$]   *Associative law*, i.e. for any a, b, c in G, we have $(ab)c = a(bc)$.

[G$_2$]   *Identity element*, i.e. there exists an element e in G such that $ae = ea = a$ for any element a in G.

[G$_3$]   *Inverses*, i.e. for each a in G, there exists an element a^{-1} (the *inverse* of a) in G such that $aa^{-1} = a^{-1}a = e$.

A group G is said to be *abelian* (or *commutative*) if the *commutative law* holds, i.e. if $ab = ba$ for every $a, b \in G$.

When the binary operation is denoted by juxtaposition as above, the group G is said to be written *multiplicatively*.   Sometimes, when G is abelian, the binary operation is denoted by $+$ and G is said to be written *additively*.   In such a case the identity element is denoted by 0 and is called the *zero* element; and the inverse is denoted by $-a$ and is called the *negative* of a.

The number of elements in a group G, denoted by $|G|$, is called the *order* of G, and G is called a *finite group* if its order is finite.   If A and B are subsets of G then we write

$$AB = \{ab : a \in A, b \in B\} \qquad \text{or} \qquad A + B = \{a + b : a \in A, b \in B\}$$

EXAMPLE 9.3.

(a)   The set $\mathbf{Z}$ of integers is an abelian group under addition.   The identity element is 0 and $-a$ is the additive inverse of a in $\mathbf{Z}$.

(b)   The nonzero rational numbers $Q \setminus \{0\}$ form an abelian group under multiplication.   The number 1 is the identity element and q/p is the multiplicative inverse of the rational number p/q.

(c)   Let S be the set of 2×2 matrices with rational entries under the operation of matrix multiplication.   Then S is not a group since inverses do not always exist.   However, let G be the subset of 2×2 matrices with a nonzero determinant.   Then G is a group under matrix multiplication.   The identity element is

$$I = \begin{pmatrix} 1 & 0 \\ 0 & 1 \end{pmatrix}$$

The inverse of $A = \begin{pmatrix} a & b \\ c & d \end{pmatrix}$ is $A^{-1} = \begin{pmatrix} d/|A| & -b/|A| \\ -c/|A| & a/|A| \end{pmatrix}$. This is an example of a nonabelian group since matrix multiplication is noncommutative.

EXAMPLE 9.4. A one-to-one mapping σ of the set $\{1, 2, \ldots, n\}$ onto itself is called a *permutation*. We denote the permutation by

$$\sigma = \begin{pmatrix} 1 & 2 & \cdots & n \\ j_1 & j_2 & \cdots & j_n \end{pmatrix}$$

where $j_i = \sigma(i)$.

The set of such permutations is denoted by S_n, and there are $n! = 1 \cdot 2 \cdots \cdot n$ of them. We note that the composition of permutations in S_n belongs to S_n, the identity function ϵ belongs to S_n, and the inverses of permutations in S_n belong to S_n. Thus S_n forms a group under composition of functions, called the *symmetric group of degree n*. We investigate S_3 here; its elements are

$$\epsilon = \begin{pmatrix} 1 & 2 & 3 \\ 1 & 2 & 3 \end{pmatrix} \qquad \sigma_2 = \begin{pmatrix} 1 & 2 & 3 \\ 3 & 2 & 1 \end{pmatrix} \qquad \phi_1 = \begin{pmatrix} 1 & 2 & 3 \\ 2 & 3 & 1 \end{pmatrix}$$

$$\sigma_1 = \begin{pmatrix} 1 & 2 & 3 \\ 1 & 3 & 2 \end{pmatrix} \qquad \sigma_3 = \begin{pmatrix} 1 & 2 & 3 \\ 2 & 1 & 3 \end{pmatrix} \qquad \phi_2 = \begin{pmatrix} 1 & 2 & 3 \\ 3 & 1 & 2 \end{pmatrix}$$

The multiplication table of S_3 follows:

	ϵ	σ_1	σ_2	σ_3	ϕ_1	ϕ_2
ϵ	ϵ	σ_1	σ_2	σ_3	ϕ_1	ϕ_2
σ_1	σ_1	ϵ	ϕ_1	ϕ_2	σ_2	σ_3
σ_2	σ_2	ϕ_2	ϵ	ϕ_1	σ_3	σ_1
σ_3	σ_3	ϕ_1	ϕ_2	ϵ	σ_1	σ_2
ϕ_1	ϕ_1	σ_3	σ_1	σ_2	ϕ_2	ϵ
ϕ_2	ϕ_2	σ_2	σ_3	σ_1	ϵ	ϕ_1

9.5 SUBGROUPS AND NORMAL SUBGROUPS

A subset H of a group G is called a *subgroup* of G if H itself forms a group under the operation of G. One can show that H is a subgroup if H has the following three properties: (i) The identity element e belongs to H. (ii) H is closed under the operation of G, i.e. if a, $b \in H$ then $ab \in H$. (iii) H is closed under inverses, i.e. if $a \in H$ then $a^{-1} \in H$. Every group G has the subgroups $\{e\}$ and G itself. Any other subgroup of G is called a *nontrivial subgroup*.

If H is a subgroup of G and $a \in G$, then the set

$$Ha = \{ha : h \in H\}$$

is called a *right coset* of H. (Analogously, aH is called a left coset of H.) We have the following important results (proved in Problems 9.13 and 9.15).

Theorem 9.3:   Let H be a subgroup of a group G. Then the right cosets Ha form a partition of G.

Theorem 9.4 (Lagrange):   Let H be a subgroup of a finite group G. Then the order of H divides the order of G.

One can actually show that the number of right cosets of H in G, called the *index of H in G*, is equal to the number of left cosets of H in G; and both numbers are equal to $|G|$ divided by $|H|$.

Definition: A subgroup H of G is called a *normal* subgroup if $a^{-1}Ha \subset H$ for every $a \in G$. Equivalently, H is normal if $aH = Ha$ for every $a \in G$, i.e. if the right and left cosets of H coincide.

Note that every subgroup of an abelian group is normal.

Theorem 9.5:   Let H be a normal subgroup of G. Then the cosets of H in G form a group under coset multiplication. This group is called the *quotient group* and is denoted by G/H.

EXAMPLE 9.5.   Consider the group **Z** of integers under addition. Let H denote the set of multiples 5, i.e.

$$H = \{\ldots, -10, -5, 0, 5, 10, \ldots\}$$

Then H is a subgroup (necessarily normal) of **Z**. The cosets of H in **Z** follow:

$$\bar{0} = 0 + H = H = \{\ldots, -10, -5, 0, 5, 10, \ldots\}$$
$$\bar{1} = 1 + H = \{\ldots, -9, -4, 1, 6, 11, \ldots\}$$
$$\bar{2} = 2 + H = \{\ldots, -8, -3, 2, 7, 12, \ldots\}$$
$$\bar{3} = 3 + H = \{\ldots, -7, -2, 3, 8, 13, \ldots\}$$
$$\bar{4} = 4 + H = \{\ldots, -6, -1, 4, 9, 14, \ldots\}$$

For any other integer $n \in \mathbf{Z}$, $\bar{n} = n + H$ coincides with one of the above cosets. Thus by the above theorem, $\mathbf{Z}/H = \{0, 1, 2, 3, 4\}$ forms a group under coset addition; its addition table follows:

+	$\bar{0}$	$\bar{1}$	$\bar{2}$	$\bar{3}$	$\bar{4}$
$\bar{0}$	$\bar{0}$	$\bar{1}$	$\bar{2}$	$\bar{3}$	$\bar{4}$
$\bar{1}$	$\bar{1}$	$\bar{2}$	$\bar{3}$	$\bar{4}$	$\bar{0}$
$\bar{2}$	$\bar{2}$	$\bar{3}$	$\bar{4}$	$\bar{0}$	$\bar{1}$
$\bar{3}$	$\bar{3}$	$\bar{4}$	$\bar{0}$	$\bar{1}$	$\bar{2}$
$\bar{4}$	$\bar{4}$	$\bar{0}$	$\bar{1}$	$\bar{2}$	$\bar{3}$

This quotient group $\mathbf{Z}/H$ is referred to as the integers modulo 5 and is frequently denoted by $\mathbf{Z}_5$. Analogously, for any positive integer n, there exists the quotient group $\mathbf{Z}_n$ called the integers modulo n.

EXAMPLE 9.6.

(a) Consider the permutation group S_3 of degree 3 which is investigated in Example 9.4. The set $H = \{\epsilon, \sigma_1\}$ is a subgroup of S_3. Its right and left cosets are

Right Cosets	Left Cosets
$H = \{\epsilon, \sigma_1\}$	$H = \{\epsilon, \sigma_1\}$
$H\phi_1 = \{\phi_1, \sigma_2\}$	$\phi_1 H = \{\phi_1, \sigma_3\}$
$H\phi_2 = \{\phi_2, \sigma_3\}$	$\phi_2 H = \{\phi_2, \sigma_2\}$

Observe that the right cosets and the left cosets are distinct; hence H is not a normal subgroup of S_3.

(b) Consider the group G of 2×2 matrices with rational entries and nonzero determinant. [See Example 9.3(c).] Let H be the subset of G consisting of matrices whose upper-right entry is zero; i.e. matrices of the form

$$\begin{pmatrix} a & 0 \\ c & d \end{pmatrix}$$

Then H is a subgroup of G since H is closed under multiplication and inverses and $I \in H$. However, H is not a normal subgroup since, for example,

$$\begin{pmatrix} 1 & 2 \\ 1 & 3 \end{pmatrix}^{-1} \begin{pmatrix} 1 & 0 \\ 1 & 1 \end{pmatrix} \begin{pmatrix} 1 & 2 \\ 1 & 3 \end{pmatrix} = \begin{pmatrix} -1 & -4 \\ 1 & 3 \end{pmatrix}$$

does not belong to H.

On the other hand, let K be the subset of G consisting of matrices with determinant 1. One can show that K is also a subgroup of G. Moreover, for any matrix X in G and any matrix A in K, we have

$$\det(X^{-1}AX) = 1$$

Hence $X^{-1}AX$ belongs to K, so K is a normal subgroup of G.

(c) Let G be any group and let a be any element of G. As usual, we define $a^0 = e$ and $a^{n+1} = a^n \cdot a$. Clearly, $a^m \cdot a^n = a^{m+n}$ and $(a^m)^n = a^{mn}$, for any integers m and n. All the powers of a,

$$\ldots, a^{-3}, a^{-2}, a^{-1}, e, a, a^2, a^3, \ldots$$

form a subgroup of G called the *cyclic group* generated by a, and will be denoted by $gp(a)$. Suppose that the powers of a are not distinct, say $a^r = a^s$ with, say, $r > s$. Then $a^{r-s} = e$ where $r - s > 0$. The smallest positive integer m such that $a^m = e$ is called the *order* of a and will be denoted by $|a|$. If $|a| = m$, then its cyclic subgroup $gp(a)$ has m elements given by

$$gp(a) = \{e, a, a^2, a^3, \ldots, a^{m-1}\}$$

For example, consider the element ϕ_1 in the symmetric group S_3 discussed in Example 9.4. We have

$$\phi_1^1 = \phi_1, \quad \phi_1^2 = \phi_2, \quad \phi_1^3 = \phi_2 \cdot \phi_1 = \epsilon$$

Hence $|\phi_1| = 3$ and $gp(\phi_1) = \{\epsilon, \phi_1, \phi_2\}$. Observe that $|\phi_1|$ divides the order of S_3. This is true in general; that is, for any element a in a group G we have that $|a|$ equals the order of $gp(a)$ which divides $|G|$ by Lagrange's Theorem 9.4. We also remark that a group G is said to be *cyclic* if it has an element a such that $G = gp(a)$.

A mapping f from a group G into a group G' is called a *homomorphism* if

$$f(ab) = f(a) f(b)$$

for every a, b in G. In addition, if f is one-to-one and onto, then f is called an *isomorphism* and G and G' are said to be *isomorphic*, written $G \simeq G'$.

If $f : G \to G'$ is a *homomorphism*, then the *kernel* of f, written $\mathrm{Ker}\, f$, is the set of elements of G whose image is the identity element e' of G':

$$\mathrm{Ker}\, f = \{a \in G : f(a) = e'\}$$

Recall that the image of f, written $f(G)$ or $\mathrm{Im}\, f$, consists of the images of elements under f:

$$\mathrm{Im}\, f = \{b \in G' : \text{ there exist } a \in G \text{ for which } f(a) = b\}$$

The following theorem (proved in Problem 9.17) is fundamental to group theory.

Theorem 9.6: Let $f : G \to G'$ be a homomorphism with kernel K. Then K is a normal subgroup of G, and the quotient group G/K is isomorphic to the image of f.

EXAMPLE 9.7.

(a) Let G be the group of real numbers under addition, and let G' be the group of positive real numbers under multiplication. The mapping $f : G \to G'$ defined by $f(a) = 2^a$ is a homomorphism because

$$f(a + b) = 2^{a+b} = 2^a 2^b = f(a) f(b)$$

In fact, f is also one-to-one and onto; hence G and G' are isomorphic.

(b) Let G be the group of nonzero complex numbers under multiplication, and let G' be the group of nonzero real numbers under multiplication. The mapping $f : G \to G'$ defined by $f(z) = |z|$ is a homomorphism because

$$f(z_1 z_2) = |z_1 z_2| = |z_1| \, |z_2| = f(z_1) f(z_2)$$

The kernel K of f consists of those complex numbers z on the unit circle, i.e. for which $|z| = 1$. Thus G/K is isomorphic to the image of f, i.e. to the group of positive real numbers under multiplication.

(c) Let a be any element in a group G. The function $f : \mathbf{Z} \to G$ defined by $f(n) = a^n$ is a homomorphism since

$$f(m + n) = a^{m+n} = a^m \cdot a^n = f(m) \cdot f(n)$$

The image of f is $gp(a)$, the cyclic subgroup generated by a. By Theorem 9.6,

$$gp(a) \simeq \mathbf{Z}/K$$

where K is the kernel of F. If $K = \{0\}$, then $gp(a) \simeq \mathbf{Z}$. On the other hand, if m is the order of a, then $K = \{\text{multiples of } m\}$, and so $gp(a) \simeq \mathbf{Z}_m$. In other words, any cyclic group is isomorphic to either the integers $\mathbf{Z}$ under addition, or to $\mathbf{Z}_m$, the integers under addition modulo m.

9.6 RINGS, INTEGRAL DOMAINS AND FIELDS

Let R be a nonempty set with two binary operations, an operation of addition (denoted by +) and an operation of multiplication (denoted by juxtaposition). Then R is called a *ring* if the following axioms are satisfied:

[R_1]   For any $a, b, c \in R$, we have $(a + b) + c = a + (b + c)$.

[R_2]   There exists an element $0 \in R$, called the *zero* element, such that $a + 0 = 0 + a = a$ for every $a \in R$.

[R_3]   For each $a \in R$ there exists an element $-a \in R$, called the *negative* of a, such that $a + (-a) = (-a) + a = 0$.

[R_4]   For any $a, b \in R$, we have $a + b = b + a$.

[R_5]   For any $a, b, c \in R$, we have $(ab)c = a(bc)$.

[R_6]   For any $a, b, c \in R$, we have:

(i) $a(b + c) = ab + ac$, and (ii) $(b + c)a = ba + ca$.

Observe that the axioms [R_1] through [R_4] may be summarized by saying that R is an abelian group under addition.

Subtraction is defined in R by $a - b \equiv a + (-b)$.

It can be shown (see Problem 9.22) that $a \cdot 0 = 0 \cdot a = 0$ for every $a \in R$.

R is called a *commutative ring* if $ab = ba$ for every $a, b \in R$. We also say that R is a *ring with an identity element* if there exists a nonzero element $1 \in R$ such that $a \cdot 1 = 1 \cdot a = a$ for every $a \in R$.

A nonempty subset J of a ring R is called an *ideal* in R if: (i) $a - b \in J$ whenever $a, b \in J$, and (ii) $ra, ar \in J$ whenever $r \in R$, $a \in J$. Note first that J is a subring of R. Also, J is a subgroup (necessarily normal) of the additive group of R. Thus we can form the collection of cosets

$$\{a + J : a \in R\}$$

which form a partition of R.

Theorem 9.7:   Let J be an ideal in a ring R. Then the cosets $\{a + J : a \in R\}$ form a ring under the coset operations

$$(a + J) + (b + J) = a + b + J \quad \text{and} \quad (a + J)(b + J) = ab + J$$

This ring is denoted by R/J and is called the *quotient ring*.

Now let R be a commutative ring with an identity element. For any $a \in R$, the set $(a) = \{ra : r \in R\}$ is an ideal; it is called the *principal ideal* generated by a. If every ideal in R is a principal ideal, then R is called a *principal ideal ring*.

A nonzero element a in a commutative ring R is called a *zero divisor* if there exists a nonzero element b such that $ab = 0$.

Definition: A commutative ring R with an identity element is called an *integral domain* if R has no zero divisors, i.e. if $ab = 0$ implies $a = 0$ or $b = 0$.

Definition: A commutative ring R with an identity element 1 (not equal to 0) is called a *field* if every nonzero $a \in R$ has a multiplicative inverse, i.e. there exists an element $a^{-1} \in R$ such that $aa^{-1} = a^{-1}a = 1$.

A field is necessarily an integral domain; for if $ab = 0$ and $a \neq 0$, then

$$b = 1 \cdot b = a^{-1}ab = a^{-1} \cdot 0 = 0$$

We remark that a field may also be viewed as a commutative ring in which the nonzero elements form a group under multiplication.

EXAMPLE 9.8.

(a) The set $\mathbf{Z}$ of integers with the usual operations of addition and multiplication is the classical example of an integral domain (with an identity element). Every ideal J in $\mathbf{Z}$ is a principal ideal, i.e. $J = (m)$ for some integer m. The quotient ring $\mathbf{Z}_m = \mathbf{Z}/(m)$ is called the *ring of integers modulo m*. If m is prime, then $\mathbf{Z}_m$ is a field. On the other hand, if m is not prime then $\mathbf{Z}_m$ has zero divisors. For example, in the ring $\mathbf{Z}_6$ we have

$$\bar{2} \cdot \bar{3} = \bar{0} \quad \text{but} \quad \bar{2} \neq \bar{0} \quad \text{and} \quad \bar{3} \neq \bar{0}$$

(b) The rational numbers $\mathbf{Q}$ and the real numbers $\mathbf{R}$ each form a field with respect to the usual operations of addition and multiplication.

(c) Let $\mathbf{C}$ denote the set of ordered pairs of real numbers with addition and multiplication defined by

$$(a, b) + (c, d) = (a + c, b + d)$$
$$(a, b) \cdot (c, d) = (ac - bd, ad + bc)$$

Then $\mathbf{C}$ satisfies all the required properties of a field. In fact, $\mathbf{C}$ is just the field of complex numbers.

(d) The set M of all 2 by 2 matrices with real entries forms a noncommutative ring with zero divisors under the operations of matrix addition and matrix multiplication.

(e) Let R be any ring. Then the set $R[x]$ of all polynomials over R forms a ring with respect to the usual operations of addition and multiplication of polynomials. Moreover, if R is an integral domain then $R[x]$ is also an integral domain.

Now let D be an integral domain. We say that b *divides* a in D if $a = bc$ for some $c \in D$. An element $u \in D$ is called a *unit* if u divides 1, i.e. if u has a multiplicative inverse. An element $b \in D$ is called an *associate* of $a \in D$ if $b = ua$ for some unit $u \in D$. A nonunit $p \in D$ is said to be *irreducible* if $p = ab$ implies a or b is a unit.

An integral domain D is called a *unique factorization domain* if every nonunit $a \in D$ can be written uniquely (up to associates and order) as a product of irreducible elements.

EXAMPLE 9.9.

(a) The ring $\mathbf{Z}$ of integers is the classical example of a unique factorization domain. The units of $\mathbf{Z}$ are 1 and −1. The only associates of $n \in \mathbf{Z}$ are n and $-n$. The irreducible elements of $\mathbf{Z}$ are the prime numbers.

(b) The set $D = \{a + b\sqrt{13} : a, b \text{ integers}\}$ is an integral domain. The units of D are ± 1, $18 \pm 5\sqrt{13}$ and $-18 \pm 5\sqrt{13}$. The elements 2, $3 - \sqrt{13}$ and $-3 - \sqrt{13}$ are irreducible in D. Observe that $4 = 2 \cdot 2 = (3 - \sqrt{13})(-3 - \sqrt{13})$. Thus D is not a unique factorization domain. (See Problem 9.81.)

Solved Problems

OPERATIONS AND SEMIGROUPS

9.1. Consider the set $\mathbf{N}$ of positive integers, and let $*$ be the operation of least common multiple (l.c.m.) on $\mathbf{N}$.

    (a) Find $4 * 6$, $3 * 5$, $9 * 18$ and $1 * 6$.

    (b) Is $(\mathbf{N}, *)$ a semigroup? Is it commutative?

    (c) Find the identity element of $*$.

    (d) Which elements in $\mathbf{N}$, if any, have inverses and what are they?

(a) Since $x * y$ means the least common multiple of x and y, we have:

$$4 * 6 = 12, \quad 3 * 5 = 15, \quad 9 * 18 = 18, \quad 1 * 6 = 6$$

(b) One proves in number theory that $(a * b) * c = a * (b * c)$, i.e. that the operation of l.c.m. is associative, and that $a * b = b * a$, i.e. that the operation of l.c.m. is commutative. Hence $(\mathbf{N}, *)$ is a commutative semigroup.

(c) The integer 1 is the identity element since the l.c.m. of 1 and any positive integer a is a, i.e. $1 * a = a * 1 = a$ for any $a \in \mathbf{N}$.

(d) Since l.c.m. $(a, b) = 1$ if and only if $a = 1$ and $b = 1$, the only number which has an inverse is 1 and it is its own inverse.

9.2. Consider the set $\mathbf{Q}$ of rational numbers, and let $*$ be the operation on $\mathbf{Q}$ defined by

$$a * b = a + b - ab$$

(a) Find $3 * 4$, $2 * (-5)$ and $7 * \frac{1}{2}$.

(b) Is $(\mathbf{Q}, *)$ a semigroup? Is it commutative?

(c) Find the identity element for $*$.

(d) Do any of the elements in $\mathbf{Q}$ have an inverse? What is it?

(a) $\quad 3 * 4 = 3 + 4 - 3 \cdot 4 = 3 + 4 - 12 = -5$

$\quad 2 * (-5) = 2 + (-5) - 2 \cdot (-5) = 2 - 5 + 10 = 7$

$\quad 7 * \frac{1}{2} = 7 + \frac{1}{2} - 7(\frac{1}{2}) = 4$

(b) We have:

$$
\begin{aligned}
(a * b) * c &= (a + b - ab) * c \\
&= (a + b - ab) + c - (a + b - ab)c \\
&= a + b - ab + c - ac - bc + abc \\
&= a + b + c - ab - ac - bc + abc
\end{aligned}
$$

$$
\begin{aligned}
a * (b * c) &= a * (b + c - bc) \\
&= a + (b + c - bc) - a(b + c - bc) \\
&= a + b + c - bc - ab - ac + abc
\end{aligned}
$$

Hence $*$ is associative and $(\mathbf{Q}, *)$ is a semigroup. Also,

$$a * b = a + b - ab = b + a - ba = b * a$$

Hence $(\mathbf{Q}, *)$ is a commutative semigroup.

(c) An element e is an identity element if $a * e = a$ for every $a \in \mathbf{Q}$. Compute as follows:

$$a * e = a, \quad a + e - ae = a, \quad e - ea = 0, \quad e(1 - a) = 0, \quad e = 0$$

Accordingly, 0 is the identity element.

(d) In order for a to have an inverse x, we must have $a * x = 0$ since 0 is the identity element by Part (c). Compute as follows:

$$a * x = 0, \quad a + x - ax = 0, \quad a = ax - x, \quad a = x(a - 1), \quad x = a/(a - 1)$$

Thus if $a \neq 1$, then a has an inverse and it is $a/(a - 1)$.

9.3. Let S be a semigroup with identity e and let b and b' be inverses of a. Show that $b = b'$, that is, that inverses are unique if they exist.

We have:

$$b * (a * b') = b * e = b \quad \text{and} \quad (b * a) * b' = e * b' = b'$$

Since S is associative, $(b * a) * b' = b * (a * b')$; hence $b = b'$.

9.4. State whether or not each of the following subsets of the positive integers **N** is closed under the operation of multiplication:

(a) $A = \{0, 1\}$ (d) $D = \{2, 4, 6, \ldots\} = \{x : x \text{ is even}\}$

(b) $B = \{1, 2\}$ (e) $E = \{1, 3, 5, \ldots\} = \{x : x \text{ is odd}\}$

(c) $C = \{x : x \text{ is prime}\}$ (f) $F = \{2, 4, 8, \ldots\} = \{x : x = 2^n, n \in \mathbf{N}\}$

(a) We have:
$$0 \cdot 0 = 0, \quad 0 \cdot 1 = 0, \quad 1 \cdot 0 = 0 \quad \text{and} \quad 1 \cdot 1 = 1$$
Hence A is closed under multiplication.

(b) Since $2 \cdot 2 = 4$, which does not belong to B, the set B is not closed under multiplication.

(c) Note that 2 and 3 are prime but $2 \cdot 3 = 6$ is not prime; hence C is not closed under multiplication.

(d) The product of even numbers is even; hence D is closed under multiplication.

(e) The product of odd numbers is odd; hence E is closed under multiplication.

(f) Since $2^r \cdot 2^s = 2^{r+s}$, F is closed under multiplication.

9.5. State whether or not each of the sets in Problem 9.4 is closed under the operation of addition.

Since the sum of two even integers is even, the set D is closed under addition. However, each of the other sets is not closed under addition since, for example,

$$1 + 1 = 2 \notin A \qquad 1 + 3 = 4 \notin E$$
$$1 + 2 = 3 \notin B \qquad 2 + 4 = 6 \notin F$$
$$3 + 5 = 8 \notin C$$

LANGUAGES AND GRAMMARS

9.6. Suppose a grammar G has the following productions:

(a) $S \to aA$, $A \to aAB$, $B \to b$, $A \to a$

(b) $S \to aAB$, $AB \to bB$, $B \to b$, $A \to aB$

(c) $S \to aAB$, $AB \to c$, $A \to b$, $B \to AB$

(d) $S \to aB$, $B \to bA$, $B \to b$, $B \to a$, $A \to aB$, $A \to a$

What type of grammar is G?

(a) Since each production is of the form $A \to \alpha$, i.e. a variable is on the left, G is a context-free or type 2 grammar.

(b) Each production is of the form $\alpha A \beta \to \alpha \gamma \beta$, i.e. we can replace the variable A by γ providing A lies between α and β. For example, by the second production we can replace the variable A by b only if A is followed by B. Hence G is of type 1.

(c) The production $AB \to c$ means that G is of type 0.

(d) G is a regular or type 3 grammar since each production is of the form $A \to a$ or $A \to aB$.

9.7. Consider a context-free grammar G with the following productions:

$$S \to aAB, \quad A \to Bba, \quad B \to bB, \quad B \to c$$

The word $W = acbabc$ can be derived from S as follows:

$$S \Rightarrow aAB \Rightarrow a(Bba)B \Rightarrow acbaB \Rightarrow acba(bB) \Rightarrow acbabc$$

Draw the *derivation tree T* of W.

The derivation tree T of W is the ordered rooted tree in Fig. 9-2(e). Note that S is the root of T and that the ordered leaves of T produce the word W. Also, any nonleaf of T is a variable, say A, and the immediate successors of A form a word α where $A \to \alpha$ is a production of G used in the derivation of W from S. Figure 9-2 shows how to construct T. (We note that derivation trees are useful only for type 2 and type 3 grammars.)

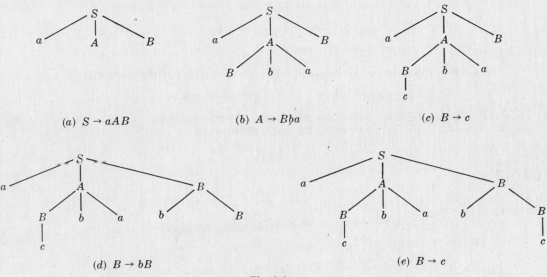

(a) $S \to aAB$ (b) $A \to Bba$ (c) $B \to c$

(d) $B \to bB$ (e) $B \to c$

Fig. 9-2

9.8. Find the language $L(G)$ generated by the grammar G with variables S, A, B, terminals a, b and productions $S \to aB$, $B \to b$, $B \to bA$, $A \to aB$. What type of language is $L(G)$?

Observe that we can only use the first production once since the start symbol S does not appear anywhere else. Also, we can only obtain a terminal word by finally using the second production. Otherwise we alternately add a's and b's using the third and fourth productions. In other words,

$$L(G) = \{(ab)^n = ababab\ldots ab : n \in \mathbf{N}\}$$

$L(G)$ is a type 3, i.e. regular language since the productions in the grammar are of the form $A \to a$ or $A \to aB$.

9.9. Let L be the set of all words in a and b with an even number of a's. Find a regular grammar G which will generate L. (Compare with Problem 7.7.)

We claim that the grammar G with the following productions will generate L:

$$S \to aA, \ S \to bB, \ B \to bB, \ B \to aA, \ A \to aB, \ A \to bA, \ A \to a, \ B \to b$$

Observe that the sum of the a's and A's in any word α either remains the same or is increased by 2 when any production is applied to α. Thus any word W in the terminals a and b which is derived from S must contain an even number of a's. In other words $L(G) \subset L$. On the other hand, it is clear which productions should be used to print any word V in L; that is, we use $S \to aA$ or $S \to bB$ according as V begins with a or b, and we use $A \to aB$ or $B \to aA$ if any subsequent letter is an a and we use $A \to bA$ or $B \to bB$ if any subsequent letter is a b. For the last letter of V we use $A \to a$ or $B \to b$. Thus $L(G) = L$.

9.10. Consider the following two languages on the letters a and b:

$$L = \{a^r b^s : r, s > 0\} \quad \text{and} \quad L' = \{a^n b^n : n > 0\}$$

Observe that both languages consist of words which begin with a's and end with b's, but in L' the number of a's must equal the number of b's.

(a) Find a regular grammar G which will generate L.

(b) Find a context-free grammar G' which will generate L'.

(a) Clearly the grammar G with the following productions will generate L:

$$S \rightarrow aA, \ A \rightarrow aA, \ A \rightarrow b, \ A \rightarrow bB, \ B \rightarrow bB, \ B \rightarrow b$$

Note that G is a regular grammar.

(b) Clearly the grammar G' with the following productions will generate L':

$$S \rightarrow ab, \ S \rightarrow aSb$$

Note that G' is a context-free grammar. We emphasize that L' is not a regular language; the proof of this fact lies beyond the scope of this text.

GROUPS

9.11. Consider the symmetric group S_3 given in Example 9.4.

(a) Find the order and the group generated by each element of S_3. (Use the multiplication table of S_3 in Example 9.4.)

(b) Find the number and all subgroups of S_3.

(c) Let $A = \{\sigma_1, \sigma_2\}$ and $B = \{\phi_1, \phi_2\}$. Find AB, $\sigma_3 A$ and $A\sigma_3$.

(d) Let $H = gp(\sigma_1)$ and $K = gp(\sigma_2)$. Show that HK is not a subgroup of G. (Compare with Problem 9.20.)

(e) Is S_3 cyclic?

(a) $\epsilon^1 = \epsilon$, so $|\epsilon| = 1$ and $gp(\epsilon) = \{\epsilon\}$. $\sigma_1^1 = \sigma_1$, $\sigma_1^2 = \epsilon$; so $|\sigma_1| = 2$ and $gp(\sigma_1) = \{\sigma_1, \epsilon\}$. Similarly, $|\sigma_2| = 2$, $gp(\sigma_2) = \{\sigma_2, \epsilon\}$; and $|\sigma_3| = 2$, $gp(\sigma_3) = \{\sigma_3, \epsilon\}$. By Example 9.6(c) $|\phi_1| = 3$, and $gp(\phi_1) = \{\epsilon, \phi_1, \phi_2\}$. Also, $\phi_2^1 = \phi_2$, $\phi_2^2 = \phi_1$, $\phi_2^3 = \phi_1 \cdot \phi_2 = \epsilon$; hence $|\phi_2| = 3$ and $gp(\phi_2) = \{\epsilon, \phi_2, \phi_1\}$.

(b) First of all, $H_1 = \{\epsilon\}$ and $H_2 = S_3$ are subgroups of S_3. Any other subgroup of S_3 must have order 2 or 3 since its order must divide $|S_3| = 6$. Since 2 and 3 are prime numbers, these subgroups must be cyclic (Problem 9.52) and hence must appear in Part (a). Thus the other subgroups of S_3 are

$$H_3 = \{\epsilon, \sigma_1\}, \ H_4 = \{\epsilon, \sigma_2\}, \ H_5 = \{\epsilon, \sigma_3\}, \ H_6 = \{\epsilon, \phi_1, \phi_2\}$$

Accordingly, S_3 has six subgroups.

(c) Multiply each element of A by each element of B:

$$\sigma_1 \phi_1 = \sigma_2, \ \ \sigma_1 \phi_2 = \sigma_3, \ \ \sigma_2 \phi_1 = \sigma_3, \ \ \sigma_2 \phi_2 = \sigma_1$$

Hence $AB = \{\sigma_1, \sigma_2, \sigma_3\}$.

Multiply σ_3 by each element of A:

$$\sigma_3 \sigma_1 = \phi_1, \ \ \sigma_3 \sigma_2 = \phi_2, \quad \text{hence} \quad \sigma_3 A = \{\phi_1, \phi_2\}$$

Multiply each element of A by σ_3:

$$\sigma_1 \sigma_3 = \phi_2, \ \ \sigma_2 \sigma_3 = \phi_1, \quad \text{hence} \quad A\sigma_3 = \{\phi_1, \phi_2\}$$

(d) $H = \{\epsilon, \sigma_1\}$, $K = \{\epsilon, \sigma_2\}$ and then $HK = \{\epsilon, \sigma_1, \sigma_2, \phi_1\}$, which is not a subgroup of S_3 since HK has four elements.

(e) S_3 is not cyclic since S_3 is not generated by any of its elements.

9.12. Consider the group $G = \{1, 2, 3, 4, 5, 6\}$ under multiplication modulo 7.

(a) Find the multiplication table of G.

(b) Find 2^{-1}, 3^{-1}, 6^{-1}.

(c) Find the orders and subgroups generated by 2 and 3.

(d) Is G cyclic?

(a) To find $a * b$ in G, find the remainder when the product ab is divided by 7. For example, $5 \cdot 6 = 30$ which yields a remainder of 2 when divided by 7; hence $5 * 6 = 2$ in G. The multiplication table of G follows:

*	1	2	3	4	5	6
1	1	2	3	4	5	6
2	2	4	6	1	3	5
3	3	6	2	5	1	4
4	4	1	5	2	6	3
5	5	3	1	6	4	2
6	6	5	4	3	2	1

(b) Note first that 1 is the identity element of G. Recall that a^{-1} is that element of G such that $aa^{-1} = 1$. Hence $2^{-1} = 4$, $3^{-1} = 5$ and $6^{-1} = 6$.

(c) We have $2^1 = 2$, $2^2 = 4$, but $2^3 = 1$. Hence $|2| = 3$ and $gp(2) = \{1, 2, 4\}$. We have $3^1 = 3$, $3^2 = 2$, $3^3 = 6$, $3^4 = 4$, $3^5 = 5$, $3^6 = 1$. Hence $|3| = 6$ and $gp(3) = G$.

(d) G is cyclic since $G = gp(3)$.

9.13. Prove Theorem 9.3: Let H be a subgroup of a group G. Then the right cosets Ha form a partition of G.

Since $e \in H$, $a = ea \in Ha$; hence every element belongs to a coset. In fact, $a \in Ha$. Now suppose Ha and Hb are not disjoint. Say $c \in Ha \cap Hb$. The proof is complete if we show that $Ha = Hb$.

Since c belongs to both Ha and Hb, we have $c = h_1 a$ and $c = h_2 b$ where $h_1, h_2 \in H$. Then $h_1 a = h_2 b$, and so $a = h_1^{-1} h_2 b$. Let $x \in Ha$. Then

$$x = h_3 a = h_3 h_1^{-1} h_2 b$$

where $h_3 \in H$. Since H is a subgroup, $h_3 h_1^{-1} h_2 \in H$; hence $x \in Hb$. Since x was any element of Ha, we have $Ha \subset Hb$. Similarly, $Hb \subset Ha$. Both inclusions imply $Ha = Hb$, and the theorem is proved.

9.14. Let H be a finite subgroup of G. Show that H and any coset Ha have the same number of elements.

Let $H = \{h_1, h_2, \ldots, h_k\}$, where H has k elements. Then $Ha = \{h_1 a, h_2 a, \ldots, h_k a\}$. However, $h_i a = h_j a$ implies $h_i = h_j$; hence the k elements listed in Ha are distinct. Thus H and Ha have the same number of elements.

9.15. Prove Theorem 9.4 (Lagrange): Let H be a subgroup of a group G. Then the order of H divides the order of G.

Suppose H has r elements and there are s right cosets; say

$$Ha_1, \ Ha_2, \ \ldots, \ Ha_s$$

By Theorem 9.3, the cosets partition G and by Problem 9.14, each coset has r elements. Therefore G has rs elements, and so the order of H divides the order of G.

9.16. Suppose $f: G \to G'$ is a group homomorphism.

Prove: (a) $f(e) = e'$, and (b) $f(a^{-1}) = f(a)^{-1}$.

(a) Since $e = ee$ and f is a homomorphism, we have

$$f(e) \ = \ f(ee) \ = \ f(e) \, f(e)$$

Multiplying both sides by $f(e)^{-1}$ gives us our result.

(b) Using part (a) and that $aa^{-1} = a^{-1}a = e$, we have

$$e' \ = \ f(e) \ = \ f(aa^{-1}) \ = \ f(a) \, f(a^{-1}) \qquad \text{and} \qquad e' \ = \ f(e) \ = \ f(a^{-1}a) \ = \ f(a^{-1}) \, f(a)$$

Hence $f(a^{-1})$ is the inverse of $f(a)$; that is, $f(a^{-1}) = f(a)^{-1}$.

9.17. Prove Theorem 9.6: Let $f: G \to G'$ be a homomorphism with kernel K. Then K is a normal subgroup of G, and the quotient group G/K is isomorphic to the image of f.

Proof that K is normal. By Problem 9.16, $f(e) = e'$, so $e \in K$. Now suppose $a, b \in K$ and $g \in G$. Then $f(a) = e'$ and $f(b) = e'$. Hence

$$
\begin{aligned}
f(ab) \ &= \ f(a) \, f(b) \ = \ e'e' \ = \ e' \\
f(a^{-1}) \ &= \ f(a)^{-1} \ = \ e'^{-1} \ = \ e' \\
f(gag^{-1}) \ &= \ f(g) \, f(a) \, f(g^{-1}) \ = \ f(g) e' \, f(g)^{-1} \ = \ e'
\end{aligned}
$$

Hence ab, a^{-1} and gag^{-1} belong to K, so K is a normal subgroup.

Proof that $G/K \cong H$, where H is the image of f. Let $\phi: G/K \to H$ be defined by

$$\phi(Ka) \ = \ f(a)$$

We show that ϕ is well-defined, i.e. if $Ka = Kb$ then $\phi(Ka) = \phi(Kb)$. Suppose $Ka = Kb$. Then $ab^{-1} \in K$ (Problem 9.48). Then $f(ab^{-1}) = e'$, and so

$$f(a) \, f(b)^{-1} \ = \ f(a) \, f(b^{-1}) \ = \ f(ab^{-1}) \ = \ e'$$

Hence $f(a) = f(b)$, and so $\phi(Ka) = \phi(Kb)$. Thus ϕ is well-defined. We next show that ϕ is a homomorphism:

$$\phi(KaKb) \ = \ \phi(Kab) \ = \ f(ab) \ = \ f(a) \, f(b) \ = \ \phi(Ka) \, \phi(Kb)$$

Thus ϕ is a homomorphism. We next show that ϕ is one-to-one. Suppose $\phi(Ka) = \phi(Kb)$. Then

$$f(a) \ = \ f(b) \qquad \text{or} \qquad f(a) \, f(b)^{-1} \ = \ e' \qquad \text{or} \qquad f(a) \, f(b^{-1}) \ = \ e' \qquad \text{or} \qquad f(ab^{-1}) \ = \ e'$$

Thus $ab^{-1} \in K$, and by Problem 9.48 we have $Ka = Kb$. Thus ϕ is one-to-one. We next show that ϕ is onto. Let $h \in H$. Since H is the image of f, there exists $a \in G$ such that $f(a) = h$. Thus $\phi(Ka) = f(a) = h$, and so ϕ is onto. Consequently $G/K \cong H$ and the theorem is proved.

9.18. Let G be a group and let $g \in G$. Define a function $\hat{g}: G \to G$ by $\hat{g}(x) = gxg^{-1}$. Show that $\hat{g}$ is an isomorphism of G onto G, i.e. show that: (a) $\hat{g}$ is a homomorphism, (b) $\hat{g}$ is one-to-one, (c) $\hat{g}$ is onto.

(a) We have $\hat{g}(xy) = gxyg^{-1} = (gxg^{-1})(gyg^{-1}) = \hat{g}(x) \, \hat{g}(y)$. Hence g is a homomorphism.

(b) Suppose $\hat{g}(x) = \hat{g}(y)$. Then $gxg^{-1} = gyg^{-1}$. By cancellation, $x = y$. Thus $\hat{g}$ is one-to-one.

(c) Suppose $z \in G$. Then $\hat{g}(g^{-1}zg) = gg^{-1}zgg^{-1} = z$. Hence $\hat{g}$ is onto.

9.19. Prove: Every subgroup of a cyclic group G is cyclic.

Since G is cyclic, there is an element a in G such that $G = gp(a)$. Let H be a subgroup of G. If $H = \{e\}$, then $H = gp(e)$ and is cyclic. Otherwise, H contains a nonzero power of a. Since H is a subgroup, it must be closed under inverses and so contains positive powers of a. Let m be the smallest power of a such that a^m belongs to H. We claim that $b = a^m$ generates H. Let x be any other element of H; since x belongs to G we have $x = a^n$ for some integer n. Dividing n by m we get a quotient q and a remainder r, i.e.

$$n = mq + r$$

where $0 \leqq r < m$. Then

$$a^n = a^{mq+r} = a^{mq} \cdot a^r = b^q \cdot a^r \quad \text{so} \quad a^r = b^{-q}a^n$$

But $a^n, b \in H$. Since H is a subgroup, $b^{-q}a^n \in H$, which means $a^r \in H$. However, m was the smallest positive power of a belonging to H. Therefore $r = 0$. Hence $a^n = b^q$. Thus b generates H, and so H is cyclic.

9.20. Let H be a subgroup and let K be a normal subgroup of a group G. Prove that HK is a subgroup of G.

We must show that $e \in HK$ and that HK is closed under multiplication and inverses. Since H and K are subgroups, $e \in H$ and $e \in K$. Hence $e = e \cdot e$ belongs to HK. Suppose $x, y \in HK$. Then $x = h_1 k_1$ and $y = h_2 k_2$ where $h_1, h_2 \in H$ and $k_1, k_2 \in K$. Then

$$xy = h_1 k_1 h_2 k_2 = h_1 h_2 (h_2^{-1} k_1 h_2) k_2$$

Since K is normal, $h_2^{-1} k_1 h_2 \in K$; and since H and K are subgroups, $h_1 h_2 \in H$ and $(h_2^{-1} k_2 h_2) k_2 \in K$. Thus $xy \in HK$, and so HK is closed under multiplication. We also have that

$$x^{-1} = (h_1 k_1)^{-1} = k_1^{-1} h_1^{-1} = h_1^{-1} (h_1 k_1^{-1} h_1^{-1})$$

Since K is a normal subgroup $h_1 k_1^{-1} h_1^{-1}$ belongs to K. Also h_1^{-1} belongs to H. Therefore $x^{-1} \in HK$, and hence HK, is closed under inverses. Consequently, HK is a subgroup.

RINGS, INTEGRAL DOMAINS, FIELDS

9.21. Consider the ring $\mathbf{Z}_{10} = \{0, 1, 2, \ldots, 9\}$ of integers modulo 10.

(a) Find the units of $\mathbf{Z}_{10}$.

(b) Find -3, -8, and 3^{-1}.

(c) Let $f(x) = 2x^2 + 4x + 4$. Find the roots of $f(x)$ over $\mathbf{Z}_{10}$.

(a) By Problem 9.71, those integers relatively prime to the modulus $m = 10$ are the units in $\mathbf{Z}_{10}$. Hence the units are 1, 3, 7 and 9.

(b) By $-a$ in a ring R we mean that element such that $a + (-a) = (-a) + a = 0$. Hence $-3 = 7$ since $3 + 7 = 7 + 3 = 0$ in $\mathbf{Z}_{10}$. Similarly $-8 = 2$. By a^{-1} in a ring R we mean that element such that $a \cdot a^{-1} = a^{-1} \cdot a = 1$. Hence $3^{-1} = 7$ since $3 \cdot 7 = 7 \cdot 3 = 1$ in $\mathbf{Z}_{10}$.

(c) Substitute each of the ten elements of $\mathbf{Z}_{10}$ into $f(x)$ to see which elements yield 0. We have:

$$f(0) = 4, \quad f(2) = 0, \quad f(4) = 2, \quad f(6) = 0, \quad f(8) = 4$$

$$f(1) = 0, \quad f(3) = 4, \quad f(5) = 4, \quad f(7) = 0, \quad f(9) = 2$$

Thus the roots are 1, 2, 6 and 7. (This example shows that a polynomial of degree n can have more than n roots over an arbitrary ring. This cannot happen if the ring is a field.)

9.22. Prove that in a ring R we have:

(i) $a \cdot 0 = 0 \cdot a = 0$, (ii) $a(-b) = (-a)b = -ab$, (iii) $(-1)a = -a$ if R has an identity element 1.

(i) Since $0 = 0 + 0$, we have

$$a \cdot 0 = a(0 + 0) = a \cdot 0 + a \cdot 0$$

Adding $-(a \cdot 0)$ to both sides yields $0 = a \cdot 0$. Similarly $0 \cdot a = 0$.

(ii) Using $b + (-b) = (-b) + b = 0$, we have

$$ab + a(-b) \;=\; a(b + (-b)) \;=\; a \cdot 0 \;=\; 0$$

$$a(-b) + ab \;=\; a((-b) + b) \;=\; a \cdot 0 \;=\; 0$$

Hence $a(-b)$ is the negative of ab; that is, $a(-b) = -ab$. Similarly, $(-a)b = -ab$.

(iii) We have

$$a + (-1)a \;=\; 1 \cdot a + (-1)a \;=\; (1 + (-1))a \;=\; 0 \cdot a \;=\; 0$$

$$(-1)a + a \;=\; (-1)a + 1 \cdot a \;=\; ((-1) + 1)a \;=\; 0 \cdot a \;=\; 0$$

Hence $(-1)a$ is the negative of a; that is, $(-1)a = -a$.

9.23. In an integral domain D, show that if $ab = ac$ with $a \neq 0$ then $b = c$.

Since $ab = ac$ we have

$$ab - ac = 0 \quad \text{and so} \quad a(b - c) = 0$$

Since $a \neq 0$, we must have $b - c = 0$, since D has no zero divisors. Hence $b = c$.

9.24. Suppose J and K are ideals in a ring R. Prove that $J \cap K$ is an ideal in R.

Since J and K are ideal, $0 \in J$ and $0 \in K$. Hence $0 \in J \cap K$. Now let $a, b \in J \cap K$ and let $r \in R$. Then $a, b \in J$ and $a, b \in K$. Since J and K are ideals,

$$a - b, \; ra, \; ar \in J \quad \text{and} \quad a - b, \; ra, \; ar \in K$$

Hence $a - b, \; ra, \; ar \in J \cap K$. Therefore $J \cap K$ is an ideal.

9.25. Let J be an ideal in a ring R with an identity element 1. Prove:

(a) If $1 \in J$ then $J = R$. (b) If any unit $u \in J$ then $J = R$.

(a) If $1 \in J$ then for any $r \in R$ we have $r \cdot 1 \in J$ or $r \in J$. Hence $J = R$.

(b) If $u \in J$ then $u^{-1} \cdot u \in J$ or $1 \in J$. Hence $J = R$ by Part (a).

9.26. Prove: (a) A finite integral domain D is a field.

(b) $\mathbf{Z}_p$ is a field where p is a prime number.

(c) (Fermat) If p is prime, then $a^p \equiv a \pmod{p}$ for any integer a.

(a) Suppose D has n elements, say $D = \{a_1, a_2, \ldots, a_n\}$. Let a be any nonzero element of D. Consider the n elements

$$aa_1, \; aa_2, \; \ldots, \; aa_n$$

Since $a \neq 0$, we have $aa_i = aa_j$ implies $a_i = a_j$ (Problem 9.23). Thus the n elements above are distinct, and so they must be a rearrangement of the elements of D. One of them, say aa_k, must equal the identity element 1 of D; that is, $aa_k = 1$. Thus a_k is the inverse of a. Since a was any nonzero element of D, we have that D is a field.

(b) Recall $\mathbf{Z}_p = \{0, 1, 2, \ldots, p - 1\}$. We show that $\mathbf{Z}_p$ has no zero divisors. Suppose $a * b = 0$ in $\mathbf{Z}_p$; that is, $ab \equiv 0 \pmod{p}$. Then p divides ab. Since p is prime, p divides a or p divides b. Thus $a \equiv 0 \pmod{p}$ or $b \equiv 0 \pmod{p}$; that is, $a \equiv 0$ or $b = 0$ in $\mathbf{Z}_p$. Accordingly, $\mathbf{Z}_p$ has no zero divisors and so is an integral domain. By Part (a), $\mathbf{Z}_p$ is a field.

(c) If p divides a, then $a \equiv 0 \pmod{p}$ and so $a^p \equiv a \equiv 0 \pmod{p}$. Suppose p does not divide a, then a may be viewed as a nonzero element of $\mathbf{Z}_p$. Since $\mathbf{Z}_p$ is a field, its nonzero elements form a group G under multiplication of order $p - 1$. By Problem 9.47(f), $a^{p-1} = 1$ in $\mathbf{Z}_p$. In other words, $a^{p-1} \equiv 1 \pmod{p}$. Multiplying by a gives $a^p \equiv a \pmod{p}$, and the theorem is proved.

Supplementary Problems

OPERATIONS AND SEMIGROUPS

9.27. Let $*$ be the operation on the real numbers defined by

$$a * b = a + b + 2ab$$

(a) Find $2 * 3$, $3 * (-5)$ and $7 * \frac{1}{2}$.

(b) Is $(R, *)$ a semigroup? Is it commutative?

(c) Find the identity element.

(d) Which elements have inverses and what are they?

9.28. Let A be a nonempty set with the operation $*$ defined by $a * b = a$. Assume A has more than one element. (a) Is A a semigroup? (b) Is A commutative? (c) Does A have an identity element? (d) Which elements, if any, have inverses and what are they?

9.29. Let $A = \{a, b\}$. Find the number of operations on A, and exhibit one which is neither associative nor commutative.

9.30. Let S be the set $\mathbf{Q} \times \mathbf{Q}$ of ordered pairs of rational numbers with the operation $*$ defined by

$$(a, b) * (x, y) = (ax, ay + b)$$

(a) Find $(3, 4) * (1, 2)$ and $(-1, 3) * (5, 2)$.

(b) Is S a semigroup? Is it commutative?

(c) Find the identity element of S.

(d) Which elements, if any, have inverses and what are they?

9.31. Let $A = \{\ldots, -9, -6, -3, 0, 3, 6, \ldots\}$, i.e. the multiples of three. Is A closed under (a) addition, (b) multiplication, (c) subtraction, (d) division (except by zero)?

9.32. Find a set A of three real numbers which is closed under (a) multiplication, (b) addition.

9.33. Let S be an infinite set. Let A be the collection of finite subsets of S and let B be the collection of infinite subsets of S.

(a) Is A closed under (i) union, (ii) intersection, (iii) complements?

(b) Is B closed under (i) union, (ii) intersection, (iii) complements?

9.34. Let S be a set with an operation $*$. Note that the product $a * b * c$ of three elements can be defined in two ways, $(a * b) * c$ and $a * (b * c)$. (a) How many ways can the product $a * b * c * d$ of four elements be defined? (b) Prove that if $*$ is associative then the product of n elements, $a_1 * a_2 * \cdots * a_n$, is the same for any parenthetical arrangement.

LANGUAGES AND GRAMMARS

9.35. Consider the words $U = a^2 ba^3 b^2$ and $V = bab^2 a$ on $A = \{a, b\}$. Find UV, VU and V^2.

9.36. Consider a countable alphabet $A = \{a_1, a_2, a_3, \ldots\}$. Let B_k be the set of words on A such that the sum of the subscripts of the letters of each word is k. For example, $W = a_2 a_3 a_3 a_6 a_4$ belongs to B_{18} since $2 + 3 + 3 + 6 + 4 = 18$. (a) Find B_4. (b) Prove each B_k is finite. (c) Prove that the set $W(A)$ of all words on A is countable. (d) Prove that any language L on A is countable.

9.37. Find a regular grammar G which generates the language L which consists of all the words on a and b with exactly one b, i.e.

$$L = \{b, a^r b, ba^s, a^r ba^s : r, s > 0\}$$

9.38. Find a regular grammar G which generates the language L which consists of all the words on a and b such that no two a's appear next to each other.

9.39. Find a regular grammar G which generates the language L which consists of all words of the form $a^r b^s c^t$, $r, s, t > 0$, i.e. a's followed by b's followed by c's.

9.40. Find a context-free grammar G which generates the language L which consists of all words on a and b with twice as many a's as b's.

9.41. Consider the context-free grammar G with productions $S \to (a, aAS)$ and $A \to bS$. Find the derivation tree of the word $W = abaabaa$.

9.42. Consider a context-free grammar G, and let Fig. 9-3 give the derivation tree of a word W. (a) Find W. (b) Which terminals, variables and productions must lie in G?

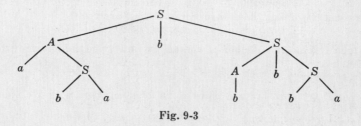

Fig. 9-3

GROUPS

9.43. Consider $\mathbf{Z}_{20} = \{0, 1, \ldots, 19\}$ under addition modulo 20. Let H be the group generated by 5. (a) Find the elements and order of H. (b) Find the cosets of H in $\mathbf{Z}_{20}$.

9.44. Consider $G = \{1, 5, 7, 11, 13, 17\}$ under multiplication modulo 18.

(a) Construct the multiplication table of G.

(b) Find 5^{-1}, 7^{-1} and 17^{-1}.

(c) Find the order and group generated by: (i) 5, (ii) 13.

(d) Is G cyclic?

9.45. Consider $G = \{1, 5, 7, 11\}$ under multiplication modulo 12. (a) Find the **orders of each element**. (b) Is G cyclic? (c) Find **all subgroups**.

9.46. Consider the symmetric group S_4. (See Example 9.4.) Let

$$\alpha = \begin{pmatrix} 1 & 2 & 3 & 4 \\ 3 & 4 & 2 & 1 \end{pmatrix} \quad \text{and} \quad \beta = \begin{pmatrix} 1 & 2 & 3 & 4 \\ 2 & 4 & 3 & 1 \end{pmatrix}$$

(a) Find $\alpha\beta$, $\beta\alpha$, α^2 and α^{-1}.

(b) Find the orders of α, β and $\alpha\beta$.

9.47. Prove the following results for a group G.

(a) The identity element e is unique.

(b) Each a in G has a unique inverse a^{-1}.

(c) $(a^{-1})^{-1} = a$, $(ab)^{-1} = b^{-1}a^{-1}$ and, more generally, $(a_1 a_2 \cdots a_k)^{-1} = a_k^{-1} \cdots a_2^{-1} a_1^{-1}$.

(d) $ab = ac$ implies $b = c$, and $ba = ca$ implies $b = c$.

(e) For any integers r and s, we have $a^r a^s = a^{r+s}$ and $(a^r)^s = a^{rs}$.

(f) $a^{|G|} = e$ when G is finite.

(g) G is abelian if and only if $(ab)^2 = a^2 b^2$ for all $a, b \in G$.

9.48. Let H be a subgroup of G. Prove:
(a) $H = Ha$ if and only if $a \in H$. (b) $Ha = Hb$ if and only if $ab^{-1} \in H$.

9.49. Suppose H is a subset of a group G. Show that H is a subgroup if:
(a) $e \in H$, (b) for all $a, b \in H$ we have $ab, a^{-1} \in H$.

9.50. Prove that the intersection of any number of subgroups (normal subgroups) of G is also a subgroup (normal subgroup) of G.

9.51. Suppose G is an abelian group. Show that any factor group G/H is also abelian.

9.52. Suppose $|G| = p$ where p is a prime. Prove: (a) G has no subgroups except G and $\{e\}$. (b) G is cyclic and every element $a \neq e$ generates G.

9.53. Suppose G has no subgroups except G and $\{e\}$. Show that G is cyclic of prime order.

9.54. Show that $G = \{1, -1, i, -i\}$ is a group under multiplication, and show that $G \cong \mathbf{Z}_4$ by giving an explicit isomorphism $f : G \to \mathbf{Z}_4$.

9.55. Let H be a subgroup of G with only two right cosets. Prove H is normal.

9.56. Let G be an abelian group and let n be a fixed integer. Show that the function $f : G \to G$ defined by $f(a) = a^n$ is a homomorphism.

9.57. Let G be the multiplicative group of complex numbers z such that $|z| = 1$, and let $\mathbf{R}$ be the additive group of real numbers. Prove $G \cong \mathbf{R}/\mathbf{Z}$.

9.58. Let H and K be groups. Let G be the product set $H \times K$ with the operation

$$(h, k) * (h', k') \;=\; (hh', kk')$$

(a) Show that G is a group (called the *direct product* of H and K).

(b) Let $H' = H \times \{e\}$. Show that (i) $H' \cong H$, (ii) H' is a normal subgroup of G, (iii) $G/H' \cong K$.

9.59. Show that there are only two nonisomorphic groups of order 4: $\mathbf{Z}_4$ and $\mathbf{Z}_2 \times \mathbf{Z}_2$.

9.60. Suppose H and N are subgroups of G with N normal. Prove that $H \cap N$ is normal in H and that $H/(H \cap N) \cong HN/N$.

RINGS

9.61. Consider the ring $\mathbf{Z}_{12} = \{0, 1, \ldots, 11\}$ of integers modulo 12.

(a) Find the units of $\mathbf{Z}_{12}$.

(b) Find the roots of $f(x) = x^2 + 4x + 4$ over $\mathbf{Z}_{12}$.

(c) Find the associates of 2.

9.62. Consider the ring $\mathbf{Z}_{30} = \{0, 1, \ldots, 29\}$ of integers modulo 30.
(a) Find -2, -7 and -11. (b) Find 7^{-1}, 11^{-1} and 26^{-1}.

9.63. Show that in a ring R:
(a) $(-a)(-b) = ab$, (b) $(-1)(-1) = 1$, if R has an identity element 1.

9.64. Suppose $a^2 = a$ for every $a \in R$. (Such a ring is called a *Boolean ring*.) Prove that R is commutative.

9.65. Let R be a ring with an identity element. We make R into another ring R' by defining $a \oplus b = a + b + 1$ and $a \cdot b = ab + a + b$. (a) Verify that R' is a ring. (b) Determine the 0-element and 1-element of R'.

9.66.   Let G be any (additive) abelian group.   Define a multiplication in G by $a \cdot b = 0$ for every $a, b \in G$. Show that this makes G into a ring.

9.67.   Let J and K be ideals in a ring R.   Prove that $J + K$ and $J \cap K$ are also ideals.

9.68.   Let R be a ring with an identity element.   Show that $(a) = \{ra : r \in R\}$ is the smallest ideal containing a.

9.69.   Show that R and $\{0\}$ are ideals of any ring R.

9.70.   Show that every nonzero ideal J of $\mathbf{Z}$ is of the form (m) where m is the smallest positive integer in J.

9.71.   Prove: (a) The units of a ring R form a group under multiplication.

(b) The units in $\mathbf{Z}_m$ are those integers which are relatively prime to m.

9.72.   Prove Theorem 9.7: Let J be an ideal in a ring R.   Then the cosets $\{a + J : a \in R\}$ form a ring under the coset operations

$$(a + J) + (b + J) \;=\; a + b + J \qquad \text{and} \qquad (a + J)(b + J) \;=\; ab + J$$

9.73.   Let R and R' be rings. A mapping $f : R \to R'$ is called a *homomorphism* (or *ring homomorphism*) if

$$\text{(i)} \quad f(a + b) \;=\; f(a) + f(b) \qquad \text{and} \qquad \text{(ii)} \quad f(ab) \;=\; f(a)\,f(b)$$

for every $a, b \in R$. Prove that if $f : R \to R'$ is a homomorphism, then the set $K = \{r \in R : f(r) = 0\}$ is an ideal in R. (The set K is called the *kernel* of f.)

9.74.   For any positive integer m, verify that $m\mathbf{Z} = \{rm : r \in \mathbf{Z}\}$ is a ring.   Prove that $2\mathbf{Z}$ and $3\mathbf{Z}$ are not isomorphic. (Rings R and R' are isomorphic if there exists a homomorphism $f : R \to R'$ which is one-to-one and onto.)

INTEGRAL DOMAINS AND FIELDS

9.75.   Prove that if $x^2 = 1$ in an integral domain D then $x = 0$ or $x = 1$.

9.76.   Let R be a finite commutative ring with no zero divisors.   Show that R is an integral domain, i.e. that R has an identity element 1.

9.77.   Prove that $F = \{a + b\sqrt{2} : a, b \text{ rational}\}$ is a field.

9.78.   Prove that $D = \{a + b\sqrt{2} : a, b \text{ integers}\}$ is an integral domain but not a field.

9.79.   A complex number $a + bi$ where a, b are integers is called a *Gaussian integer*. Show that the set G of Gaussian integers is an integral domain.   Also show that the units are ± 1 and $\pm i$.

9.80.   Let D be an integral domain and let I be an ideal in D.   Prove that the factor ring D/I is an integral domain if and only if I is a prime ideal.   (An ideal I is *prime* if $ab \in I$ implies $a \in I$ or $b \in I$.)

9.81.   Consider the integral domain $D = \{a + b\sqrt{13} : a, b \text{ integers}\}$ [see Example 9.9(b)]. If $\alpha = a + b\sqrt{13}$, we define $N(\alpha) = a^2 - 13b^2$. Prove: (i) $N(\alpha\beta) = N(\alpha)\,N(\beta)$; (ii) α is a unit if and only if $N(\alpha) = \pm 1$; (iii) the units of D are ± 1, $18 \pm 5\sqrt{13}$; and $-18 \pm 5\sqrt{13}$; (iv) the numbers 2, $3 - \sqrt{13}$ and $-3 - \sqrt{13}$ are irreducible.

COMPUTER PROGRAMMING PROBLEMS

An operation $*$ on the set $S = \{a, b, c, d\}$ is given by a 4×4 matrix with elements in S.

9.82.   Write a program which prints the identity element of $*$ if it exists.

9.83. Write a program which decides whether or not $*$ is commutative. If not, then the program should print one pair of elements x, y such that $xy \neq yx$.

9.84. Write a program which decides whether or not $*$ is associative. If not, then the program should print one triplet of elements x, y, z such that $(xy)z \neq x(yz)$.

Test the above three programs with the following data:

$$\text{(i)} \quad \begin{pmatrix} d & a & b & c \\ a & b & c & d \\ b & c & d & a \\ c & d & a & b \end{pmatrix} \qquad \text{(ii)} \quad \begin{pmatrix} a & b & c & d \\ a & b & c & d \\ a & b & c & d \\ a & b & c & d \end{pmatrix} \qquad \text{(iii)} \quad \begin{pmatrix} a & a & b & b \\ a & a & b & b \\ c & c & d & d \\ c & c & d & d \end{pmatrix}$$

9.85. Write a program which decides whether or not a subset of $\mathbf{Z}_{24} = \{0, 1, 2, \ldots, 23\}$ is closed under multiplication modulo 24. Test the program with the following input data:

(i)   0, 6, 18, 1, 12   (iv)   0, 1, 17, 23, 5

(ii)   0, 2, 4, 8, 16   (v)   1, 5, 7, 11, 13, 17, 19, 23

(iii)   1, 5, 7, 11   (vi)   1, 3, 9, 7, 15

Answers to Supplementary Problems

9.27. (a) 17, -32, $29/2$, (b) Yes, yes, (c) Zero, (d) If $a \neq \frac{1}{2}$ then a has an inverse which is $-a/(1 + 2a)$.

9.28. (a) Yes, (b) No, (c) No, (d) It is meaningless to talk about an inverse when no identity element exists.

9.29. Sixteen, since there are two choices, a or b, for each of the four products aa, ab, ba and bb. In the table at right, $ab \neq ba$. Also, $(aa)b = bb = a$, but $a(ab) = aa = b$.

$*$	a	b
a	b	a
b	b	a

9.30. (a) $(3, 10)$, $(-5, 1)$, (b) Yes, no, (c) $(1, 0)$, (d) The element (a, b) has an inverse if $a \neq 0$, and its inverse is $(1/a, -b/a)$.

9.31. (a) Yes, (b) Yes, (c) Yes, (d) No

9.32. (a) $\{1, -1, 0\}$, (b) There is no set.

9.33. (a) Yes, yes, no, (b) Yes, no, no

9.34. (a) Five: $(ab)(cd)$, $a((bc)d)$, $a(b(cd))$, $((ab)c)d$ and $(a(bc))d$

9.35. $UV = a^2ba^3b^3ab^2a$, $VU = bab^2a^3ba^3b^2$, $V^2 = bab^2abab^2a$

9.36. (a) a_1^4, $a_1^2a_2$, $a_1a_2a_1$, $a_2a_1^2$, a_1a_3, a_3a_1, a_2^2, a_4

(b) No word in B_k can have more than k letters and no subscript can be larger than k; hence B_k is finite.

(c) $W(A) = \cup_k B_k$ and the B_k are finite.

(d) L is a subset of the countable set $W(A)$.

9.37. $S \to (b, aA)$, $A \to (b, aA, bB)$, $B \to (a, aB)$

9.38. $S \to (a, b, aB, bA)$, $A \to (bA, ab, a, b)$, $B \to (b, bA)$

9.39. $S \to aA$, $A \to (aA, bB)$, $B \to (bB, c, cC)$, $C \to (c, cC)$

9.40. $S \to (AAB, ABA, BAA)$, $A \to (a, BAAA, ABAA, AABA, AAAB)$,
$B \to (b, BBAA, BABA, BAAB, ABAB, AABB)$

9.41. See Fig. 9-4.

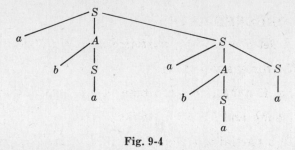

Fig. 9-4

9.42. (a) $ababbbba$, (b) a, b must be terminals, A, S must be variables, and $S \to (AbS, ba)$, $A \to (aS, b)$ must be productions.

9.43. (a) $H = \{0, 5, 10, 15\}$ and $|H| = 4$

(b) H, $1 + H = \{1, 6, 11, 16\}$, $2 + H = \{2, 7, 12, 17\}$, $3 + H = \{3, 8, 13, 18\}$, $4 + H = \{4, 9, 14, 19\}$

9.44. (a) See table below.

(b) $5^{-1} = 11$, $7^{-1} = 13$, $17^{-1} = 17$

(c) (i) $gp(5) = G$, $|5| = 6$, (ii) $gp(13) = \{1, 7, 13\}$, $|13| = 3$

(d) Yes, since $G = gp(5)$.

1	5	7	11	13	17
5	7	17	1	11	13
7	17	13	5	1	11
11	1	5	13	17	7
13	11	1	17	7	5
17	13	11	7	5	1

9.45. (a) $|1| = 1$, $|5| = 2$, $|7| = 2$, $|11| = 2$, (b) No, (c) G, $\{1\}$, $\{1, 7\}$, $\{1, 5\}$, $\{1, 11\}$

9.46. (a) $\alpha\beta = 3142$, $\beta\alpha = 4123$, $\alpha^2 = 2143$, $\alpha^{-1} = 4312$

(b) $|\alpha| = 4$, $|\beta| = 3$, $|\alpha\beta| = 4$

9.54. $f(1) = 0$, $f(i) = 1$, $f(-1) = 2$, $f(-i) = 3$

9.61. (a) 1, 5, 7, 11, (b) 4, 10, (c) $\{2, 10\}$

9.62. (a) $-2 = 28$, $-7 = 23$, $-11 = 19$

(b) $7^{-1} = 13$, $11^{-1} = 11$, 26^{-1} does not exist since 26 is not a unit.

9.64. Show $-a = a$ using $a + a = (a + a)^2$. Then show $ab = -ba$ by $a + b = (a + b)^2$.

9.65. (b) -1 is the 0-element, and 0 is the 1-element.

Chapter 10

Posets and Lattices

10.1 PARTIALLY ORDERED SETS

A relation R on a set S is called a *partial order* on S if R is:

(1) reflexive, i.e. $a\,R\,a$ for every a in S;

(2) antisymmetric, i.e. if $a\,R\,b$ and $b\,R\,a$ then $a = b$; and

(3) transitive, i.e. if $a\,R\,b$ and $b\,R\,c$ then $a\,R\,c$.

A set S together with a partial order on S is called a *partially ordered set* or *poset*.

EXAMPLE 10.1.

(a) Let $\mathcal{S}$ be any class of sets. The relation of set inclusion $\subset$ is a partial order on $\mathcal{S}$. For $A \subset A$ for any set in $\mathcal{S}$, if $A \subset B$ and $B \subset A$ then $A = B$, and if $A \subset B$ and $B \subset C$ then $A \subset C$.

(b) Consider the positive integers **N**. We say that "a divides b", written $a \mid b$, if there is an integer c such that $ac = b$. For example, $2 \mid 4$, $3 \mid 12$, $7 \mid 21$, and so on. This relation of divisibility is a partial order on **N**.

(c) The relation $\leqq$ is also a partial order on the positive integers **N**. (In fact $\leqq$ is a partial order on any subset of the real numbers.) This relation is sometimes called the *usual order* on **N**.

We usually denote a partial order relation by $\precsim$, and

$$a \precsim b$$

is read "a precedes b". With this notation, we also use the following additional notation:

$a \prec b$ means $a \precsim b$ and $a \neq b$; read "a strictly precedes b".

$b \succsim a$ means $a \precsim b$; read "b succeeds a".

$b \succ a$ means $a \prec b$; read "b strictly succeeds a".

$\not\precsim, \not\prec, \not\succsim$ and $\not\succ$ are self-explanatory.

When there is no ambiguity, we will frequently use the symbols $\leqq$, $<$, $>$ and $\geqq$ instead of, respectively, $\precsim$, $\prec$, $\succsim$ and $\succ$. We also note that if $\precsim$ is a partial order on a set S, then the inverse relation is $\succsim$ and it is also a partial order on S called the *inverse order*.

Two elements a and b in a poset are said to be *comparable* if

$$a \precsim b \quad \text{or} \quad b \precsim a$$

that is, if one precedes the other; otherwise a and b are said to be *noncomparable*. For example, the integers 3 and 5 in Example 10.1(b) are noncomparable since neither divides the other.

The word "partial" is used in defining a partially ordered set A because some elements of A need not be comparable. If, on the other hand, every pair of elements in a poset A are comparable, then A is said to be *totally ordered* or *linearly ordered* and A is called a *chain*. For example, the positive integers **N** with the usual ordering $\leqq$ is a linearly ordered set.

If S and T are totally ordered sets, then the product set $S \times T$ can be totally ordered as follows:

$$(a, b) \prec (a', b') \quad \text{if} \quad a \prec a' \quad \text{or if} \quad a = a' \quad \text{and} \quad b \prec b'$$

This order is called the *lexicographical order* on $S \times T$ since it is similar to the way words are arranged in a dictionary.

Any subset A of a poset S is partially ordered by the relation on S, i.e. for any a, b in A we let $a \preceq b$ as elements of A according as $a \preceq b$ as elements of S. We note that subsets of S may be linearly ordered although S is not. For example, $\{2, 4, 16, 64\}$ is a linearly ordered subset of the nonlinearly ordered set $\mathbf{N}$ of positive integers ordered by divisibility.

10.2 DIAGRAM OF A POSET

Let S be a partially ordered set. We say a in S is an *immediate predecessor* of b in S, or b is an *immediate successor* of a, written

$$a \ll b$$

if $a < b$ but no element of S lies between a and b, i.e. there is no x in S such that $a < x < b$.

Suppose S is a finite poset. Then the order on S is completely known once we know all pairs a, b in S such that $a \ll b$, i.e. the relation $\ll$ on S. This follows from the fact that $x < y$ if and only if there exists elements

$$x = a_0, a_1, \ldots, a_m = y \quad \text{such that} \quad a_{i-1} \ll a_i \quad \text{for} \quad i = 1, \ldots, m$$

By the *diagram* of a finite poset S we mean the directed graph whose vertices are the elements of S and in which there is an arc from a to b if $a \ll b$ in S. (Instead of drawing an arrow from a to b, we sometimes place b higher than a in the diagram and draw a line from a to b which slants upward.) In the diagram of S, there is a (directed) path from a vertex x to a vertex y if and only if $x < y$. Also, there cannot be any cycles in the diagram of S since the order relation is antisymmetric.

EXAMPLE 10.2.

(a) Let $A = \{1, 2, 3, 4, 6, 8, 9, 12, 18, 24\}$ be ordered by the relation "x divides y". The diagram of A is given in Fig. 10-1(a) (Unlike rooted trees, the direction of a line in the diagram of a poset is always upward.)

(b) Let $B = \{a, b, c, d, e\}$. The diagram in Fig. 10-1(b) defines a partial order on B in the natural way. That is, $d \preceq b$, $d \preceq a$, $e \preceq c$ and so on.

(c) The diagram of a finite linearly ordered set, i.e. a finite chain, consists simply of one path. For example, Fig. 10-1(c) shows the diagram of a chain with five elements.

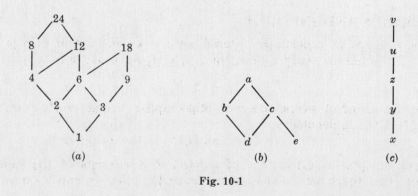

$\qquad\qquad$ (a) $\qquad\qquad\qquad\qquad\qquad\qquad$ (b) $\qquad\qquad\qquad\qquad$ (c)

Fig. 10-1

EXAMPLE 10.3. A *partition* of a positive integer m is a set of positive integers whose sum is m. For example, there are 7 partitions of $m = 5$:

$$5, \ 3\text{-}2, \ 2\text{-}2\text{-}1, \ 1\text{-}1\text{-}1\text{-}1\text{-}1, \ 4\text{-}1, \ 3\text{-}1\text{-}1, \ 2\text{-}1\text{-}1\text{-}1$$

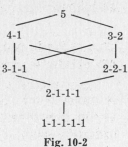

Fig. 10-2

We order the partitions of an integer m as follows. A partition P_1 precedes a partition P_2 if the integers in P_1 can be added to obtain the integers in P_2 or, equivalently, if the integers in P_2 can be further subdivided to obtain the integers in P_1. For example,

$$2\text{-}2\text{-}1 \qquad \text{precedes} \qquad 3\text{-}2$$

since $2 + 1 = 3$. On the other hand, 3-1-1 and 2-2-1 are noncomparable. Figure 10-2 gives the diagram of the partitions of $m = 5$.

Frequently we want to assign positive integers to the elements of a finite partially ordered set S in such a way that the order in S is preserved. That is, we seek a function $f : S \to \mathbf{N}$ so that if $a \preceq b$ then $f(a) \leqq f(b)$. Such a function is called a *consistent enumeration* of S. The fact that this can always be done is the content of the next theorem.

Theorem 10.1: There exists a consistent enumeration for any finite poset S.

We emphasize that such an enumeration need not be unique. For example, the following are two such enumerations for the poset in Fig. 10-1(b):

(i) $f(d) = 1, \ f(e) = 2, \ f(b) = 3, \ f(c) = 4, \ f(a) = 5$

(ii) $g(e) = 1, \ g(d) = 2, \ g(c) = 3, \ g(b) = 4, \ g(a) = 5$

However the chain in Fig. 10-1(c) admits only one consistent enumeration if we map the set into $\{1, 2, 3, 4, 5\}$. Specifically, we assign:

$$h(x) = 1, \quad h(y) = 2, \quad h(z) = 3, \quad h(u) = 4, \quad h(v) = 5$$

An element a in a poset S is said to be *maximal* if no element succeeds a, i.e. if $a \preceq x$ implies $x = a$. Analogously, an element b in a poset S is said to be *minimal* if no element precedes b, i.e. if $y \preceq b$ implies $y = b$. There can be more than one maximal and more than one minimal element. In Fig. 10-1(a), 18 and 24 are maximal elements but only 1 is a minimal element. In Fig. 10-1(b), d and e are minimal elements but only a is a maximal element. In Fig. 10-1(c), x is the only minimal element and y is the only maximal element.

An infinite poset may have neither maximal nor minimal elements. For example, the set of integers $\mathbf{Z} = \{\ldots, -2, -1, 0, 1, 2, \ldots\}$ with the usual ordering has no maximal and no minimal element. However, Theorem 10.1 shows that a finite poset S must have at least one minimal and at least one maximal element. Specifically, if a in S is the element assigned the smallest (largest) integer by any consistent enumeration, then a is a minimal (maximal) element.

10.3 SUPREMUM AND INFIMUM

Let A be a subset of a partially ordered set S. An element M in S is called an *upper bound* of A if M succeeds every element of A, i.e. if, for every x in A, we have

$$x \preceq M$$

If an upper bound of A precedes every other upper bound of A, then it is called the *supremum* of A and is denoted by

$$\sup{(A)}$$

We also write $\sup(a_1, \ldots, a_n)$ instead of $\sup(A)$ if A consists of the elements $a_1, \ldots, a_n$. We emphasize that there can be at most one $\sup(A)$; however, $\sup(A)$ may not exist. For example, in Fig. 10-2, the partitions 4-1, 3-2 and 5 are upper bounds for 3-1-1 and 2-2-1; but sup (3-1-1, 2-2-1) does not exist, as no upper bound precedes the other two.

Analogously, an element m in a poset S is called a *lower bound* of a subset A of S if m precedes every element of A, i.e. if, for every y in A, we have

$$m \lesssim y$$

If a lower bound of A succeeds every other lower bound of A, then it is called the *infimum* of A and is denoted by

$$\inf (A) \quad \text{or} \quad \inf (a_1, \ldots, a_n)$$

if A consists of the elements $a_1, \ldots, a_n$. There can be at most one inf (A) although inf (A) may not exist. For example, in Fig. 10-2, the partitions 4-1 and 3-2 have four lower bounds but inf {4-1, 3-2} does not exist since no lower bound succeeds the other three.

Some texts use the term *least upper bound* instead of supremum and *greatest lower bound* instead of infimum, and then write lub (A) for sup (A) and glb (A) for inf (A).

We now give important examples of posets where sup (a, b) and inf (a, b) do exist for any pair of elements a and b.

EXAMPLE 10.4.

(a) Let the positive integers **N** be partially ordered by the relation of divisibility, and let a, b be in **N**. The *greatest common divisor* of a and b, denoted by

$$\gcd (a, b)$$

is the largest integer which divides a and b. The *least common multiple* of a and b, denoted by

$$\text{lcm} (a, b)$$

is the smallest integer divisible by both a and b.

An important theorem in number theory says that every common divisor of a and b divides gcd (a, b). One can also prove that lcm (a, b) divides every multiple of a and b. Thus

$$\gcd (a, b) = \inf (a, b) \quad \text{and} \quad \text{lcm} (a, b) = \sup (a, b)$$

In other words, inf (a, b) and sup (a, b) do exist for any pair of elements of **N** ordered by divisibility.

(b) For any positive integer m, we will let D_m denote the set of divisors of m ordered by divisibility. The diagram of

$$D_{36} = \{1, 2, 3, 4, 6, 9, 12, 18, 36\}$$

appears in Fig. 10-3. Again, inf $(a, b) =$ gcd (a, b) and sup $(a, b) =$ lcm (a, b) exist.

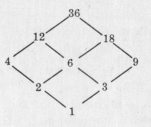

Fig. 10-3

10.4 LATTICES

Let L be a nonempty set closed under two binary operations called *meet* and *join*, denoted respectively by $\wedge$ and $\vee$. Then L is called a *lattice* if the following axioms hold where a, b, c are any elements in L:

[L₁] Commutative Law:

    (1a) $a \wedge b = b \wedge a$ (1b) $a \vee b = b \vee a$

[L₂] Associative Law:

    (2a) $(a \wedge b) \wedge c = a \wedge (b \wedge c)$ (2b) $(a \vee b) \vee c = a \vee (b \vee c)$

[L₃] Absorption Law:

    (3a) $a \wedge (a \vee b) = a$ (3b) $a \vee (a \wedge b) = a$

We will sometimes denote the lattice by $(L, \wedge, \vee)$ when we want to show which operations are involved.

The *dual* of any statement in a lattice $(L, \wedge, \vee)$ is defined to be the statement that is obtained by interchanging $\wedge$ and $\vee$. For example, the dual of

$$a \wedge (b \vee a) = a \vee a \quad \text{is} \quad a \vee (b \wedge a) = a \wedge a$$

Notice that the dual of each axiom of a lattice is also an axiom. Accordingly, the principle of duality holds; that is:

Theorem 10.2 (Principle of Duality): The dual of any theorem in a lattice is also a theorem.

This follows from the fact that the dual theorem can be proven by using the dual of each step of the proof of the original theorem.

An important property of lattices follows directly from the absorption laws.

Theorem 10.3 (Idempotent Law):   (i) $a \wedge a = a$,   (ii) $a \vee a = a$.

The proof of (i) requires only two lines:

$$a \wedge a = a \wedge (a \vee (a \wedge b)) \quad \text{(using (3}b\text{))}$$
$$= a \quad \text{(using (3}a\text{))}$$

The proof of (ii) follows from the above principle of duality (or can be proved in a similar manner).

Given a lattice L, we can define a partial order on L as follows:

$$a \precsim b \quad \text{if} \quad a \wedge b = a$$

Analogously, we could define

$$a \precsim b \quad \text{if} \quad a \vee b = b$$

We state these results in a theorem.

Theorem 10.4:   Let L be a lattice. Then:

(i) $a \wedge b = a$ if and only if $a \vee b = b$.

(ii) The relation $a \precsim b$ (defined by $a \wedge b = a$ or $a \vee b = b$) is a partial order on L.

Now that we have a partial order on any lattice L, we can picture L by a diagram as was done for partially ordered sets in general.

EXAMPLE 10.5. Let C be a collection of sets closed under intersection and union. Then $(C, \cap, \cup)$ is a lattice. In this lattice, the partial order relation is the same as the set inclusion relation. Figure 10-4 shows the diagram of the lattice L of all subsets of $\{a, b, c\}$.

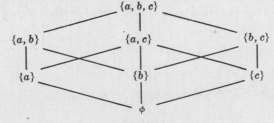

Fig. 10-4

We have shown how to define a partial order on a lattice L. The next theorem tells us when we can define a lattice on a partially ordered set P such that the lattice will give back the original order on P.

Theorem 10.5: Let P be a poset such that the $\inf(a, b)$ and $\sup(a, b)$ exist for any a, b in P. Letting

$$a \wedge b = \inf(a, b) \quad \text{and} \quad a \vee b = \sup(a, b)$$

we have that $(P, \wedge, \vee)$ is a lattice. Furthermore, the partial order on P induced by the lattice is the same as the original partial order on P.

The converse of the above theorem is also true. That is, let L be a lattice and let $\preceq$ be the induced partial order on L. Then $\inf(a, b)$ and $\sup(a, b)$ exist for any pair a, b in L and the lattice obtained from the poset $(L, \preceq)$ is the original lattice. Accordingly, we have the following:

Alternate Definition: A lattice is a partially ordered set in which

$$a \wedge b = \inf(a, b) \quad \text{and} \quad a \vee b = \sup(a, b)$$

exist for any pair of elements a and b.

We note first that any linearly ordered set is a lattice since $\inf(a, b) = a$ and $\sup(a, b) = b$ whenever $a \preceq b$. By Example 10.4, the positive integers $\mathbf{N}$ and the set D_m of divisors of m are lattices under the relation of divisibility.

Suppose M is a nonempty subset of a lattice L. We say M is a *sublattice* of L if M itself is a lattice (with respect to the operations of L). We note that M is a sublattice of L if and only if M is closed under the operations of $\wedge$ and $\vee$ of L. For example, the set D_m of divisors of m is a sublattice of the positive integers N under divisibility.

Two lattices L and L' are said to be *isomorphic* if there is a one-to-one correspondence $f : L \to L'$ such that

$$f(a \wedge b) = f(a) \wedge f(b) \quad \text{and} \quad f(a \vee b) = f(a) \vee f(b)$$

for any elements a, b in L.

10.5 BOUNDED LATTICES

A lattice L is said to have a *lower bound* 0 if for any element x in L we have $0 \preceq x$. Analogously, L is said to have an *upper bound* I if for any x in L we have $x \preceq I$. We say L is *bounded* if L has both a lower bound 0 and an upper bound I. In such a lattice we have the identities

$$a \vee I = I, \quad a \wedge I = a, \quad a \vee 0 = a, \quad a \wedge 0 = 0$$

for any element a in L.

The nonnegative integers with the usual ordering,

$$0 < 1 < 2 < 3 < 4 < \cdots$$

have 0 as a lower bound but have no upper bound. On the other hand, the lattice $P(U)$ of all subsets of any universal set U is a bounded lattice with U as an upper bound and the empty set $\varnothing$ as a lower bound.

Suppose $L = \{a_1, a_2, \ldots, a_n\}$ is a finite lattice. Then

$$a_1 \vee a_2 \vee \cdots \vee a_n \quad \text{and} \quad a_1 \wedge a_2 \wedge \cdots \wedge a_n$$

are upper and lower bounds for L respectively. Thus we have

Theorem 10.6: Every finite lattice L is bounded.

10.6 DISTRIBUTIVE LATTICES

A lattice L is said to be *distributive* if for any elements a, b, c in L we have the following:

[L₄]   Distributive Law:

$$(4a) \quad a \wedge (b \vee c) = (a \wedge b) \vee (a \wedge c) \qquad (4b) \quad a \vee (b \wedge c) = (a \vee b) \wedge (a \vee c)$$

Otherwise, L is said to be *nondistributive*. We note that by the principle of duality the condition $(4a)$ holds if and only if $(4b)$ holds.

Figure 10-5(a) is a nondistributive lattice since

$$a \vee (b \wedge c) = a \vee 0 = a$$

but

$$(a \vee b) \wedge (a \vee c) = I \wedge c = c$$

Figure 10-5(b) is also a nondistributive lattice. In fact, we have the following characterization of such lattices.

Theorem 10.7:   A lattice L is nondistributive if and only if it contains a sublattice isomorphic to Fig. 10-5(a) or (b).

The proof of this theorem lies beyond the scope of this text.

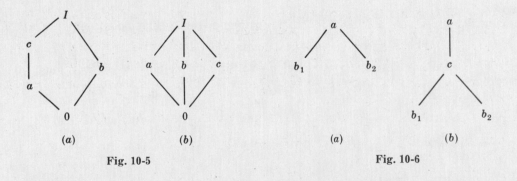

<table>
<tr><td>(a)</td><td>(b)</td><td>(a)</td><td>(b)</td></tr>
<tr><td colspan="2">Fig. 10-5</td><td colspan="2">Fig. 10-6</td></tr>
</table>

Let L be a lattice with a lower bound 0. An element a in L is said to be *join irreducible* if $a = x \vee y$ implies $a = x$ or $a = y$. (Prime numbers under multiplication have this property, i.e. if $p = ab$ then $p = a$ or $p = b$ where p is prime.) Clearly 0 is join irreducible. If a has at least two immediate predecessors, say b_1 and b_2 as in Fig. 10-6(a), then $a = b_1 \vee b_2$, and so a is not join irreducible. On the other hand, if a has a unique immediate predecessor c, then $a \neq \sup (b_1, b_2) = b_1 \vee b_2$ for any other elements b_1 and b_2 because c would lie between the b's and a as in Fig. 10-6(b). In other words, $a \neq 0$ is join irreducible if and only if a has a unique immediate predecessor. Those elements which immediately succeed 0, called *atoms*, are join irreducible. However, lattices can have other join-irreducible elements. For example, the element c in Fig. 10-5(a) is not an atom but is join-irreducible since a is its only immediate predecessor.

If an element a in a finite lattice L is not join irreducible, then we can write $a = b_1 \vee b_2$. Then we can write b_1 and b_2 as the join of other elements if they are not join irreducible; and so on. Since L is finite we finally have

$$a = d_1 \vee d_2 \vee \cdots \vee d_n$$

where the d's are join irreducible. If d_i precedes d_j then $d_i \vee d_j = d_j$; so we can delete the d_i from the expression. In other words, we can assume that the d's are *irredundant*, i.e.

no d precedes any other d.   We emphasize that such an expression need not be unique, e.g. $I = a \vee b$ and $I = b \vee c$ in both lattices in Fig. 10-5. We now state the main theorem of this section (proved in Problem 10.15).

Theorem 10.8:   Let L be a finite distributive lattice.   Then every a in L can be written uniquely (except for order) as the join of irredundant join-irreducible elements.

Actually this theorem can be generalized to lattices with *finite length*, i.e. where all linearly ordered subsets are finite. (Problem 10.10 gives an infinite lattice with finite length.)

10.7 COMPLEMENTED LATTICES

Let L be a bounded lattice with lower bound 0 and upper bound I. Let a be an element of L. An element x in L is called a *complement* of a if

$$a \vee x = I \quad \text{and} \quad a \wedge x = 0$$

Complements need not exist and need not be unique.   For example, the elements a and c are both complements of b in Fig. 10-5(a).   Also, the elements y, z and u in the chain in Fig. 10-1 have no complements.   We have the following result.

Theorem 10.9:   Let L be a bounded distributive lattice.   Then complements are unique if they exist.

Proof.   Suppose x and y are complements of any element a in L.   Then

$$a \vee x = I, \quad a \vee y = I, \quad a \wedge x = 0, \quad a \wedge y = 0$$

Using distributivity,

$$x = x \vee 0 = x \vee (a \wedge y) = (x \vee a) \wedge (x \vee y) = I \wedge (x \vee y) = x \vee y$$

Similarly,

$$y = y \vee 0 = y \vee (a \wedge x) = (y \vee a) \wedge (y \vee x) = I \wedge (y \vee x) = y \vee x$$

Thus

$$x = x \vee y = y \vee x = y$$

and the theorem is proved.

A lattice L is said to be *complemented* if L is bounded and every element in L has a complement.   Figure 10-5(b) shows a complemented lattice where complements are not unique.   On the other hand, the lattice $P(U)$ of all subsets of a universal set U is complemented, and each subset A of U has the unique complement $A^c = U \setminus A$.

Theorem 10.10:   Let L be a complemented lattice with unique complements. Then the join-irreducible elements of L, other than 0, are its atoms.

Combining this theorem and Theorems 10.8 and 10.9, we get an important result.

Theorem 10.11:   Let L be a finite complemented distributive lattice.   Then every element a in L is the join of a unique set of atoms.

Remark: Some texts define a lattice L to be complemented if each a in L has a unique complement.   Theorem 10.10 is then stated differently.

Solved Problems

ORDERED SETS AND SUBSETS

10.1. Suppose the positive integers $N = \{1, 2, 3, \ldots\}$ are ordered by divisibility (see Example 10.1(b)).

(1) Insert the correct symbol, $<$, $>$, or $\parallel$ (not comparable) between each pair of numbers:

  (a) 2___8,   (b) 18___24,   (c) 9___3,   (d) 5___15

(2) State whether or not each of the following subsets of N are linearly ordered.

  (a) $\{24, 2, 6\}$,   (b) $\{3, 15, 5\}$,   (c) $\{15, 5, 30\}$,   (d) $\{2, 8, 32, 4\}$,   (e) $\{1, 2, 3, \ldots\}$,   (f) $\{7\}$

(1) (a) Since 2 divides 8, 2 precedes 8, i.e. $2 < 8$.

  (b) 18 does not divide 24, and 24 does not divide 18; hence $18 \parallel 24$.

  (c) Since 9 is divisible by 3, $9 > 3$.

  (d) Since 5 divides 15, $5 < 15$.

(2) (a) Since 2 divides 6 which divides 24, the set is totally ordered.

  (b) Since 3 and 5 are not comparable, the set is not totally ordered.

  (c) The set is totally ordered since 5 divides 15 which divides 30.

  (d) The set is totally ordered since $2 < 4 < 8 < 32$.

  (e) The set is not totally ordered since 2 and 3 are not comparable.

  (f) Any set consisting of one element is totally ordered.

10.2. Let $V = \{a, b, c, d, e\}$ be ordered by the diagram in Fig. 10-7.

Insert the correct symbol, $<$, $>$, or $\parallel$ (not comparable) between each pair of elements:

(a) a___e (c) d___a

(b) b___c (d) c___d

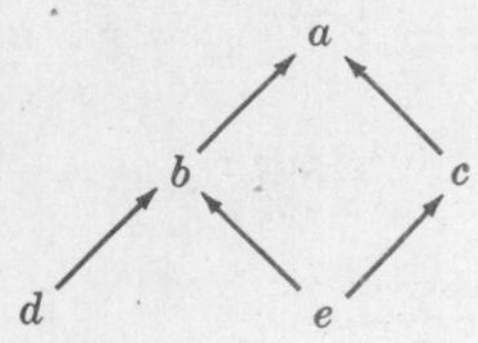

Fig. 10-7

(a) Since there is a "path" from e to c to a, e precedes a; hence $a > e$.

(b) There is no path from b to c, or vice versa; hence $b \parallel c$.

(c) There is a path from d to b to a; hence $d < a$.

(d) Neither $d < c$ nor $c < d$; hence $c \parallel d$.

10.3. Let $N \times N$ be ordered lexicographically. Insert the correct symbol, $<$ or $>$, between each of the following pairs of elements of $N \times N$:

(a) $(5, 78)$___$(7, 1)$,   (b) $(4, 6)$___$(4, 2)$,   (c) $(5, 5)$___$(4, 23)$,   (d) $(1, 3)$___$(1, 2)$

Note that according to the lexicographical ordering,

$$(a, b) < (a', b') \quad \text{if } a < a' \quad \text{or} \quad \text{if } a = a' \text{ but } b < b'$$

(a) $(5, 78) < (7, 1)$, since $5 < 7$.

(b) $(4,6) \succ (4,2)$,   since $4 = 4$ but $6 > 2$.

(c) $(5,5) \succ (4,23)$, since $5 > 4$.

(d) $(1,3) \succ (1,2)$,   since $1 = 1$ but $3 > 2$.

10.4. Let $A = \{2,3,4,\ldots\} = \mathbf{N} \setminus \{1\}$ be ordered by "x divides y". (a) Find all minimal elements. (b) Find all maximal elements.

(a) If p is a prime number, then only p divides p (since $1 \notin A$); hence all the prime numbers are minimal elements. Furthermore, if $a \in A$ is not a prime number, then there is an integer $b \in A$ other than a which divides a. Thus the only minimal elements of A are the prime numbers.

(b) There are no maximal elements in A because, for any element $a \in A$, we have that a divides, for example, $2a$.

10.5. Let $B = \{1,2,3,4,5\}$ be ordered by the diagram in Fig. 10-8. (a) Find all the minimal elements of B. (b) Find all the maximal elements of B.

(a) No element strictly precedes 4 or 5; hence 4 and 5 are the minimal elements of B.

(b) The only maximal element is 1.

Fig. 10-8

10.6. Let $W = \{1,2,\ldots,7,8\}$ be ordered as shown in Fig. 10-9. Consider the subset $V = \{4,5,6\}$ of W. (a) Find the set of upper bounds of V. (b) Find the set of lower bounds of V. (c) Does $\sup(V)$ exist? Does $\inf(V)$ exist?

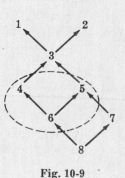

(a) Each of the elements in $\{1,2,3\}$, and only these elements, succeeds every element in V and hence is an upper bound of V.

(b) Only 6 and 8 precede every element of V; hence $\{6,8\}$ is the set of lower bounds of V. Note that 7 is not a lower bound of V since 7 does not precede 4 or 6.

(c) Since 3 precedes all upper bounds of V, $\sup(V) = 3$. Note that 3 does not belong to V. Since 6 succeeds each lower bound of V, $\inf(V) = 6$. Note that 6 does belong to V.

Fig. 10-9

10.7. Let $D = \{1,2,3,4,5,6\}$ be ordered as shown in Fig. 10-10. Consider the subset $E = \{2,3,4\}$ of D. (a) Find the upper bounds of E. (b) Find the lower bounds of E. (c) Does $\sup(E)$ exist? Does $\inf(E)$ exist?

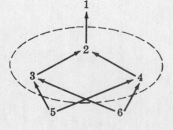

(a) Both 1 and 2, and no other elements, succeed every element of E, and hence are the upper bounds of E.

(b) Only 5 and 6 precede every element of E, and so are the lower bounds of E.

(c) Since 2 precedes all upper bounds of E, we have $\sup(E) = 2$. No lower bound succeeds all the lower bounds; hence $\inf(E)$ does not exist.

Fig. 10-10

10.8. Prove Theorem 10-1:   There exists a consistent enumeration for any finite poset S.

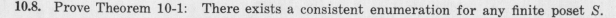

The proof is by induction on the number of elements in S. If $S = \{s\}$ has only one element then obviously $f : S \to \mathbf{N}$, defined by $f(s) = 1$, is a consistent enumeration of S. Now suppose S has

$n > 1$ elements and the theorem holds for posets with less than n elements. Suppose $b \in S$, and consider the subset $T = S \setminus \{b\}$ of S. Since T has $n-1$ elements, there exists a consistent enumeration $g : T \to \mathbf{N}$ of T. Then the function $h : T \to \mathbf{N}$ defined by $h(x) = 2g(x)$ is also a consistent enumeration. (Prove!) Note that the image of h only contains even numbers. Let $f : S \to \mathbf{N}$ be defined as follows:

$$f(x) = h(x) \quad \text{if} \quad x \neq b$$

$$f(b) = \begin{cases} f(a) + 1 & \text{if } a \text{ strictly precedes } b \\ 1 & \text{if no element precedes } b \end{cases}$$

Then f is a consistent enumeration of S.

LATTICES

10.9. Write the dual of each statement:

(a) $(a \wedge b) \vee c = (b \vee c) \wedge (c \vee a)$,   (b) $(a \wedge b) \vee a = a \wedge (b \vee a)$

Replace $\vee$ by $\wedge$ and $\wedge$ by $\vee$ in each statement to obtain the dual statement:

(a) $(a \vee b) \wedge c = (b \wedge c) \vee (c \wedge a)$

(b) $(a \vee b) \wedge a = a \vee (b \wedge a)$

10.10. Give an example of an infinite lattice L with finite length.

Let

$$L = \{0, 1, a_1, a_2, a_3, \ldots\}$$

and let L be ordered as in Fig. 10-11; that is, for each $n \in \mathbf{N}$ we have

$$0 < a_n < 1$$

Then L has finite length since L has no infinite linearly ordered subset.

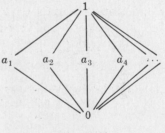

Fig. 10-11

10.11. Prove Theorem 10.4: Let L be a lattice. Then: (i) $a \wedge b = a$ if and only if $a \vee b = b$. (ii) The relation $a \precsim b$ (defined by $a \wedge b = a$ or $a \vee b = b$) is a partial order on L.

(a) Suppose $a \wedge b = a$. Using the absorption law in the first step we have:

$$b = b \vee (b \wedge a) = b \vee (a \wedge b) = b \vee a = a \vee b$$

Now suppose $a \vee b = b$. Again using the absorption law in the first step we have:

$$a = a \wedge (a \vee b) = a \wedge b$$

Thus $a \wedge b = a$ if and only if $a \vee b = b$.

(b) For any a in L, we have $a \wedge a = a$ by idempotency. Hence $a \precsim a$, and so $\precsim$ is reflexive.

Suppose $a \precsim b$ and $b \precsim a$. Then $a \wedge b = a$ and $b \wedge a = b$. Therefore, $a = a \wedge b = b \wedge a = b$, and so $\precsim$ is antisymmetric.

Lastly, suppose $a \precsim b$ and $b \precsim c$. Then $a \wedge b = a$ and $b \wedge c = b$. Thus

$$a \wedge c = (a \wedge b) \wedge c = a \wedge (b \wedge c) = a \wedge b = a$$

Therefore $a \precsim c$, and so $\precsim$ is transitive. Accordingly, $\precsim$ is a partial order on L.

10.12. Which of the partially ordered sets in Fig. 10-12 are lattices?

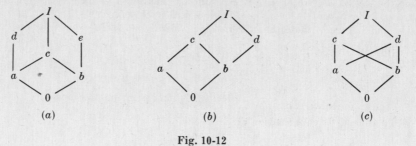

$$(a) \qquad\qquad\qquad (b) \qquad\qquad\qquad (c)$$

Fig. 10-12

A partially ordered set is a lattice if and only if $\sup(x, y)$ and $\inf(x, y)$ exist for each pair x, y in the set. Only (c) is not a lattice since $\{a, b\}$ has three upper bounds, c, d and I, and no one of them precedes the other two, i.e. $\sup(a, b)$ does not exist.

10.13. Consider the lattice L in Fig. 10-12(a).

(a) Which nonzero elements are join irreducible?

(b) Which elements are atoms?

(c) Which of the following are sublattices of L:

$$L_1 = \{0, a, b, I\} \qquad L_3 = \{a, c, d, I\}$$
$$L_2 = \{0, a, e, I\} \qquad L_4 = \{0, c, d, I\}$$

(d) Is L distributive?

(e) Find complements, if they exist, for the elements a, b and c.

(f) Is L a complemented lattice?

(a) Those nonzero elements with a unique immediate predecessor are join irreducible. Hence a, b, d and e are join irreducible.

(b) Those elements which immediately succeed 0 are atoms, hence a and b are the atoms.

(c) A subset L' is a sublattice if it is closed under $\wedge$ and $\vee$. L_1 is not a sublattice since $a \vee b = c$, which does not belong to L_1. The set L_4 is not a sublattice since $c \wedge d = a$ does not belong to L_4. The other two sets, L_2 and L_3, are sublattices.

(d) L is not distributive since $M = \{0, a, d, e, I\}$ is a sublattice which is isomorphic to the nondistributive lattice in Fig. 10-5(a).

(e) We have $a \wedge e = 0$ and $a \vee e = I$, so a and e are complements. Also b and d are complements. However, c has no complement.

(f) L is not a complemented lattice since c has no complement.

10.14. Consider the lattice M in Fig. 10-12(b).

(a) Find the nonzero join-irreducible elements and atoms of M.

(b) Is M distributive?

(c) Is M complemented?

(a) The nonzero elements with a unique predecessor are a, b and d, and of these three only a and b are atoms since their unique predecessor is 0.

(b) M is distributive since M does not have a sublattice which is isomorphic to one of the lattices in Fig. 10-5.

(c) M is not complemented since b has no complement.   Note a is the only solution to $b \wedge x = 0$ but $b \vee a = c \neq I$.

10.15. Prove Theorem 10.8:   Let L be a finite distributive lattice.   Then every a in L can be written uniquely (except for order) as the join of irredundant join-irreducible elements.

Since L is finite we can write a as the join of irredundant join-irreducible elements as discussed in Section 10.6. Thus we need to prove uniqueness. Suppose

$$a = b_1 \vee b_2 \vee \cdots \vee b_r = c_1 \vee c_2 \vee \cdots \vee c_s$$

where the b's are irredundant and join irreducible and the c's are irredundant and irreducible.   For any given i we have

$$b_i \lesssim (b_1 \vee b_2 \vee \cdots \vee b_r) = (c_1 \vee c_2 \vee \cdots \vee c_s)$$

Hence

$$b_i = b_i \wedge (c_1 \vee c_2 \vee \cdots \vee c_s) = (b_i \wedge c_1) \vee (b_i \wedge c_2) \vee \cdots \vee (b_i \wedge c_s)$$

Since b_i is join irreducible, there exists a j such that $b_i = b_i \wedge c_j$, and so $b_i \lesssim c_j$.   By a similar argument, for c_j there exists a b_k such that $c_j \lesssim b_k$.   Therefore

$$b_i \lesssim c_j \lesssim b_k$$

which gives $b_i = c_j = b_k$ since the b's are irredundant.   Accordingly, the b's and c's may be paired off. Thus the representation for a is unique except for order.

10.16. Prove Theorem 10.10:   Let L be a complemented lattice with unique complements. Then the join-irreducible elements of L, other than 0, are its atoms.

Suppose a is join irreducible and is not an atom.   Then a has a unique immediate predecessor $b \neq 0$. Let b' be the complement of b.   Since $b \neq 0$ we have $b' \neq I$. If a precedes b', then $b \lesssim a \lesssim b'$, and so $b \wedge b' = b'$, which is impossible since $b \wedge b' = I$.   Thus a does not precede b', and so $a \wedge b'$ must strictly precede a.   Since b is the unique immediate predecessor of a, we also have that $a \wedge b'$ precedes b as in Fig. 10-13.   But $a \wedge b'$ precedes b'.   Hence

$$a \wedge b' \lesssim \inf (b, b') = b \wedge b' = 0$$

Thus $a \wedge b' = 0$. Since $a \vee b = a$, we also have that

$$a \vee b' = (a \vee b) \vee b' = a \vee (b \vee b') = a \vee I = I$$

Therefore b' is a complement of a. Since complements are unique, $a = b$. This contradicts the assumption that b is an immediate predecessor of a. Thus the only join-irreducible elements of L are its atoms.

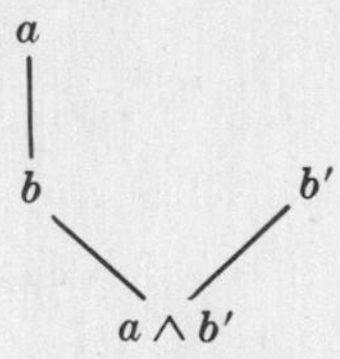

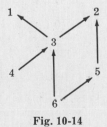

Fig. 10-13

Supplementary Problems

ORDERED SETS AND SUBSETS

10.17. Let

$$V = \{1, 2, 3, 4, 5, 6\}$$

be ordered as in Fig. 10-14.

(a) Find all minimal and maximal elements of V.

(b) Find all linearly ordered subsets of V, each of which contains at least three elements.

Fig. 10-14

10.18. Let $M = \{2, 3, 4, \ldots\}$ and let $M \times M$ be ordered as follows:

$$(a, b) \preceq (c, d)$$

if a divides c *and* if b is less than or equal to d. Find all minimal and maximal elements of $M \times M$.

10.19. Let $A = (\mathbf{N}, \leqq)$, the positive integers with the usual order, let $B = (\mathbf{N}, \geqq)$, the positive integers with the inverse order, and let $A \times B$ be ordered lexicographically. Insert the correct symbol, $\prec$ or $\succ$, between each of the following pairs of elements $A \times B$.

     (*a*) (1, 3)____(1, 5),   (*b*) (4, 1)____(2, 18),   (*c*) (4, 30)____(4, 4),   (*d*) (2, 2)____(15, 15)

10.20. Let $W = \{1, 2, \ldots, 7, 8\}$ be ordered as in Fig. 10-15.

  (1)   Consider the subset $A = \{4, 5, 7\}$ of W.

       (*a*) Find the set of upper bounds of A.   (*c*) Does sup (A) exist?
       (*b*) Find the set of lower bounds of A.   (*d*) Does inf (A) exist?

  (2)   Consider the subset $B = \{2, 3, 6\}$ of W.

       (*a*) Find the set of upper bounds of B (*c*) Does sup (B) exist?
       (*b*) Find the set of lower bounds of B.   (*d*) Does inf (B) exist?

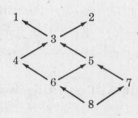

  (3)   Consider the subset $C = \{1, 2, 4, 7\}$ of W.

       (*a*) Find the set of upper bounds of C.   (*c*) Does sup (C) exist?
       (*b*) Find the set of lower bounds of C.   (*d*) Does inf (C) exist?

  (4)   Find the number of linearly ordered subsets of W with

       (*a*) five elements,   (*b*) six elements.   **Fig. 10-15**

10.21. Draw the diagram of the partitions of m (see Example 10.3) where (*a*) $m = 4$, (*b*) $m = 6$.

10.22. Let D_m denote the positive divisors of m ordered by divisibility. Draw the diagram of (*a*) D_{12}, (*b*) D_{15}, (*c*) D_{16}, (*d*) D_{17}.

10.23. Let

$$B = \{a, b, c, d, e, f\}$$

be ordered as in Fig. 10-16.

(*a*) Find minimal and maximal elements of B.

(*b*) List two and find the number of consistent enumerations of B into the set

$$\{1, 2, 3, 4, 5, 6\}$$

Fig. 10-16

10.24. Let $S = \{a, b, c, d, e, f\}$ be a poset and suppose there are exactly six pairs of elements such that the first immediately precedes the second:

$$f \ll a, \quad f \ll d, \quad e \ll b, \quad c \ll f, \quad e \ll c, \quad b \ll f$$

(*a*) Find all maximal elements of S. (*b*) Find all minimal elements of S. (*c*) Find all pairs of elements, if any, which are noncomparable.

LATTICES

10.25. Consider the lattices L in Fig. 10-17. (*a*) Find all sublattices with five elements. (*b*) Find all (i) join-irreducible elements, (ii) atoms. (*c*) Find complements of a and b, if they exist. (*d*) Is L distributive? Complemented?

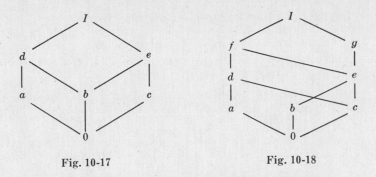

Fig. 10-17   Fig. 10-18

10.26.   Consider the lattice M in Fig. 10-18. (a) Find the join-irreducible elements. (b) Find the atoms.
(c) Find complements of a and b, if they exist. (d) Express each x in M as the join of irredundant
join-irreducible elements. (e) Is M distributive? Complemented?

10.27.   Consider the lattice $D_{60} = \{1, 2, 3, 4, 5, 6, 10, 12, 15, 20, 30, 60\}$, the divisors of 60 ordered by divisibility. (a) Draw the diagram of D_{60}. (b) Which elements are join irreducible? Atoms? (c) Find
complements of 2 and 10, if they exist. (d) Express each number x as the join of a minimum
number of irredundant join-irreducible elements.

10.28.   Consider the lattice $\mathbf{N}$ of positive integers ordered by divisibility. (a) Which elements are join
irreducible? (b) Which elements are atoms?

10.29.   Show that the following "weak" distributive laws hold for any lattice:

(a) $a \vee (b \wedge c) \;\leqq\; (a \vee b) \wedge (a \vee c)$

(b) $a \wedge (b \vee c) \;\geqq\; (a \wedge b) \vee (a \wedge c)$

10.30.   Use Theorem 10.7 to prove: (a) Every linearly ordered set is a distributive lattice. (b) Every sublattice of a distributive lattice is also distributive.

10.31.   A lattice M is said to be modular if whenever $a \leqq c$ we have the law

$$a \vee (b \wedge c) \;=\; (a \vee b) \wedge c$$

(a) Prove that every distributive lattice is modular.

(b) Verify that the nondistributive lattice in Fig. 10-5(b) is modular; hence the converse of (a) is
not true.

(c) Prove that the nondistributive lattice in Fig. 10-5(a) is nonmodular. [In fact, one can prove
that every nonmodular lattice contains a sublattice isomorphic to Fig. 10-5(a).]

10.32.   An element a in a lattice L is said to be *meet irreducible* if $a = x \wedge y$ implies $a = x$ or $a = y$.
Find all meet-irreducible elements in (a) Fig. 10-17, (b) Fig. 10-18, (c) D_{60} (see Problem 10.27).

10.33.   Let R be a ring. Let L be the collection of all ideals of R. For any ideal J and K of R we define

$$J \vee K \;=\; J + K \qquad \text{and} \qquad J \wedge K \;=\; J \cap K$$

Prove that L is a bounded lattice.

10.34.   Let $S = \{1, 2, 3, 4\}$. Three partitions of S are:

$$P_1 = \{\overline{1, 2}, \overline{3}, \overline{4}\} = [\{1, 2\}, \{3\}, \{4\}], \quad P_2 = \{\overline{1, 2}, \overline{3, 4}\}, \quad P_3 = \{\overline{1, 3}, \overline{2}, \overline{4}\}$$

(a) Find the other nine partitions of S.

(b) Let L be the collection of twelve partitions S ordered by *refinement*, i.e. $P_i \precsim P_j$ if each cell of P_i is a subset of a cell of P_j. For example $P_1 \precsim P_2$, but P_2 and P_3 are noncomparable. Show that L is a bounded lattice and draw its diagram.

COMPUTER PROGRAMMING PROBLEMS

Let $S = \{1, 2, 3, 4, 5, 6\}$ be partially ordered by an immediate predecessor relation $\lll$, where $\lll$ contains six elements. Suppose $\lll$ is punched into a deck of six cards where each card contains an element of $\lll$.

10.35. Write a program which prints (a) all maximal and (b) all minimal elements of S.

10.36. Write a program which decides for any elements m, n in S whether or not $m \prec n$.

10.37. Write a program which prints a consistent enumeration of S into $\{1, 2, \ldots, 6\}$.

Test the above three programs with the data:

(i) $3 \lll 2,\ 6 \lll 4,\ 2 \lll 5,\ 3 \lll 1,\ 4 \lll 5,\ 6 \lll 1$

(ii) $2 \lll 5,\ 3 \lll 6,\ 5 \lll 4,\ 5 \lll 1,\ 2 \lll 3,\ 5 \lll 6$

(iii) $6 \lll 1,\ 6 \lll 4,\ 5 \lll 2,\ 3 \lll 6,\ 5 \lll 3,\ 2 \lll 6$

10.38. Write a subprogram called LGCD such that LGCD (M, N) computes the greatest common divisor of M and N, and a subprogram called LCM such that LCM(M, N) computes the least common multiple of M and N.

Answers to Supplementary Problems

10.17. (a) 6 is minimal, and 1 and 2 are maximal.

(b) $\{1, 3, 4\}, \{1, 3, 6\}, \{2, 3, 4\}, \{2, 3, 6\}, \{2, 5, 6\}$

10.18. Any ordered pair $(p, 2)$ where p is a prime, is a minimal element. There are no maximal elements.

10.19. (a) $(1, 3) \succ (1, 5)$, (b) $(4, 1) \succ (2, 18)$, (c) $(4, 30) \prec (4, 4)$, (d) $(2, 2) \prec (15, 15)$

10.20. (1) (a) $\{1, 2, 3\}$, (b) $\{8\}$, (c) $\sup (A) = 3$, (d) $\inf (A) = 8$

(2) (a) $\{2\}$, (b) $\{6, 8\}$, (c) $\sup (B) = 2$, (d) $\inf (B) = 6$

(3) (a) $\emptyset$. There are no upper bounds. (b) $\{8\}$, (c) No, (d) $\inf (C) = 8$

10.21. See Fig. 10-19.

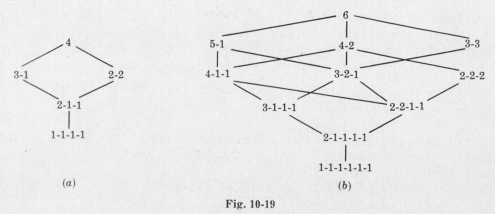

(a)   (b)

Fig. 10-19

10.22. See Fig. 10-20.

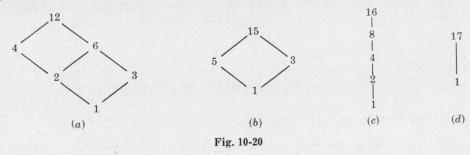

Fig. 10-20

10.23. (a) a is maximal, and d and f are minimal.

(b) There are eleven: *dfebca, dfecba, dfceba, fdebca, fdecba, fdceba, fedbca, fedcba, fcdeba, fecdba, fcedba*

10.24. (a) a and d, (b) e, (c) (b, c), (a, d)

10.25. (a) There are six: $0abdI$, $0acdI$, $0adeI$, $0bceI$, $0aceI$, $0cdeI$.

(b) (i) $a, b, e, 0$, (ii) a, b, c

(c) c and e are complements of a. b has no complement.

(d) No. No.

10.26. (a) $a, b, c, g, 0$

(b) a, b, c

(c) g is a unque complement of a. There is no complement of b.

(d) $I = a \vee g$, $f = a \vee b = a \vee c$, $e = b \vee c$, $d = a \vee c$
Other elements are join irreducible.

(e) No. No.

10.27. (a) See Fig. 10-21.

(b) 1, 2, 3, 4, 5   The atoms are 2, 3 and 5.

(c) 2 has no complement; 3 is a complement of 10.

(d) $60 = 4 \vee 3 \vee 5$, $30 = 2 \vee 3 \vee 5$, $20 = 4 \vee 5$,
$15 = 3 \vee 5$, $12 = 3 \vee 4$, $10 = 2 \vee 5$, $6 = 2 \vee 3$

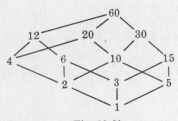

Fig. 10-21

10.28. (a) Powers of prime numbers and 1.   (b) primes

10.31. (a) If $a \leqq c$ then $a \vee c = c$.   Hence
$$a \vee (b \wedge c) = (a \vee b) \wedge (a \vee c) = (a \vee b) \wedge c$$

(c) Here $a \leqq c$. But $a \vee (b \wedge c) = a \vee 0 = a$ and $(a \vee b) \wedge c = I \wedge c = c$; hence
$a \vee (b \wedge c) \neq (a \vee b) \wedge c$.

10.32. Geometrically, an element $a \neq I$ is meet irreducible if and only if a has only one immediate successor.   (a) a, c, d, e, I, (b) a, b, d, f, g, I, (c) 4, 6, 10, 12, 15, 60

10.34. (a) $\{\overline{1}, \overline{2}, \overline{3}, \overline{4}\}$, $\{\overline{1, 4}, \overline{2}, \overline{3}\}$, $\{\overline{1, 3}, \overline{2, 4}\}$, $\{\overline{1, 4}, \overline{2, 3}\}$, $\{\overline{1, 2}, \overline{3}, \overline{4}\}$,
$\{\overline{1, 2, 4}, \overline{3}\}$, $\{\overline{1, 3, 4}, \overline{2}\}$, $\{\overline{2, 3, 4}, \overline{1}\}$, $\{\overline{1, 2, 3, 4}\}$

Chapter 11

Proposition Calculus

11.1 STATEMENTS AND COMPOUND STATEMENTS

Statements (or *verbal assertions*) will be denoted by the letters

$$p, q, r$$

(with or without subscripts). The fundamental property of a statement is that it is either *true* or *false*, but not both. The truthfulness or falsity of a statement is called its *truth value*. Some statements are *composite*, that is, composed of *substatements* and various connectives which we discuss subsequently. Such composite statements are called *compound statements*.

EXAMPLE 11.1.

(a) "Roses are red and violets are blue" is a compound statement with substatements "Roses are red" and "Violets are blue".

(b) "He is intelligent or studies every night" is, implicitly, a compound statement with substatements "He is intelligent" and "He studies every night".

(c) "Where are you going?" is not a statement since it is neither true nor false.

The fundamental property of a compound statement is that its truth value is completely determined by the truth values of its substatements together with the way in which they are connected to form the compound statement. We begin with a study of some of these connectives.

11.2 CONJUNCTION, $p \wedge q$

Any two statements can be combined by the word "and" to form a compound statement called the *conjunction* of the original statements. Symbolically,

$$p \wedge q$$

denotes the conjunction of the statements p and q, read "p and q".

The truth value of the compound statement $p \wedge q$ is given by the following table:

p	q	$p \wedge q$
T	T	T
T	F	F
F	T	F
F	F	F

Here, the first line is a short way of saying that if p is true and q is true then $p \wedge q$ is true. The other lines have analogous meaning. We regard this table as defining precisely the truth value of the compound statement $p \wedge q$ as a function of the truth values of p and of q. Observe that $p \wedge q$ is true only in the case that both substatements are true.

EXAMPLE 11.2. Consider the following four statements:

(i)   Paris is in France and $2 + 2 = 4$.

(ii) Paris is in France and $2 + 2 = 5$.

(iii) Paris is in England and $2 + 2 = 4$.

(iv) Paris is in England and $2 + 2 = 5$.

Only the first statement is true. Each of the other statements is false since at least one of its substatements is false.

11.3   DISJUNCTION, $p \vee q$

Any two statements can be combined by the word "or" (in the sense of "and/or") to form a new statement which is called the *disjunction* of the original two statements. Symbolically,

$$p \vee q$$

denotes the disjunction of the statements p and q and is read "p or q".

The truth value of $p \vee q$ is given by the following table, which we regard as defining $p \vee q$:

p	q	$p \vee q$
T	T	T
T	F	T
F	T	T
F	F	F

Observe that $p \vee q$ is false only when both substatements are false.

EXAMPLE 11.3. Consider the following four statements:

(i)   Paris is in France or $2 + 2 = 4$.

(ii) Paris is in France or $2 + 2 = 5$.

(iii) Paris is in England or $2 + 2 = 4$.

(iv) Paris is in England or $2 + 2 = 5$.

Only (iv) is false. Each of the other statements is true since at least one of its substatements is true.

Remark: The English word "or" is commonly used in two distinct ways. Sometimes it is used in the sense of "p or q or both", i.e. at least one of the two alternates occurs, as above, and sometimes it is used in the sense of "p or q but not both", i.e. exactly one of the two alternatives occurs. For example, the sentence "He will go to Harvard or to Yale" uses "or" in the latter sense, called the *exclusive disjunction*. Unless otherwise stated, "or" shall be used in the former sense. This discussion points out the precision we gain from our symbolic language: $p \vee q$ is defined by its truth table and *always* means "p and/or q".

11.4   NEGATION, $\sim p$

Given any statement p, another statement, called the *negation* of p, can be formed by writing "It is false that..." before p or, if possible, by inserting in p the word "not". Symbolically,

$$\sim p$$

denotes the negation of p (read "not p").

The truth value of $\sim p$ is given by the following table:

p	$\sim p$
T	F
F	T

In other words, if p is true then $\sim p$ is false, and if p is false then $\sim p$ is true. Thus the truth value of the negation of any statement is always the opposite of the truth value of the original statement.

EXAMPLE 11.4.   Consider the following statements.

(a) Paris is in France.

(b) It is false that Paris is in France.

(c) Paris is not in France.

(d) $2 + 2 = 5$

(e) It is false that $2 + 2 = 5$.

(f) $2 + 2 \neq 5$

Then (b) and (c) are each the negation of (a); and (e) and (f) are each the negation of (d). Since (a) is true, the statements (b) and (c) are false; and since (d) is false, the statements (e) and (f) are true.

Remark: The logical notation for the connectives "and", "or", and "not" is not completely standard. For example, some texts use

$$p \,\&\, q, \, p \cdot q \text{ or } pq \quad \text{for} \quad p \wedge q$$

$$p + q \qquad\qquad \text{for} \quad p \vee q$$

$$p', \bar{p} \text{ or } \rceil p \quad\quad \text{for} \quad \sim p$$

11.5   PROPOSITIONS AND TRUTH TABLES

By repetitive use of the logical connectives ($\wedge$, $\vee$, $\sim$ and others discussed subsequently), we can construct compound statements that are more involved. In the case where the substatements $p, q, \ldots$ of a compound statement $P(p, q, \ldots)$ are variables, we call the compound statement a *proposition*.

Now the truth value of a proposition depends exclusively upon the truth values of its variables, that is, the truth value of a proposition is known once the truth values of its variables are known. A simple concise way to show this relationship is through a *truth table*. The truth table, for example, of the proposition $\sim(p \wedge \sim q)$ is constructed as follows:

p	q	$\sim q$	$p \wedge \sim q$	$\sim(p \wedge \sim q)$
T	T	F	F	T
T	F	T	T	F
F	T	F	F	T
F	F	T	F	T

Observe that the first columns of the table are for the variables $p, q, \ldots$ and that there are enough rows in the table to allow for all possible combinations of T and F for these *variables*. (For 2 variables, as above, 4 rows are necessary; for 3 variables, 8 rows are necessary; and, in general, for n variables, 2^n rows are required.) There is then a column for each "elementary" stage of the construction of the proposition, the truth value at each step being determined from the previous stages by the definitions of the connectives $\wedge$, $\vee$, $\sim$. Finally we obtain the truth value of the proposition, which appears in the last column.

Remark: The truth table of the above proposition consists precisely of the columns under the variables and the column under the proposition:

p	q	~(p ∧ ~q)
T	T	T
T	F	F
F	T	T
F	F	T

The other columns were merely used in the construction of the truth table.

Another way to construct the above truth table for ~(p ∧ ~q) is as follows. First construct the following table:

p	q	~	(p	∧	~	q)
T	T					
T	F					
F	T					
F	F					
		Step				

Observe that the proposition is written on the top row to the right of its variables, and that there is a column under each variable or connective in the proposition. Truth values are then entered into the truth table in various steps as follows:

p	q	~	(p	∧	~	q)
T	T		T			T
T	F		T			F
F	T		F			T
F	F		F			F
		Step	1			1

(a)

p	q	~	(p	∧	~	q)
T	T		T		F	T
T	F		T		T	F
F	T		F		F	T
F	F		F		T	F
		Step	1		2	1

(b)

p	q	~	(p	∧	~	q)
T	T		T	F	F	T
T	F		T	T	T	F
F	T		F	F	F	T
F	F		F	F	T	F
		Step	1	3	2	1

(c)

p	q	~	(p	∧	~	q)
T	T	T	T	F	F	T
T	F	F	T	T	T	F
F	T	T	F	F	F	T
F	F	T	F	F	T	F
		Step 4	1	3	2	1

(d)

The truth table of the proposition then consists of the original columns under the variables and the last column entered into the table, i.e. the last step.

11.6 TAUTOLOGIES AND CONTRADICTIONS

Some propositions $P(p, q, \ldots)$ contain only T in the last column of their truth tables, i.e. are true for any truth values of their variables. Such propositions are called *tautologies*. Similarly, a proposition $P(p, q, \ldots)$ is called a *contradiction* if it contains only

F in the last column of its truth table, i.e. is false for any truth values of its variables. For example, the proposition "p or not p", i.e. $p \vee \sim p$, is a tautology and the proposition "p and not p", i.e. $p \wedge \sim p$, is a contradiction. This is verified by constructing their truth tables.

p	$\sim p$	$p \vee \sim p$		p	$\sim p$	$p \wedge \sim p$
T	F	T		T	F	F
F	T	T		F	T	F

We note that the negation of a tautology is a contradiction since it is always false, and the negation of a contradiction is a tautology since it is always true.

Now let $P(p, q, \ldots)$ be a tautology, and let $P_1(p, q, \ldots), P_2(p, q, \ldots), \ldots$ be any propositions. Since $P(p, q, \ldots)$ does not depend upon the particular truth values of its variables $p, q, \ldots$, we can substitute P_1 for p, P_2 for q, $\ldots$ in the tautology $P(p, q, \ldots)$ and still have a tautology. In other words:

Theorem 11.1 (Principle of Substitution):   If $P(p, q, \ldots)$ is a tautology, then $P(P_1, P_2, \ldots)$ is a tautology for any propositions $P_1, P_2, \ldots$.

11.7   LOGICAL EQUIVALENCE

Two propositions $P(p, q, \ldots)$ and $Q(p, q, \ldots)$ are said to be *logically equivalent*, or simply *equivalent* or *equal*, denoted by

$$P(p, q, \ldots) \equiv Q(p, q, \ldots)$$

if they have identical truth tables. For example, consider the truth tables of $\sim(p \wedge q)$ and $\sim p \vee \sim q$:

p	q	$p \wedge q$	$\sim(p \wedge q)$		p	q	$\sim p$	$\sim q$	$\sim p \vee \sim q$
T	T	T	F		T	T	F	F	F
T	F	F	T		T	F	F	T	T
F	T	F	T		F	T	T	F	T
F	F	F	T		F	F	T	T	T

Since the truth tables are the same, i.e. both propositions are false in the first case and true in the other three cases, the propositions $\sim(p \wedge q)$ and $\sim p \vee \sim q$ are logically equivalent and we can write:

$$\sim(p \wedge q) \equiv \sim p \vee \sim q$$

Consider now the statement,

"It is false that roses are red and violets are blue."

This statement can be written in the form $\sim(p \wedge q)$ where p is "roses are red" and q is "violets are blue". However, by the above truth tables, $\sim(p \wedge q)$ is logically equivalent to $\sim p \vee \sim q$. Thus the given statement has the same meaning as the statement

"Roses are not red, or violets are not blue."

11.8   ALGEBRA OF PROPOSITIONS

Propositions satisfy various laws which are listed in Table 11.1. (In the table, t and f are restricted to the truth values "true" and "false" respectively.) We state this result formally.

Theorem 11.2:   Propositions satisfy the laws of Table 11.1.

Table 11.1 Laws of the Algebra of Propositions

Idempotent Laws

1a. $p \vee p \equiv p$ 1b. $p \wedge p \equiv p$

Associative Laws

2a. $(p \vee q) \vee r \equiv p \vee (q \vee r)$ 2b. $(p \wedge q) \wedge r \equiv p \wedge (q \wedge r)$

Commutative Laws

3a. $p \vee q \equiv q \vee p$ 3b. $p \wedge q \equiv q \wedge p$

Distributive Laws

4a. $p \vee (q \wedge r) \equiv (p \vee q) \wedge (p \vee r)$ 4b. $p \wedge (q \vee r) \equiv (p \wedge q) \vee (p \wedge r)$

Identity Laws

5a. $p \vee f \equiv p$ 5b. $p \wedge t \equiv p$

6a. $p \vee t \equiv t$ 6b. $p \wedge f \equiv f$

Complement Laws

7a. $p \vee {\sim}p \equiv t$ 7b. $p \wedge {\sim}p \equiv f$

8a. ${\sim}t \equiv f$ 8b. ${\sim}f \equiv t$

Involution Law

9. ${\sim}{\sim}p \equiv p$

DeMorgan's Laws

10a. ${\sim}(p \vee q) \equiv {\sim}p \wedge {\sim}q$ 10b. ${\sim}(p \wedge q) \equiv {\sim}p \vee {\sim}q$

11.9 CONDITIONAL AND BICONDITIONAL STATEMENTS

Many statements, particularly in mathematics, are of the form "If p then q". Such statements are called *conditional* statements and are denoted by

$$p \rightarrow q$$

The conditional $p \rightarrow q$ is frequently read "p implies q" or "p only if q".

Another common statement is of the form "p if and only if q". Such statements are called *biconditional* statements and are denoted by

$$p \leftrightarrow q$$

The truth values of $p \rightarrow q$ and $p \leftrightarrow q$ are given in the following tables:

p	q	$p \rightarrow q$
T	T	T
T	F	F
F	T	T
F	F	T

p	q	$p \leftrightarrow q$
T	T	T
T	F	F
F	T	F
F	F	T

Observe that the conditional $p \rightarrow q$ is false only when the first part p is true and the second part q is false. In case p is false, the conditional $p \rightarrow q$ is true regardless of the truth value of q. Observe also that $p \leftrightarrow q$ is true when p and q have the same truth values and false otherwise.

Now consider the truth table of the proposition ${\sim}p \vee q$:

p	q	${\sim}p$	${\sim}p \vee q$
T	T	F	T
T	F	F	F
F	T	T	T
F	F	T	T

Observe that the above truth table is identical to the truth table of $p \to q$. Hence $p \to q$ is logically equivalent to the proposition $\sim p \lor q$:

$$p \to q \equiv \sim p \lor q$$

In other words, the conditional statement "If p then q" is logically equivalent to the statement "Not p or q" which only involves the connectives $\lor$ and $\sim$ and thus was already a part of our language. We may regard $p \to q$ as an abbreviation for an oft-recurring statement.

11.10 ARGUMENTS

An *argument* is an assertion that a given set of propositions $P_1, P_2, \ldots, P_n$, called *premises*, yields (has as a consequence) another proposition Q, called the *conclusion*. Such an argument is denoted by

$$P_1, P_2, \ldots, P_n \vdash Q$$

The notion of a "logical argument" or "valid argument" is formalized as follows:

An argument $P_1, P_2, \ldots, P_n \vdash Q$ is said to be *valid* if Q is true whenever all the premises $P_1, P_2, \ldots, P_n$ are true.

An argument which is not valid is called a *fallacy*.

EXAMPLE 11.5.

(a) The following argument is valid:

$$p, p \to q \vdash q \qquad \text{(Law of Detachment)}$$

The proof of this rule follows from the following truth table.

p	q	$p \to q$
T	T	T
T	F	F
F	T	T
F	F	T

For p is true in Cases (lines) 1 and 2, and $p \to q$ is true in Cases 1, 3 and 4; hence p and $p \to q$ are true simultaneously in Case 1. Since in this case q is true, the argument is valid.

(b) The following argument is a fallacy:

$$p \to q, q \vdash p$$

For $p \to q$ and q are both true in Case (line) 3 in the above truth table, but in this case p is false.

Now the propositions $P_1, P_2, \ldots, P_n$ are true simultaneously if and only if the proposition $P_1 \land P_2 \land \cdots \land P_n$ is true. Thus the argument $P_1, P_2, \ldots, P_n \vdash Q$ is valid if and only if Q is true whenever $P_1 \land P_2 \land \cdots \land P_n$ is true or, equivalently, if the proposition $(P_1 \land P_2 \land \cdots \land P_n) \to Q$ is a tautology. We state this result formally.

Theorem 11.3: The argument $P_1, P_2, \ldots, P_n \vdash Q$ is valid if and only if the proposition $(P_1 \land P_2 \land \cdots \land P_n) \to Q$ is a tautology.

We apply this theorem in the next example.

EXAMPLE 11.6. A fundamental principle of logical reasoning states:

"If p implies q and q implies r, then p implies r."

that is, the following argument is valid:

$$p \to q, \, q \to r \, \vdash \, p \to r \quad \text{(Law of Syllogism)}$$

This fact is verified by the following truth table which shows that the proposition

$$[(p \to q) \wedge (q \to r)] \to (p \to r)$$

is a tautology:

p	q	r	[$(p$	$\to$	$q)$	$\wedge$	$(q$	$\to$	$r)]$	$\to$	$(p$	$\to$	$r)$
T	T	T	T	T	T	T	T	T	T	T	T	T	T
T	T	F	T	T	T	F	T	F	F	T	T	F	F
T	F	T	T	F	F	F	F	T	T	T	T	T	T
T	F	F	T	F	F	F	F	T	F	T	T	F	F
F	T	T	F	T	T	T	T	T	T	T	F	T	T
F	T	F	F	T	T	F	T	F	F	T	F	T	F
F	F	T	F	T	F	T	F	T	T	T	F	T	T
F	F	F	F	T	F	T	F	T	F	T	F	T	F
Step			1	2	1	3	1	2	1	4	1	2	1

Equivalently, the argument is valid since the premises $p \to q$ and $q \to r$ are true simultaneously only in cases (lines) 1, 5, 7 and 8, and in these cases the conclusion $p \to r$ is also true. (Observe that the truth table required $2^3 = 8$ lines since there are three variables p, q and r.)

We now apply the above theory to arguments involving specific statements. We emphasize that the validity of an argument does not depend upon the truth values nor the content of the statements appearing in the argument, but upon the particular form of the argument. This is illustrated in the following examples.

EXAMPLE 11.7. Consider the following argument:

S_1: If a man is a bachelor, he is unhappy.

S_2: If a man is unhappy, he dies young.

....

S: Bachelors die young.

Here the statement S below the line denotes the conclusion of the argument, and the statements S_1 and S_2 above the line denote the premises. We claim that the argument $S_1, S_2 \vdash S$ is valid. For the argument is of the form

$$p \to q, \, q \to r \, \vdash \, p \to r$$

where p is "He is a bachelor", q is "He is unhappy" and r is "He dies young"; and by Example 11.6 this argument (Law of Syllogism) is valid.

11.11 LOGICAL IMPLICATION

A proposition $P(p, q, \ldots)$ is said to *logically imply* a proposition $Q(p, q, \ldots)$, written

$$P(p, q, \ldots) \Rightarrow Q(p, q, \ldots)$$

if $Q(p, q, \ldots)$ is true whenever $P(p, q, \ldots)$ is true.

EXAMPLE 11.8.   We claim that p logically implies $p \lor q$.   For consider the truth tables of p and $p \lor q$ in the table below.   Observe that p is true in Cases (lines) 1 and 2, and in these cases $p \lor q$ is also true. In other words, p logically implies $p \lor q$.

p	q	$p \lor q$
T	T	T
T	F	T
F	T	T
F	F	F

Now if $Q(p, q, \ldots)$ is true whenever $P(p, q, \ldots)$ is true, then the argument

$$P(p, q, \ldots) \;\vdash\; Q(p, q, \ldots)$$

is valid; and conversely.   Furthermore, the argument $P \vdash Q$ is valid if and only if the conditional statement $P \to Q$ is always true, i.e. a tautology.   We state this result formally.

Theorem 11.4:   For any propositions $P(p, q, \ldots)$ and $Q(p, q, \ldots)$, the following three state-ments are equivalent:

    (i) $P(p, q, \ldots)$ logically implies $Q(p, q, \ldots)$.

    (ii)   The argument $P(p, q, \ldots) \vdash Q(p, q, \ldots)$ is valid.

    (iii)   The proposition $P(p, q, \ldots) \to Q(p, q, \ldots)$ is a tautology.

We note that some logicians and many texts use the word "implies" in the same sense as we use "logically implies", and so they distinguish between "implies" and "if . . . then". These two distinct concepts are, of course, intimately related as seen in the above theorem.

Solved Problems

STATEMENT AND COMPOUND STATEMENTS

11.1.   Let p be "It is cold" and let q be "It is raining".   Give a simple verbal sentence which describes each of the following statements:

    (1) $\sim p$,   (2) $p \land q$,   (3) $p \lor q$,   (4) $q \lor \sim p$,   (5) $\sim p \land \sim q$,   (6) $\sim \sim q$

    In each case, translate $\land$, $\lor$ and $\sim$ to read "and", "or" and "It is false that" or "not", respec-tively, and then simplify the English sentence.

    (1) It is not cold.

    (2) It is cold and raining.

    (3) It is cold or it is raining.

    (4) It is raining or it is not cold.

    (5) It is not cold and it is not raining.

    (6) It is not true that it is not raining.

11.2.   Let p be "He is tall" and let q be "He is handsome".   Write each of the following statements in symbolic form using p and q.

    (1) He is tall and handsome.

    (2) He is tall but not handsome.

    (3) It is false that he is short or handsome.

(4) He is neither tall or handsome.

(5) He is tall, or he is short and handsome.

(6) It is not true that he is short or not handsome.

(Assume that "He is short" means "He is not tall", i.e. $\sim p$.)

(1) $p \wedge q$	(3) $\sim(\sim p \vee q)$	(5) $p \vee (\sim p \wedge q)$
(2) $p \wedge \sim q$	(4) $\sim p \wedge \sim q$	(6) $\sim(\sim p \vee \sim q)$

PROPOSITIONS AND THEIR TRUTH TABLES

11.3. Find the truth table of $\sim p \wedge q$.

p	q	$\sim p$	$\sim p \wedge q$
T	T	F	F
T	F	F	F
F	T	T	T
F	F	T	F

Method 1

p	q	$\sim$	p	$\wedge$	q
T	T	F	T	F	T
T	F	F	T	F	F
F	T	T	F	T	T
F	F	T	F	F	F
Step		2	1	3	1

Method 2

11.4. Find the truth table of $\sim(p \vee q)$.

p	q	$p \vee q$	$\sim(p \vee q)$
T	T	T	F
T	F	T	F
F	T	T	F
F	F	F	T

Method 1

p	q	$\sim$	$(p$	$\vee$	$q)$
T	T	F	T	T	T
T	F	F	T	T	F
F	T	F	F	T	T
F	F	T	F	F	F
Step		3	1	2	1

Method 2

11.5. Find the truth table of $\sim(p \vee \sim q)$.

p	q	$\sim q$	$p \vee \sim q$	$\sim(p \vee \sim q)$
T	T	F	T	F
T	F	T	T	F
F	T	F	F	T
F	F	T	T	F

Method 1

p	q	$\sim$	$(p$	$\vee$	$\sim$	$q)$
T	T	F	T	T	F	T
T	F	F	T	T	T	F
F	T	T	F	F	F	T
F	F	F	F	T	T	F
Step		4	1	3	2	1

Method 2

TAUTOLOGIES AND CONTRADICTIONS

11.6. Verify that the proposition $p \vee \sim(p \wedge q)$ is a tautology.

Construct the truth table of $p \vee \sim(p \wedge q)$:

p	q	$p \wedge q$	$\sim(p \wedge q)$	$p \vee \sim(p \wedge q)$
T	T	T	F	T
T	F	F	T	T
F	T	F	T	T
F	F	F	T	T

Since the truth value of $p \vee \sim(p \wedge q)$ is T for all values of p and q, it is a tautology.

11.7. Verify that the proposition $(p \wedge q) \wedge \sim(p \vee q)$ is a contradiction.

Construct the truth table of $(p \wedge q) \wedge \sim(p \vee q)$:

p	q	$p \wedge q$	$p \vee q$	$\sim(p \vee q)$	$(p \wedge q) \wedge \sim(p \vee q)$
T	T	T	T	F	F
T	F	F	T	F	F
F	T	F	T	F	F
F	F	F	F	T	F

Since the truth value of $(p \wedge q) \wedge \sim(p \vee q)$ is F for all values of p and q, it is a contradiction.

LOGICAL EQUIVALENCE

11.8. Prove that disjunction distributes over conjunction; that is, prove the Distributive Law: $p \vee (q \wedge r) \equiv (p \vee q) \wedge (p \vee r)$.

Construct the required truth tables.

p	q	r	$q \wedge r$	$p \vee (q \wedge r)$	$p \vee q$	$p \vee r$	$(p \vee q) \wedge (p \vee r)$
T	T	T	T	T	T	T	T
T	T	F	F	T	T	T	T
T	F	T	F	T	T	T	T
T	F	F	F	T	T	T	T
F	T	T	T	T	T	T	T
F	T	F	F	F	T	F	F
F	F	T	F	F	F	T	F
F	F	F	F	F	F	F	F

Since the truth tables are identical, the propositions are equivalent.

11.9. Prove that the operation of disjunction can be written in terms of the operations of conjunction and negation. Specifically, $p \vee q \equiv \sim(\sim p \wedge \sim q)$.

Construct the required truth tables:

p	q	$p \vee q$	$\sim p$	$\sim q$	$\sim p \wedge \sim q$	$\sim(\sim p \wedge \sim q)$
T	T	T	F	F	F	T
T	F	T	F	T	F	T
F	T	T	T	F	F	T
F	F	F	T	T	T	F

Since the truth tables are identical, the propositions are equivalent.

11.10. Simplify each proposition by using the laws listed in Table 11.1.   (a) $p \vee (p \wedge q)$, (b) $\sim(p \vee q) \vee (\sim p \wedge q)$.

(a)

	Statement		Reason
(1)	$p \vee (p \wedge q) \equiv (p \wedge t) \vee (p \wedge q)$	(1)	Identity law
(2)	$\equiv p \wedge (t \vee q)$	(2)	Distributive law
(3)	$\equiv p \wedge t$	(3)	Identity law
(4)	$\equiv p$	(4)	Identity law

(b)

	Statement		Reason
(1)	$\sim(p \vee q) \vee (\sim p \wedge q) \equiv (\sim p \wedge \sim q) \vee (\sim p \wedge q)$	(1)	DeMorgan's law
(2)	$\equiv \sim p \wedge (\sim q \vee q)$	(2)	Distributive law
(3)	$\equiv \sim p \wedge t$	(3)	Complement law
(4)	$\equiv \sim p$	(4)	Identity law

NEGATION

11.11. Prove DeMorgan's laws: (a) $\sim(p \wedge q) \equiv \sim p \vee \sim q$;   (b) $\sim(p \vee q) \equiv \sim p \wedge \sim q$

In each case construct the required truth tables.

(a)

p	q	$p \wedge q$	$\sim(p \wedge q)$	$\sim p$	$\sim q$	$\sim p \vee \sim q$
T	T	T	F	F	F	F
T	F	F	T	F	T	T
F	T	F	T	T	F	T
F	F	F	T	T	T	T

(b)

p	q	$p \vee q$	$\sim(p \vee q)$	$\sim p$	$\sim q$	$\sim p \wedge \sim q$
T	T	T	F	F	F	F
T	F	T	F	F	T	F
F	T	T	F	T	F	F
F	F	F	T	T	T	T

11.12. Verify: $\sim \sim p \equiv p$

p	$\sim p$	$\sim \sim p$
T	F	T
F	T	F

11.13. Use the results of the preceding problems to simplify each of the following statements.

(a) It is not true that his mother is English or his father is French.

(b) It is not true that he studies physics but not mathematics.

(c) It is not true that sales are decreasing and prices are rising.

(d) It is not true that it is not cold or it is raining.

(a) Let p denote "His mother is English" and let q denote "His father is French". Then the given statement is $\sim(p \vee q)$. But $\sim(p \vee q) \equiv \sim p \wedge \sim q$. Hence the given statement is logically equivalent to the statement "His mother is not English and his father is not French.

(b) Let p denote "He studies physics" and let q denote "He studies mathematics". Then the given statement is $\sim(p \wedge \sim q)$. But $\sim(p \wedge \sim q) \equiv \sim p \vee \sim \sim q \equiv \sim p \vee q$. Hence the given statement is logically equivalent to the statement "He does not study physics or he studies mathematics".

(c) Since $\sim(p \wedge q) \equiv \sim p \vee \sim q$, the given statement is logically equivalent to the statement "Sales are increasing or prices are falling".

(d) Since $\sim(\sim p \vee q) \equiv p \wedge \sim q$, the given statement is logically equivalent to the statement "It is cold and it is not raining".

CONDITIONAL AND BICONDITIONAL

11.14. Rewrite the following statements without using the conditional.

(a) If it is cold, he wears a hat.

(b) If productivity increases, then wages rise.

   Recall that "If p then q" is equivalent to "Not p or q"; that is, $p \to q \equiv \sim p \vee q$.

(a) It is not cold or he wears a hat.

(b) Productivity does not increase or wages rise.

11.15. (a) Show that "p implies q and q implies p" is logically equivalent to the biconditional "p if and only if q"; that is, $(p \to q) \wedge (q \to p) \equiv p \leftrightarrow q$.

(b) Show that the biconditional $p \leftrightarrow q$ can be written in terms of the original three connectives $\vee$, $\wedge$ and $\sim$.

(a)

p	q	$p \leftrightarrow q$	$p \to q$	$q \to p$	$(p \to q) \wedge (q \to p)$
T	T	T	T	T	T
T	F	F	F	T	F
F	T	F	T	F	F
F	F	T	T	T	T

(b) Now $p \to q \equiv \sim p \vee q$ and $q \to p \equiv \sim q \vee p$; hence by

$$p \leftrightarrow q \equiv (p \to q) \wedge (q \to p) \equiv (\sim p \vee q) \wedge (\sim q \vee p)$$

11.16. Determine the truth table of $(p \to q) \to (p \wedge q)$.

p	q	$p \to q$	$p \wedge q$	$(p \to q) \to (p \wedge q)$
T	T	T	T	T
T	F	F	F	T
F	T	T	F	F
F	F	T	F	F

11.17. Determine the truth table of $(p \to q) \vee \sim(p \leftrightarrow \sim q)$.

p	q	$(p$	$\to$	$q)$	$\vee$	$\sim$	$(p$	$\leftrightarrow$	$\sim$	$q)$
T	T	T	T	T	T	T	T	F	F	T
T	F	T	F	F	F	F	T	T	T	F
F	T	F	T	T	T	F	F	T	F	T
F	F	F	T	F	T	T	F	F	T	F
Step		1	2	1	5	4	1	3	2	1

11.18. Consider the conditional proposition $p \to q$ and other simple propositions containing p and q:

$$q \to p, \quad \sim p \to \sim q \quad \text{and} \quad \sim q \to \sim p$$

These are called respectively the *converse*, *inverse* and *contrapositive* of the original conditional proposition $p \to q$. Which if any of these propositions are logically equivalent to $p \to q$?

Construct their truth tables:

p	q	$\sim p$	$\sim q$	Conditional $p \to q$	Converse $q \to p$	Inverse $\sim p \to \sim q$	Contrapositive $\sim q \to \sim p$
T	T	F	F	T	T	T	T
T	F	F	T	F	T	T	F
F	T	T	F	T	F	F	T
F	F	T	T	T	T	T	T

Only the contrapositive $\sim q \to \sim p$ is logically equivalent to the original conditional proposition $p \to q$.

ARGUMENTS AND LOGICAL IMPLICATION

11.19. Show that the following argument is valid: $p \leftrightarrow q, \ q \vdash p$.

Method 1.

Construct the truth table on the right. Now $p \leftrightarrow q$ is true in Cases (lines) 1 and 4, and q is true in Cases 1 and 3; hence $p \leftrightarrow q$ and q are true simultaneously only in Case 1 where p is also true. Thus the argument $p \leftrightarrow q, \ q \vdash p$ is valid.

p	q	$p \leftrightarrow q$
T	T	T
T	F	F
F	T	F
F	F	T

Method 2.

Construct the truth table of $[(p \leftrightarrow q) \wedge q] \to p$:

p	q	$p \leftrightarrow q$	$(p \leftrightarrow q) \wedge q$	$[(p \leftrightarrow q) \wedge q] \to p$
T	T	T	T	T
T	F	F	F	T
F	T	F	F	T
F	F	T	F	T

Since $[(p \leftrightarrow q) \wedge q] \to p$ is a tautology, the argument is valid.

11.20. Test the validity of the following argument:

> If I study, then I will not fail mathematics.
>
> If I do not play basketball, then I will study.
>
> But I failed mathematics.
>
> ....
>
> Therefore, I played basketball.

First translate the arguments into symbolic form. Let p be "I study", q be "I fail mathematics" and r be "I play basketball". Then the given argument is as follows:

$$p \to \sim q, \ \sim r \to p, \ q \ \vdash \ r$$

To test the validity of the argument, construct the truth tables of the given propositions $p \to \sim q$, $\sim r \to p$, q and r:

p	q	r	$\sim q$	$p \to \sim q$	$\sim r$	$\sim r \to p$
T	T	T	F	F	F	T
T	T	F	F	F	T	T
T	F	T	T	T	F	T
T	F	F	T	T	T	T
F	T	T	F	T	F	T
F	T	F	F	T	T	F
F	F	T	T	T	F	T
F	F	F	T	T	T	F

Now the premises $p \to \sim q$, $\sim r \to p$ and q are true simultaneously only in Case (line) 5, and in that case the conclusion r is also true; hence the argument is valid.

11.21. Show that $p \leftrightarrow q$ logically implies $p \to q$.

Method 1.

Construct the truth tables of $p \leftrightarrow q$ and $p \to q$:

p	q	$p \leftrightarrow q$	$p \to q$
T	T	T	T
T	F	F	F
F	T	F	T
F	F	T	T

Now $p \leftrightarrow q$ is true in lines 1 and 4, and in these cases $p \to q$ is also true. Hence $p \leftrightarrow q$ logically implies $p \to q$.

Method 2.

Construct the truth table of $(p \leftrightarrow q) \to (p \to q)$. It will be a tautology; hence, by Theorem 11.4, $p \leftrightarrow q$ logically implies $p \to q$.

11.22. Show that $p \leftrightarrow \sim q$ does not logically imply $p \to q$.

Method 1.

Construct the truth tables of $p \leftrightarrow \sim q$ and $p \to q$:

p	q	$\sim q$	$p \leftrightarrow \sim q$	$p \to q$
T	T	F	F	T
T	F	T	T	F
F	T	F	T	T
F	F	T	F	T

Recall that $p \leftrightarrow \sim q$ logically implies $p \to q$ if $p \to q$ is true whenever $p \leftrightarrow \sim q$ is true. But $p \leftrightarrow \sim q$ is true in Case (line) 2 in the above table, and in that Case $p \to q$ is false. Hence $p \leftrightarrow \sim q$ does not logically imply $p \to q$.

Method 2.

Construct the truth table of the proposition $(p \leftrightarrow \sim q) \to (p \to q)$. It will not be a tautology; hence, by Theorem 4.2, $p \leftrightarrow \sim q$ does not logically imply $p \to q$.

MISCELLANEOUS PROBLEMS

11.23. Let Apq denote $p \wedge q$ and let Np denote $\sim p$. Rewrite the following propositions using A and N instead of $\wedge$ and $\sim$:

(a) $p \wedge \sim q$, (b) $\sim(\sim p \wedge q)$, (c) $\sim p \wedge (\sim q \wedge r)$, (d) $\sim(p \wedge \sim q) \wedge (\sim q \wedge \sim r)$

(a) $p \wedge \sim q = p \wedge Nq = ApNq$

(b) $\sim(\sim p \wedge q) = \sim(Np \wedge q) = \sim(ANpq) = NANpq$

(c) $\sim p \wedge (\sim q \wedge r) = Np \wedge (Nq \wedge r) = Np \wedge (ANqr) = ANpANqr$

(d) $\sim(p \wedge \sim q) \wedge (\sim q \wedge \sim r) = \sim(ApNq) \wedge (ANqNr) = (NApNq) \wedge (ANqNr) = ANApNqANqNr$

Notice that there are no parentheses in the final answer when A and N are used instead of $\wedge$ and $\sim$. It has been proven that they are not needed. Furthermore, since every connective is logically equivalent to A and N, i.e. $\wedge$ and $\sim$, the above notation suffices for any development of the algebra of propositions.

11.24. Rewrite the following propositions using $\wedge$ and $\sim$ instead of A and N.

(a) $NApq$, (b) $ANpq$, (c) $ApNq$, (d) $ApAqr$, (e) $NAANpqr$, (f) $ANpAqNr$

(a) $Napq = N(p \wedge q) = \sim(p \wedge q)$ (c) $ApNq = Ap(\sim q) = p \wedge \sim q$

(b) $ANpq = A(\sim p)q = \sim p \wedge q$ (d) $ApAqr = Ap(q \wedge r) = p \wedge (q \wedge r)$

(e) $NAANpqr = NAA(\sim p)qr = NA(\sim p \wedge q)r = N[(\sim p \wedge q) \wedge r] = \sim[(\sim p \wedge q) \wedge r]$

(f) $ANpAqNr = ANpAq(\sim r) = ANp(q \wedge \sim r) = A(\sim p)(q \wedge \sim r) = \sim p \wedge (q \wedge \sim r)$

Notice that the propositions involving A and N are unraveled from right to left.

11.25. The propositional connective $\veebar$ is called the *exclusive disjunction*; $p \veebar q$ is read "p or q but not both".

(1) Construct a truth table for $p \veebar q$.

(2) Prove: $p \veebar q \equiv (p \vee q) \wedge \sim(p \wedge q)$. Hence $\veebar$ can be written in terms of the original three connectives $\vee$, $\wedge$ and $\sim$.

(1) Note that $p \veebar q$ is true if p is true but not if both p and q are true; hence the truth table of $p \veebar q$ is as follows:

p	q	$p \veebar q$
T	T	F
T	F	T
F	T	T
F	F	F

(2) Consider the following truth table.

p	q	$(p$	$\vee$	$q)$	$\wedge$	$\sim$	$(p$	$\wedge$	$g)$
T	T	T	T	T	F	F	T	T	T
T	F	T	T	F	T	T	T	F	F
F	T	F	T	T	T	T	F	F	T
F	F	F	F	F	F	T	F	F	F
Step		1	2	1	4	3	1	2	1

Since the truth tables of $p \veebar q$ and $(p \vee q) \wedge \sim(p \vee q)$ are identical, $p \veebar q \equiv (p \vee q) \wedge \sim(p \wedge q)$.

11.26. The propositional connective $\downarrow$ is called the *joint denial*; $p \downarrow q$ is read "Neither p nor q".

  (1) Construct a truth table for $p \downarrow q$.

  (2) Prove: The three connectives $\vee$, $\wedge$ and $\sim$ may be expressed in terms of the connective $\downarrow$ as follows:

$$(a) \ \ \sim p \equiv p \downarrow p, \ \ \ \ (b) \ \ p \wedge q \equiv (p \downarrow p) \downarrow (q \downarrow q), \ \ \ \ (c) \ \ p \vee q \equiv (p \downarrow q) \downarrow (p \downarrow q).$$

  (1) Note $p \downarrow q$ is true if neither p is true nor q is true; hence the truth table of $p \downarrow q$ is as follows:

p	q	$p \downarrow q$
T	T	F
T	F	F
F	T	F
F	F	T

  (2) *(a)*   *(b)*

p	$\sim p$	$p \downarrow p$
T	F	F
F	T	T

p	q	$p \wedge q$	$p \downarrow p$	$q \downarrow q$	$(p \downarrow p) \downarrow (q \downarrow q)$
T	T	T	F	F	T
T	F	F	F	T	F
F	T	F	T	F	F
F	F	F	T	T	F

  (c)

p	q	$p \vee q$	$p \downarrow q$	$(p \downarrow q) \downarrow (p \downarrow q)$
T	T	T	F	T
T	F	T	F	T
F	T	T	F	T
F	F	F	T	F

Supplementary Problems

STATEMENTS AND COMPOUND STATEMENTS

11.27. Let p be "Marc is rich" and let q be "Marc is happy". Write each of the following in symbolic form.

  (a) Marc is poor but happy.

  (b) Marc is neither rich nor happy.

  (c) Marc is either rich or unhappy.

  (d) Marc is poor or else he is both rich and unhappy.

11.28. Let p be "Erik reads *Newsweek*", let q be "Erik reads *The New Yorker*" and let r be "Erik reads *Time*". Write each of the following in symbolic form.

  (a) Erik reads *Newsweek* or *The New Yorker*, but not *Time*.

  (b) Erik reads *Newsweek* and *The New Yorker*, or he does not read *Newsweek* and *Time*.

(c) It is not true that Erik reads *Newsweek* but not *Time*.

(d) It is not true that Erik reads *Time* or *The New Yorker* but not *Newsweek*.

11.29. Let *p* be "Audrey speaks French" and let *q* be "Audrey speaks Danish".   Give a simple verbal sentence which describes each of the following.

(a) $p \vee q$,   (b) $p \wedge q$,   (c) $p \wedge \sim q$,   (d) $\sim p \vee \sim q$,   (e) $\sim \sim p$,   (f) $\sim (\sim p \wedge \sim q)$

PROPOSITIONS AND THEIR TRUTH TABLES

11.30. Find the truth table of each proposition:

(a) $p \vee \sim q$,   (b) $\sim p \wedge \sim q$,   (c) $\sim (\sim p \wedge q)$,   (d) $\sim (\sim p \vee \sim q)$

11.31. Find the truth table of each proposition:

(a) $(p \wedge \sim q) \vee r$,   (b) $\sim p \vee (q \wedge \sim r)$,   (c) $(p \vee \sim r) \wedge (q \vee \sim r)$

LOGICAL EQUIVALENCE

11.32. Prove the associative law for disjunction: $(p \vee q) \vee r \equiv p \vee (q \vee r)$

11.33. Prove that conjunction distributes over disjunction: $p \wedge (q \vee r) \equiv (p \wedge q) \vee (q \wedge r)$

11.34. Prove $p \vee (p \wedge q) \equiv p$ by constructing the appropriate truth tables [see Problem 11.10(a)].

11.35. Prove $\sim (p \vee q) \vee (\sim p \wedge q) \equiv \sim p$ by constructing the appropriate truth tables [see Problem 11.10(b)]

11.36. (a) Express $\vee$ in terms of $\wedge$ and $\sim$.

(b) Express $\wedge$ in terms of $\vee$ and $\sim$.

11.37. Simplify: (a) $\sim (p \wedge \sim q)$,   (b) $\sim (\sim p \vee q)$,   (c) $\sim (\sim p \wedge \sim q)$

11.38. Write the negation of each of the following statements as simply as possible.

(a) He is tall but handsome.

(b) He has blond hair or blue eyes.

(c) He is neither rich nor happy.

(d) He lost his job or he did not go to work today.

(e) Neither Marc nor Erik is unhappy.

(f) Audrey speaks Spanish or French, but not German.

11.39. Prove the following equivalences by using the laws of algebra of propositions listed in Table 11-1.

(a) $p \wedge (p \vee q) \equiv p$,   (b) $(p \wedge q) \vee \sim p \equiv \sim p \vee q$,   (c) $p \wedge (\sim p \vee q) \equiv p \wedge q$

CONDITIONAL AND BICONDITIONAL

11.40. Find the truth table of each proposition: (a) $(\sim p \vee q) \rightarrow p$,   (b) $q \leftrightarrow (\sim q \wedge p)$

11.41. Find the truth table of each proposition: (a) $(p \leftrightarrow \sim q) \rightarrow (\sim p \wedge q)$,   (b) $(\sim q \vee p) \leftrightarrow (q \rightarrow \sim p)$

11.42. Find the truth table of each proposition:

(a) $[p \wedge (\sim q \to p)] \wedge \sim[(p \leftrightarrow \sim q) \to (q \vee \sim p)]$, (b) $[q \leftrightarrow (r \to \sim p)] \vee [(\sim q \to p) \leftrightarrow r]$

11.43. Prove: (a) $(p \wedge q) \to r \equiv (p \to r) \vee (q \to r)$, (b) $p \to (q \to r) \equiv (p \wedge \sim r) \to \sim q$

11.44. Determine the equivalent contrapositive of each statement:

(a) If he has courage he will win. (b) Only if he does not tire will he win.

11.45. Find: (a) Contrapositive of $p \to \sim q$.   (c) Contrapositive of the converse of $p \to \sim q$.

(b) Contrapositive of $\sim p \to q$.   (d) Converse of the contrapositive of $\sim p \to \sim q$.

ARGUMENTS AND LOGICAL IMPLICATION

11.46. Test the validity of each argument: (a) $\sim p \to q, p \vdash \sim q$; (b) $\sim p \to q, q \vdash p$

11.47. Test the validity of each argument: (a) $p \to q, r \to \sim q \vdash r \to \sim p$; (b) $p \to \sim q, \sim r \to \sim q \vdash p \to \sim r$

11.48. Test the validity of each argument: (a) $p \to \sim q, r \to p, q \vdash \sim r$; (b) $p \to q, r \vee \sim q, \sim r \vdash \sim p$

11.49. Translate into symbolic form and test the validity of the argument:

(a) If 6 is even, then 2 does not divide 7.

Either 5 is not prime or 2 divides 7.

But 5 is prime.
....
Therefore, 6 is odd (not even).

(b) On my wife's birthday, I bring her flowers.

Either it's my wife's birthday or I work late.

I did not bring my wife flowers today.
....
Therefore, today I worked late.

(c) If I work, I cannot study.

Either I work, or I pass mathematics.

I passed mathematics.
....
Therefore, I studied.

(d) If I work, I cannot study.

Either I study, or I pass mathematics.

I worked.
....
Therefore, I passed mathematics.

11.50. Show that (a) $p \wedge q$ logically implies p, (b) $p \vee q$ does not logically imply p.

11.51. Show that   (a) q logically implies $p \to q$,   (b) $\sim p$ logically implies $p \to q$.

11.52. Show that $p \wedge (q \vee r)$ logically implies $(p \wedge q) \vee r$.

11.53. Determine those propositions which logically imply   (a) a tautology,   (b) a contradiction.

MISCELLANEOUS PROBLEMS

11.54. Let Apq denote $p \wedge q$ and let Np denote $\sim p$.   (See Problem 11.23.)   Rewrite the following propositions using A and N instead of $\wedge$ and $\sim$.

(a) $\sim p \wedge q$, (b) $\sim p \wedge \sim q$, (c) $\sim (p \wedge \sim q)$, (d) $(\sim p \wedge q) \wedge \sim r$

11.55. Rewrite the following propositions using $\wedge$ and $\sim$ instead of A and N.

(a) $NApNq$, (b) $ANApqNr$, (c) $AApNrAqNp$, (d) $ANANpANqrNp$

Answers to Supplementary Problems

11.27. (a) $\sim p \wedge q$,   (b) $\sim p \wedge \sim q$,   (c) $p \vee \sim q$,   (d) $\sim p \vee (p \wedge \sim q)$

11.28. (a) $(p \vee q) \wedge \sim r$,   (b) $(p \wedge q) \vee \sim (p \wedge r)$,   (c) $\sim (p \wedge \sim r)$,   (d) $\sim [(r \vee q) \wedge \sim p]$

11.29. (a) Audrey speaks French or Danish.

(b) Audrey speaks French and Danish.

(c) Audrey speaks French but not Danish.

(d) Audrey does not speak French or she does not speak Danish.

(e) It is not true that Audrey does not speak French.

(f) It is not true that Audrey speaks neither French nor Danish.

11.30.

p	q	$p \vee \sim q$	$\sim p \wedge \sim q$	$\sim (\sim p \wedge q)$	$\sim (\sim p \vee \sim q)$
T	T	T	F	T	T
T	F	T	F	T	F
F	T	F	F	F	F
F	F	T	T	T	F

11.31.

p	q	r	(a)	(b)	(c)
T	T	T	T	F	T
T	T	F	F	T	T
T	F	T	T	F	F
T	F	F	T	F	T
F	T	T	T	T	F
F	T	F	F	T	T
F	F	T	T	T	F
F	F	F	F	T	T

11.36. (a) $p \vee q \equiv \sim(\sim p \wedge \sim q)$,   (b) $p \wedge q \equiv \sim(\sim p \vee \sim q)$

11.37. (a) $\sim p \vee q$,   (b) $p \wedge \sim q$,   (c) $p \vee q$

11.38. (c) He is rich or happy.   (f) Audrey speaks German but neither Spanish nor French.

11.39. (a) $p \wedge (p \vee q) \equiv (p \vee f) \wedge (p \vee q) \equiv p \vee (f \wedge q) \equiv p \vee f \equiv p$

11.40. (a) TTFF, (b) FFFT

11.41.. (a) TFTT, (b) FTFT

11.42. (a) FTFF, (b) TTTFTTFT

11.43. *Hint*: Construct the appropriate truth tables,

11.44. (a) If he does not win, then he does not have courage.   (b) If he tires, then he will not win.

11.45. (a) $q \rightarrow \sim p$,   (b) $\sim q \rightarrow p$,   (c) $\sim p \rightarrow q$,   (d) $p \rightarrow q$

11.46. (a) fallacy, (b) valid

11.47. (a) valid, (b) fallacy

11.48. (a) valid, (b) valid

11.49. (a) $p \rightarrow \sim q, \sim r \vee q, r \vdash \sim p$; valid.   (c) $p \rightarrow \sim q, p \vee r, r \vdash q$; fallacy.

   (b) $p \rightarrow q, p \vee r, \sim q \vdash r$; valid.   (d) $p \rightarrow \sim q, q \vee r, p \vdash r$; valid.

11.53. (a) Every proposition logically implies a tautology.

   (b) Only a contradiction logically implies a contradiction.

11.54. (a) $ANpq$, (b) $ANpNq$, (c) $NApNq$, (d) $AANpqNr$

11.55. (a) $\sim(p \wedge \sim q)$,   (b) $\sim(p \wedge q) \wedge \sim r$,   (c) $(p \wedge \sim r) \wedge (q \wedge \sim p)$,   (d) $\sim(\sim p \wedge (\sim q \wedge r)) \wedge \sim p$

Chapter 12

Boolean Algebra

12.1 BASIC DEFINITIONS

Both sets and propositions have similar properties, i.e. satisfy identical laws. These laws are used to define an abstract mathematical structure called a *Boolean algebra,* which is named after the mathematician George Boole (1813–1864).

Let B be a nonempty set with two binary operations $+$ and $*$, a unary operation $'$, and two distinct elements 0 and 1. Then B is called a *Boolean algebra* if the following axioms hold where a, b, c are any elements in B:

[B₁] Commutative Laws:

$(1a)\quad a + b = b + a$ $(1b)\quad a * b = b * a$

[B₂] Distributive Laws:

$(2a)\quad a + (b * c) = (a + b) * (a + c)$ $(2b)\quad a * (b + c) = (a * b) + (a * c)$

[B₃] Identity Laws:

$(3a)\quad a + 0 = a$ $(3b)\quad a * 1 = a$

[B₄] Complement Laws:

$(4a)\quad a + a' = 1$ $(4b)\quad a * a' = 0$

We will sometimes designate a Boolean algebra by $\langle B, +, *, ', 0, 1 \rangle$ when we want to emphasize its six parts. We say 0 is the *zero* element, 1 is the *unit* element and a' is the *complement* of a. We will usually drop the symbol $*$ and use juxtaposition instead. Then (2b) is written $a(b + c) = ab + ac$ which is the familiar algebraic identity of rings and fields. However, (2a) becomes $a + bc = (a + b)(a + c)$, which is certainly not a usual identity in algebra.

The operations $+$, $*$ and $'$ are called sum, product and complement respectively. We adopt the usual convention that, unless we are guided by parentheses, $'$ has precedence over $*$, and $*$ has precedence over $+$. For example,

$a + b * c$ means $a + (b * c)$ and not $(a + b) * c$ $a * b'$ means $a * (b')$ and not $(a * b)'$

Of course when $a + b * c$ is written $a + bc$ then the meaning is clear.

EXAMPLE 12.1.

(a) Let B be the set with two elements $\{0, 1\}$ with binary operations $+$ and $*$ defined by

+	1	0
1	1	1
0	1	0

*	1	0
1	1	0
0	0	0

and the unary operation $'$ defined by $0' = 1$ and $1' = 0$. Then B is a Boolean algebra.

(b) Let C be a collection of sets closed under union, intersection and complement. Then C is a Boolean algebra with the empty set $\emptyset$ as the zero element and the universal set U as the unit element.

(c) Let $D_{70} = \{1, 2, 5, 7, 10, 14, 35, 70\}$, the divisors of 70. Define $+$, $*$ and $'$ on D_{70} by

$$a + b = \operatorname{lcm}(a, b), \quad a * b = \gcd(a, b), \quad a' = 70/a$$

Then D_{70} is a Boolean algebra with 1 the zero element and 70 the unit element.

Suppose C is a nonempty subset of a Boolean algebra B. We say C is a *subalgebra* of B if C itself is a Boolean algebra (with respect to the operations of B). We note that C is a subalgebra of B if and only if C is closed under the three operations of B, i.e. $+$, $*$ and $'$. For example, $\{1, 2, 35, 70\}$ is a subalgebra of D_{70} in Example 11.1(c).

Two Boolean algebras B and B' are said to be *isomorphic* if there is a one-to-one correspondence $f : B \to B'$ which preserves the three operations, i.e. such that

$$f(a + b) = f(a) + f(b), \quad f(a * b) = f(a) * f(b) \quad \text{and} \quad f(a') = f(a)'$$

for any elements a, b in B.

12.2 DUALITY

The *dual* of any statement in a Boolean algebra B is the statement obtained by interchanging the operations $+$ and $*$, and interchanging their identity elements 0 and 1 in the original statement. For example, the dual of

$$(1 + a) * (b + 0) = b \quad \text{is} \quad (0 * a) + (b * 1) = b$$

Observe the symmetry in the axioms of a Boolean algebra B. That is, the dual of the set of axioms of B is the same as the original set of axioms. Accordingly, the important principle of duality holds in B. Namely,

Theorem 12.1 (Principle of Duality):   The dual of any theorem in a Boolean algebra is also a theorem.

In other words, if any statement is a consequence of the axioms of a Boolean algebra, then the dual is also a consequence of those axioms since the dual statement can be proven by using the dual of each step of the proof of the original statement.

12.3 BASIC THEOREMS

Using the axioms [B_1] through [B_4], we prove (Problem 12.4) the following theorem.

Theorem 12.2:   Let a, b, c be any elements in a Boolean algebra B.

(i)   Idempotent Laws:

(5a) $a + a = a$ (5b) $a * a = a$

(ii) Boundedness Laws:

(6a) $a + 1 = 1$ (6b) $a * 0 = 0$

(iii) Absorption Laws:

(7a) $a + (a * b) = a$ (7b) $a * (a + b) = a$

(iv) Associative Laws:

(8a) $(a + b) + c = a + (b + c)$ (8b) $(a * b) * c = a * (b * c)$

Theorem 12.2 and our axioms still do not contain all the properties of sets listed in Table 1.1. The next two theorems give us the remaining properties.

Theorem 12.3:   Let a be any element of a Boolean algebra B.

     (i)   (Uniqueness of Complement)

          If $a + x = 1$ and $a * x = 0$, then $x = a'$.

     (ii)   (Involution Law) $(a')' = a$

     (iii) $(9a)$ $0' = 1$ $(9b)$ $1' = 0$

Theorem 12.4 (DeMorgan's laws): $(10a)$ $(a + b)' = a' * b'$ $(10b)$ $(a * b)' = a' + b'$

We prove these theorems in Problems 12.5 and 12.6.

12.4 BOOLEAN ALGEBRAS AS LATTICES

By Theorem 12.2 and axiom $[B_1]$, every Boolean algebra B satisfies the associative, commutative and absorption laws and hence is a lattice where $+$ and $*$ are the join and meet operations respectively. With respect to this lattice, $a + 1 = 1$ implies $a \leqq 1$ and $a * 0 = 0$ implies $0 \leqq a$, for any element $a \in B$. Thus B is a bounded lattice. Furthermore, axioms $[B_2]$ and $[B_4]$ show that B is also distributive and complemented. Conversely, every bounded, distributive and complemented lattice L satisfies the axioms $[B_1]$ through $[B_4]$. Accordingly, we have the following

Alternate Definition: A Boolean algebra B is a bounded, distributive and complemented lattice.

Since a Boolean algebra B is a lattice, it has a natural partial ordering (and so its diagram can be drawn). Recall (Chapter 10) that we define $a \leqq b$ when the equivalent conditions $a + b = b$ and $a * b = a$ hold. Since we are in a Boolean algebra, we can actually say much more.

Theorem 12.5:   The following are equivalent in a Boolean algebra:

     (1) $a + b = b$,   (2) $a * b = a$,   (3) $a' + b = 1$,   (4) $a * b' = 0$

Thus in a Boolean algebra we can write $a \leqq b$ whenever any of the above four conditions is known to be true.

EXAMPLE 12.2.

(a) Consider a Boolean algebra of sets. Then set A precedes set B if A is a subset of B. Theorem 12.4 states that if $A \subset B$, as illustrated in the Venn diagram in Fig. 12-1, then the following conditions hold:

    (1) $A \cup B = B$ (3) $A^c \cup B = U$

    (2) $A \cap B = A$ (4) $A \cap B^c = \emptyset$

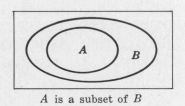

A is a subset of B

Fig. 12-1

(b) Consider the Boolean algebra of the proposition calculus. Then the proposition P precedes the proposition Q if P logically implies Q, i.e. if $P \Rightarrow Q$.

12.5 REPRESENTATION THEOREM

Let B be a finite Boolean algebra. Recall (Section 10.6) that an element a in B is an atom if a immediately succeeds 0, that is if $0 \lessdot a$. Let A be the set of atoms of B and let

$P(A)$ be the Boolean algebra of all subsets of the set A of atoms. By Theorem 10.11, each $x \neq 0$ in B can be expressed uniquely (except for order) as the sum (join) of atoms, i.e. elements of A. Say,

$$x = a_1 + a_2 + \cdots + a_r$$

is such a representation. Consider the function $f : B \to P(A)$ defined by

$$f(x) = \{a_1, a_2, \ldots, a_r\}$$

The mapping is well defined since the representation is unique.

Theorem 12.6:   The above mapping $f : B \to P(A)$ is an isomorphism.

Thus we see the intimate relationship between set theory and abstract Boolean algebras in the sense that every finite Boolean algebra is structurally the same as a Boolean algebra of sets.

If a set A has n elements, then its power set $P(A)$ has 2^n elements. Thus the above theorem gives us our next result.

Corollary 12.7:   A finite Boolean algebra has 2^n elements for some positive integer n.

EXAMPLE 12.3.   Consider the Boolean algebra $D_{70} = \{1, 2, 5, \ldots, 70\}$ of divisors of 70. [See Example 12.1(c).] Its diagram is given in Fig. 12-2(a). Note that $A = \{2, 5, 7\}$ is the set of atoms of D_{70}. We have: $10 = 2 \vee 5$, $14 = 2 \vee 7$, $35 = 5 \vee 7$, $70 = 2 \vee 5 \vee 7$, which is the unique representation of nonatoms by atoms. Figure 12-2(b) gives the diagram of the Boolean algebra of the power set $P(A)$ of A. Observe that the two diagrams are structurally the same.

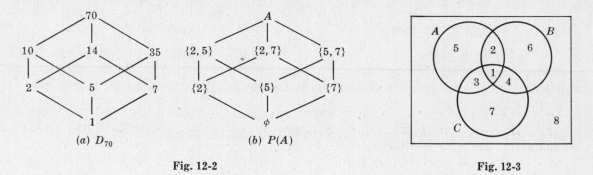

(a) D_{70} (b) $P(A)$

Fig. 12-2   Fig. 12-3

12.6 DISJUNCTIVE NORMAL FORM FOR SETS

We motivate the concept of the disjunctive normal form from an example of set theory. Consider the Venn diagram in Fig. 12-3 of three sets A, B and C. Observe that these three sets partition the rectangle (universal set) into the eight numbered sets which can be represented as follows:

(1) $A \cap B \cap C$ (3) $A \cap B^c \cap C$ (5) $A \cap B^c \cap C^c$ (7) $A^c \cap B^c \cap C$

(2) $A \cap B \cap C^c$ (4) $A^c \cap B \cap C$ (6) $A^c \cap B \cap C^c$ (8) $A^c \cap B^c \cap C^c$

Each of the eight sets is of the form $A^* \cap B^* \cap C^*$ where A^* is A or A^c, B^* is B or B^c and C^* is C or C^c. Any nonempty set expression involving A, B and C, for example,

$$[(A \cap B^c)^c \cup (A^c \cap C^c)] \cap [(B^c \cup C)^c \cap (A \cup C^c)]$$

will represent some area in Fig. 12-3 and hence will uniquely equal the union of one or more of the eight sets. This unique representation is the same as the full disjunctive normal form for Boolean algebras which we discuss below.

12.7 DISJUNCTIVE NORMAL FORM

Consider a set of variables (or letters or symbols), say $x_1, x_2, \ldots, x_n$. By a *Boolean expression* E in these variables, sometimes written $E(x_1, \ldots, x_n)$, we mean any variable or any expression built up from the variables using the Boolean operations $+$, $*$ and $'$. For example,

$$E = (x + y'z)' + (xyz' + x'y)' \quad \text{and} \quad F = ((xy'z' + y)' + x'z)'$$

are Boolean expressions in x, y and z.

A *literal* is a variable or complemented variable, e.g. x, x', y, y', and so on. By a *fundamental product* we mean a literal or a product of two or more literals in which no two literals involve the same variable. For example,

$$xz', \ xy'z, \ x, \ y', \ yz', \ x'yz$$

are fundamental products but $xyx'z$ and $xyzy$ are not.

One fundamental product P_1 is said to be *included in* or *contained in* another fundamental product P_2 if the literals of P_1 are also literals of P_2. For example, $x'z$ is included in $x'yz$ but not in $xy'z$ since x' is not a literal of the latter product. If P_1 is included in P_2 then by the absorption law $P_1 + P_2 = P_1$.

A Boolean expression E is said to be in *disjunctive normal form* (dnf) if E is a fundamental product or the sum of two or more fundamental products of which none is included in another. For example, consider the expressions

$$E_1 = xz' + y'z + xyz' \quad \text{and} \quad E_2 = xz' + x'yz' + xy'z$$

The first is not in disjunctive normal form since xz' is contained in xyz', but the second is in dnf.

Using the Boolean algebra laws, we can construct an algorithm to transform any Boolean expression E into disjunctive normal form as follows:

(1) Using DeMorgan's laws and involution, we can move the complement operation into any parenthesis until finally it only applies to variables. Then E will only consist of sums and products of literals.

(2) Using the distributive law, we can further transform E into a sum of products; and then using the commutative, idempotent and absorption laws we can finally transform E into disjunctive normal form.

For example, by (1)

$$E = ((ab)'c)'((a' + c)(b' + c'))' = ((ab)'' + c')((a' + c)' + (b' + c')')$$
$$= (ab + c')(ac' + bc)$$

and by (2)

$$E = abac' + abbc + ac'c' + bcc' = abc' + abc + ac' + 0$$
$$= ac' + abc$$

which is in dnf. Observe that ac' is included in abc', so by the absorption law

$$ac' + (ac' * b) = ac'$$

A Boolean expression $E(x_1, x_2, \ldots, x_n)$ is said to be in *full disjunctive normal form* if it is in dnf and each fundamental product involves all the variables. We can easily change any Boolean expression E in dnf into full disjunctive normal form by multiplying any fundamental product P of E by $x_i + x_i'$ if P does not involve x_i. For example, we transform $E = E(a, b, c)$ above into full disjunctive normal form by

$$E \;=\; ac' + abc \;=\; ac'(b + b') + abc \;=\; abc' + ab'c' + abc$$

We note that $x_i + x_i' = 1$, so multiplying P by $x_i + x_i'$ is permissible. The following theorem applies.

Theorem 12.8:   Every nonzero Boolean expression $E(x_i, x_2, \ldots, x_n)$ can be put into full disjunctive normal form and such a representation is unique.

Warning: The terminology in this section has not been standardized. Specifically, some texts use the expression "disjunctive normal form" where we use "full disjunctive form", and say "sum of products" where we say "disjunctive normal form". Also, other texts may use the expression "disjunctive canonical form" where we use "full disjunctive normal form". Accordingly, the reader should always look up the meaning of the expression "disjunctive normal form" in any text which contains sections on Boolean algebra.

12.8 SWITCHING CIRCUIT DESIGNS

Let $A, B, \ldots$ denote electrical switches, and let A and A' denote switches with the property that if one is on then the other is off, and vice versa. Two switches, say A and B, can be connected by wire in a series or parallel combination as shown in Fig. 12-4.

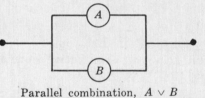

Series combination, $A \wedge B$ Parallel combination, $A \vee B$

Fig. 12-4

Let $A \wedge B$ and $A \vee B$

denote respectively that A and B are connected in series and A and B are connected in parallel.

A Boolean switching circuit design means an arrangement of wires and switches that can be constructed by repeated use of series and parallel combinations; hence it can be described by the use of connectives $\wedge$ and $\vee$.

EXAMPLE 12.4.   Circuit (1) of Fig. 12-5 can be described by $A \wedge (B \vee A')$, and circuit (2) can be described by $(A \wedge B') \vee [(A' \vee C) \wedge B]$.

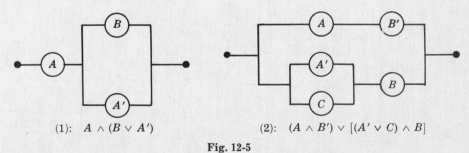

(1): $A \wedge (B \vee A')$ (2): $(A \wedge B') \vee [(A' \vee C) \wedge B]$

Fig. 12-5

With respect to a switching circuit, we will let

1   and   0

denote, respectively, that a switch or circuit is on and that a switch or circuit is off. The next two tables describe the behavior of a series circuit $A \wedge B$ and a parallel circuit $A \vee B$.

A	B	$A \wedge B$
1	1	1
1	0	0
0	1	0
0	0	0

A	B	$A \vee B$
1	1	1
1	0	1
0	1	1
0	0	0

The next table shows the relationship between a switch A and a switch A'.

A	A'
1	0
0	1

Notice that the above three tables are identical with the tables of conjunction, disjunction and negation for statements (and propositions). The only difference is that 1 and 0 are used here instead of T and F. Thus

Theorem 12.9:   The algebra of Boolean switching circuits is a Boolean algebra.

In order to find the behavior of a Boolean switching circuit, a table is constructed which is analogous to the truth tables for propositions.

EXAMPLE 12.5.

(a) Consider circuit (1) in Fig. 12-5. What is the behavior of the circuit, that is, when will the circuit be on (i.e. when will current flow) and when will the circuit be off? A "truth" table is constructed for $A \wedge (B \vee A')$ as follows:

A	B	A'	$B \vee A'$	$A \wedge (B \vee A')$
1	1	0	1	1
1	0	0	0	0
0	1	1	1	0
0	0	1	1	0

Thus current will flow only if both A and B are on.

(b) The behavior of circuit (2) in Fig. 12-5 is indicated by the following truth table for $(A \wedge B') \vee [(A' \vee C) \wedge B]$:

A	B	C	(A	$\wedge$	B')	$\vee$	[(A'	$\vee$	C)	$\wedge$	B]
1	1	1	1	0	0	1	0	1	1	1	1
1	1	0	1	0	0	0	0	0	0	0	1
1	0	1	1	1	1	1	0	1	1	0	0
1	0	0	1	1	1	1	0	0	0	0	0
0	1	1	0	0	0	1	1	1	1	1	1
0	1	0	0	0	0	1	1	1	0	1	1
0	0	1	0	0	1	0	1	1	1	0	0
0	0	0	0	0	1	0	1	1	0	0	0
Step			1	2	1	4	1	2	1	3	1

12.9 PRIME IMPLICANTS, CONSENSUS METHOD

Suppose P_1 and P_2 are fundamental products such that exactly one variable, say x_k, appears complemented in one of P_1 and P_2 and uncomplemented in the other. Then the *consensus* Q of P_1 and P_2 is the product (without repetition) of the literals of P_1 and the literals of P_2 after x_k and x_k' are deleted.   For example:

$$xyz's \text{ and } xy't \text{ have the consensus } xz'st$$

$$xy' \text{ and } y \text{ have the consensus } x$$

$$x'yz \text{ and } x'yt \text{ have no consensus}$$

$$x'yz \text{ and } xyz' \text{ have no consensus}$$

If Q is the consensus of P_1 and P_2 and if we write Q, P_1 and P_2 in full dnf, then each summand of Q will appear among the summands of P_1 or the summands of P_2. That is,

Lemma 12.10:   If Q is the consensus of P_1 and P_2 then $P_1 + P_2 + Q = P_1 + P_2$.

A fundamental product P is called a *prime implicant* of a Boolean expression E if $P + E = E$ but no other fundamental product included in P has this property. For example, suppose $E = xy' + xyz' + x'yz'$. By Problem 12.10, $xz' + E = E$, but $x + E \neq E$ and $z' + E \neq E$; hence xz' is a prime implicant of E. We next discuss the consensus method for finding the prime implicants of E.

Consider a Boolean expression $E = P_1 + P_2 + \cdots + P_n$ where the P's are fundamental products. Applying the following two steps to E will be called the *consensus method*:

Step (1):   Delete any fundamental product P_i which includes any other fundamental product P_j. (Permissible by the absorption law.)

Step (2):   Add the consensus Q of any P_i and P_j providing Q does not include any of the P's. (Permissible by Lemma 12.10.)

The following theorem gives the basic property of this method.

Theorem 12.11:   The consensus method applied to any Boolean expression E will eventually stop, and then E will be the sum of its prime implicants.

EXAMPLE 12.6.   Let $E = xyz + x'z' + xyz' + x'y'z + x'yz'$.   Then:

$$
\begin{aligned}
E &= xyz + x'z' + xyz' + x'y'z & &(x'yz' \text{ includes } x'z')\\
&= xyz + x'z' + xyz' + x'y'z + xy & &(\text{Consensus of } xyz \text{ and } xyz')\\
&= x'z' + x'y'z + xy & &(xyz \text{ and } xyz' \text{ include } xy)\\
&= x'z' + x'y'z + xy + x'y' & &(\text{Consensus of } x'z' \text{ and } x'y'z)\\
&= x'z' + xy + x'y' & &(x'y'z \text{ includes } x'y')\\
&= x'z' + xy + x'y' + yz' & &(\text{Consensus of } x'z' \text{ and } xy)
\end{aligned}
$$

Since neither step in the consensus method will now change E, the prime implicants of E are $x'z'$, xy, $x'y'$ and yz'.

12.10 MINIMAL BOOLEAN EXPRESSIONS

There are many ways of representing the same Boolean expression E. Since E may be a switching circuit, we may want a representation which is in some sense minimal. Here we define and investigate minimal disjunctive normal forms for E. Other types of minimal forms exist but their investigation lies beyond the scope of this text.

Consider a Boolean expression E in disjunctive normal form. We will let E_L denote the number of literals in E and we will let E_S denote the number of summands in E. For example, if

$$E = abc' + a'b'd + ab'c'd + a'bcd$$

then $E_L = 14$ and $E_S = 4$. Let E and F be equivalent Boolean expressions in dnf. We say that E is *simpler* then F if $E_L \leqq F_L$ and $E_S \leqq F_S$ and one of the inequalities is not equal. We say E is *minimal* if there is none which is simpler than E.

The following theorem shows the basic relationship between minimal dnf's and prime implicants.

Theorem 12.12:   A minimal disjunctive normal form of a Boolean expression E is a sum of prime implicants of E.

The consensus method can be used to express any E as a sum of all its prime implicants. One method of finding its minimal dnf is to express each prime implicant in full dnf and to delete one by one those prime implicants whose summands appear among the remaining ones. For example,

$$E = x'z' + xy + x'y' + yz'$$

is expressed as the sum of all its prime implicants (Example 12.6). Then

$$x'z' = x'z'(y + y') = x'yz' + x'y'z'$$

$$xy = xy(z + z') = xyz + xyz'$$

$$x'y' = x'y'(z + z') = x'y'z + x'y'z'$$

$$yz' = yz'(x + x') = xyz' + x'yz'$$

Note $x'z'$ can be deleted since its summands $x'yz$ and $x'y'z'$ appear among the others. Thus

$$E = xy + x'y' + yz'$$

and this is a minimal dnf for E since none of the prime implicants are *superfluous*, i.e. none can be deleted without changing E.

The above method of finding a minimal dnf is direct but inefficient. In the next section we give a geometrical method of finding minimal dnf's when the number of variables is not large. The reader is referred to other texts for other techniques for finding minimal dnf's.

12.11 KARNAUGH MAPS

Karnaugh maps are pictorial devices for finding prime implicants and minimal disjunctive normal forms for Boolean expressions involving at most six variables. We will only treat the case of two, three or four variables.

In our Karnaugh maps, fundamental products in the same variables will be represented by squares. We say that two such fundamental products P_1 and P_2 (or the squares corresponding to P_1 and P_2) are *adjacent* if P_1 and P_2 differ in exactly one literal, which must be a complemented variable in one product and uncomplemented in the other. Thus the sum of two adjacent products will be a fundamental product with one less literal. For example,

$$xyz' + xy'z' = xz'(y + y') = xz'(1) = xz'$$

$$x'yzt + x'yz't = x'yt(z + z') = x'yt(1) = x'yt$$

Note that $x'yzt$ and $xyz't$ are not adjacent. Also note that xyz' and $xyzt$ will not appear in the same Karnaugh map since they involve different variables. In the context of Karnaugh maps, we will sometimes use the terms squares and fundamental products interchangeably.

Two variables: x and y. There are four fundamental products, xy, xy', $x'y$ and $x'y'$, which correspond in the obvious way to the four squares in the Karnaugh map in Fig. 12-6(a).

Any Boolean expression $E(x, y)$ in full dnf can be represented by checks in the appropriate squares. For example,

$$E_1 = xy + xy', \qquad E_2 = xy + x'y + x'y', \qquad E_3 = xy + x'y'$$

are represented in Fig. 12-6(b), (c), (d), respectively. (The loops will be explained later.)

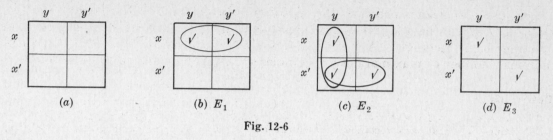

Fig. 12-6

Two squares are adjacent in Fig. 12-6(a) if they have a side in common. A prime implicant of an expression $E(x, y)$ will either be a pair of adjacent squares or an *isolated square*, i.e. a square which is not adjacent to any other square of $E(x, y)$. For example, E_1 consists of two adjacent squares designated by the loop in Fig. 12-6(b). This pair of adjacent squares represents the variable x, so x is a (the only) prime implicant of E_1 and

$$E_1 = x$$

is its minimal dnf. Observe that E_2 contains two pairs of adjacent squares (designated by the two loops) which include all the squares of E_2. The vertical pair represents y and the horizontal pair x'; so y and x' are prime implicants of E_2 and

$$E_2 = x' + y$$

is its minimal dnf. On the other hand, E_3 consists of two isolated squares which represent xy and $x'y'$; hence xy and $x'y'$ are the prime implicants of E_3 and

$$E_3 = xy + x'y'$$

is its minimal dnf.

Three variables: x, y, z. Here there are eight fundamental products,

$$xyz, \ xyz', \ xy'z, \ xy'z', \ x'yz, \ x'yz', \ x'y'z, \ x'y'z'$$

which correspond in the obvious way to the eight squares in the Karnaugh map in Fig. 12-7(a). Observe that the top of the map is labeled so that adjacent products differ by one literal. Moreover, in order for adjacent squares to have a side in common, we must identify the left edge and the right edge of the map since xyz and $xy'z$ are adjacent products and $x'yz$ and $x'y'z$ are adjacent products. In other words, if we cut out, bend and glue the map along the identified edge, then we would obtain the cylinder pictured in Fig. 12-7(b). Adjacent squares on the cylinder would then have a side in common. As before, any Boolean expression $E(x, y, z)$ in full dnf is represented on the Karnaugh map by checking the appropriate squares.

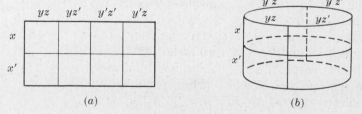

Fig. 12-7

By a *basic rectangle* in the Karnaugh map we mean a square, two adjacent squares, or four squares which form a one-by-four or a two-by-two rectangle. These basic rectangles correspond to fundamental products of three, two and one literal respectively. Moreover, a basic rectangle represents the fundamental product that is the product of those literals which appear in all the squares of the rectangle.

Consider a Boolean expression $E = E(x, y, z)$ represented by a Karnaugh map. A prime implicant of E will be *a maximal basic rectangle* in E, i.e. one which is not contained in any larger basic rectangle. A minimal dnf for E will consist of a *minimal cover* of E, i.e. a minimal number of maximal basic rectangles which include all the squares of E.

EXAMPLE 12.7.

(a) Let $E_1 = xyz + xyz' + x'yz' + x'y'z$. Figure 12-8(a) represents E_1. Observe that E_1 has three prime implicants (maximal basic rectangles) which are circled and which represent xy, yz' and $x'y'z$. Since all three are needed to cover E_1, the minimal dnf is

$$E_1 = xy + yz' + x'y'z$$

(b) Let $E_2 = xyz + xyz' + xy'z + x'yz + x'y'z$. Figure 12-8(b) represents E_2. Note that E_2 has two prime implicants, the two adjacent squares which represents xy, and the two-by-two square (overlapping the identified edge) which represents z. Since they cover E_2, we have that

$$E_2 = xy + z$$

is the minimal dnf for E_2.

(c) Let $E_3 = xyz + xyz' + x'yz' + x'y'z' + x'y'z$. Figure 12-9 represents E_3. Here E_3 has four prime implicants, xy, yz', $x'z'$ and $x'y'$. However, only one of the two dotted ones, i.e. one of yz' or $x'z'$, is needed in a minimal cover of E_3. Thus E_3 has two minimal dnf's:

$$E_3 = xy + yz' + x'y' = xy + x'z' + x'y'$$

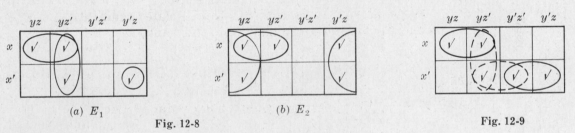

(a) E_1 (b) E_2

Fig. 12-8   Fig. 12-9

Four variables: x, y, z, t. There are sixteen fundamental products which correspond in the obvious way to the sixteen squares in the Karnaugh map in Fig. 12-10. Observe that the top line and the left side are labeled so that adjacent products differ by one literal. Again we must identify the left edge with the right edge (as we did with three variables) but we must also identify the top edge with the bottom edge. (This identification gives rise to a donut-shaped surface called a torus, and we may view our map as really being a torus.)

A basic rectangle is a square, two adjacent squares, four squares which form a one-by-four or two-by-two rectangle, or eight squares which form a two-by-four rectangle. These rectangles correspond to fundamental products with four, three, two and one literal respectively. Again, maximal basic rectangles are the prime implicants. The minimization technique for a Boolean expression $E(x, y, z, t)$ is the same as before.

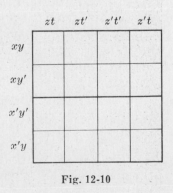

Fig. 12-10

EXAMPLE 12.8. Consider the three Boolean expressions, E_1, E_2, E_3, in variables x, y, z, t which are given by the Karnaugh maps in Fig. 12-11.

(a) The two-by-two basic rectangle represents $y'z$ since only y' and z appear in all four squares. The horizontal pair of adjacent squares represents xyz', and the adjacent squares overlapping the top and bottom edges represents $yz't'$. Hence

$$E_1 = y'z + xyz' + yz't'$$

is a minimal dnf for E_1.

(b) Only y' appears in all eight squares of the two-by-four basic rectangle, and the designated pair of adjacent squares represents xzt'. Hence

$$E_2 = y' + xzt'$$

is the minimal dnf for E_2.

(c) The four corner squares form a two-by-two basic rectangle which represents yt since only y and t appear in all the four squares. The four-by-one basic rectangle represents $x'y'$, and the two adjacent squares represent $y'zt'$. Thus

$$E_3 = yt + x'y' + y'zt'$$

is the minimal dnf for E_3.

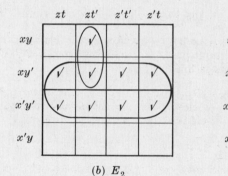 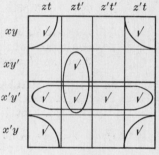

(a) E_1 (b) E_2 (c) E_3

Fig. 12-11

Solved Problems

BOOLEAN ALGEBRAS

12.1. Write the dual of each Boolean equation: (a) $(a * 1) * (0 + a') = 0$, (b) $a + a'b = a + b$.

(a) To obtain the dual equation, interchange $+$ and $*$, and interchange 0 and 1. Thus

$$(a + 0) + (1 * a') = 1$$

(b) First write the equation using $*$: $a + (a' * b) = a + b$. Then the dual is: $a * (a' + b) = a * b$, which can be written as

$$a(a' + b) = ab$$

12.2. Recall (Chapter 10) that the set D_m of divisors of m is a bounded, distributive lattice with $a + b = a \vee b = \text{lcm}\,(a, b)$ and $a * b = a \wedge b = \text{gcd}\,(a, b)$. (a) Show that D_m is a Boolean algebra if m is square free, i.e. if m is a product of distinct primes. (b) Find the atoms of D_m.

(a) We need only show that D_m is complemented. Let x be in D_m and let $x' = m/x$. Since m is a product of distinct primes, x and x' have different prime divisors. Hence $x * x' = \text{gcd}\,(x, x') = 1$ and $x + x' = \text{lcm}\,(x, x') = m$. Recall that 1 is the zero element (lower bound) of D_m, and that m is the identity element (upper bound) of D_m. Thus x' is a complement of x, and so D_m is a Boolean algebra.

(b) The atoms of D_m are the prime divisors of m.

12.3. Consider the Boolean algebra D_{210}.

(a) List its elements and draw its diagram.

(b) Find the set A of atoms.

(c) Find two subalgebras with eight elements.

(d) Is $X = \{1, 2, 6, 210\}$ a sublattice of D_{210}? A subalgebra?

(e) Is $Y = \{1, 2, 3, 6\}$ a sublattice of D_{210}? A subalgebra?

(f) Find the number of subalgebras of D_{210}.

(a) The divisors of 210 are 1, 2, 3, 5, 6, 7, 10, 14, 15, 21, 30, 35, 42, 70, 105 and 210. The diagram of D_{210} appears in Fig. 12-12.

(b) $A = \{2, 3, 5, 7\}$, the set of prime divisors of 210.

(c) $B = \{1, 2, 3, 35, 6, 70, 105, 210\}$ and $C = \{1, 5, 6, 7, 30, 35, 42, 210\}$ are subalgebras of D_{210}.

(d) X is a sublattice since it is linearly ordered. However, X is not a subalgebra since 35 is the complement of 2 in D_{210} but 35 does not belong to X. (In fact, no Boolean algebra with more than two elements is linearly ordered.)

(e) Y is a sublattice of D_{210} since it is closed under $+$ and $*$. However, Y is not a subalgebra of D_{210} since it is not closed under complements in D_{210}, e.g. $35 = 2'$ does not belong to Y. (We note that Y itself is a Boolean algebra, in fact, $Y = D_6$.)

(f) A subalgebra of D_{210} must contain two, four, eight or sixteen elements.

   (i) There can be only one two-element subalgebra which consists of the upper and lower bounds, i.e. $\{1, 210\}$.

   (ii) Since D_{210} contains sixteen elements, the only sixteen-element subalgebra is D_{210} itself.

   (iii) Any four-element subalgebra is of the form $\{1, x, x', 210\}$, i.e. consists of the upper and lower bounds and a nonbound element and its complement. There are fourteen nonbound elements in D_{210} and so there are $14/2 = 7$ pairs $\{x, x'\}$. Thus D_{210} has seven 4-element subalgebras.

   (iv) Any eight-element subalgebra S will itself contain three atoms s_1, s_2, s_3. We can choose s_1 and s_2 to be any two of the four atoms of D_{210} and then s_3 must be the product of the other two atoms, e.g. we can let $s_1 = 2$, $s_2 = 3$, $s_3 = 5 \cdot 7 = 35$ [which determines the subalgebra B in Part (c)], or we can let $s_1 = 5$, $s_2 = 7$, $s_3 = 2 \cdot 3 = 6$ [which determines the subalgebra C in Part (c)]. There are $\binom{4}{2} = 6$ ways to choose s_1 and s_2 from the four atoms of D_{210} and so D_{210} has six 8-element subalgebras.

Accordingly, D_{210} has $1 + 1 + 7 + 6 = 15$ subalgebras.

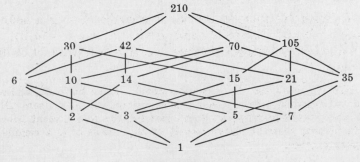

Fig. 12-12

12.4. Prove Theorem 12.2: Let a, b, c be any elements in a Boolean algebra B.

   (i)   Idempotent Laws:

         (5a) $a + a = a$ (5b) $a * a = a$

   (ii)   Boundedness Laws:

         (6a) $a + 1 = 1$ (6b) $a * 0 = 0$

   (iii)   Absorption Laws:

         (7a) $a + (a * b) = a$ (7b) $a * (a + b) = a$

   (iv)   Associative Laws:

         (8a) $(a + b) + c = a + (b + c)$ (8b) $(a * b) * c = a * (b * c)$

(5b) $a = a * 1 = a * (a + a') = (a * a) + (a * a') = (a * a) + 0 = a * a$

(5a)   Follows from (5b) and duality.

(6b) $a * 0 = (a * 0) + 0 = (a * 0) + (a * a') = a * (0 + a') = a * (a' + 0) = a * a' = 0$

(6a)   Follows from (6b) and duality.

(7b) $a * (a + b) = (a + 0) * (a + b) = a + (0 * b) = a + (b * 0) = a + 0 = a$

(7a)   Follows from (7b) and duality.

(8b)   Let $L = (a * b) * c$ and $R = a * (b * c)$. We need to prove that $L = R$. We first prove that $a + L = a + R$. Using the absorption laws in the last two steps,

$$a + L = a + ((a * b) * c) = (a + (a * b)) * (a + c) = a * (a + c) = a$$

Also, using the absorption law in the last step,

$$a + R = a + (a * (b * c)) = (a + a) * (a + (b * c)) = a * (a + (b * c)) = a$$

Thus $a + L = a + R$. Next we show that $a' + L = a' + R$. We have,

$$a' + L = a' + ((a * b) * c) = (a' + (a * b)) * (a' + c)$$
$$= ((a' + a) * (a' + b)) * (a' + c) = (1 * (a' + b)) * (a' + c)$$
$$= (a' + b) * (a' + c) = a' + (b * c)$$

Also,

$$a' + R = a' + (a * (b * c)) = (a' + a) * (a' + (b * c))$$
$$= 1 * (a' + (b * c)) = a' + (b * c)$$

Thus $a' + L = a' + R$. Consequently,

$$L = 0 + L = (a * a') + L = (a + L) * (a' + L) = (a + R) * (a' + R)$$
$$= (a * a') + R = 0 + R = R$$

(8a)   Follows from (8b) and duality.

12.5. Prove Theorem 12.3: Let a be any element of a Boolean algebra B.

   (i)   (Uniqueness of Complement) If $a + x = 1$ and $a * x = 0$, then $x = a'$.

   (ii)   (Involution Law) $(a')' = a$

   (iii)   (9a) $0' = 1$ (9b) $1' = 0$

   (i)   We have:

$$a' = a' + 0 = a' + (a * x) = (a' + a) * (a' + x) = 1 * (a' + x) = a' + x$$

Also,
$$x = x + 0 = x + (a * a') = (x + a) * (x + a') = 1 * (x + a') = x + a'$$
Hence $x = x + a' = a' + x = a'$.

(ii) By definition of complement, $a + a' = 1$ and $a * a' = 0$. By commutativity, $a' + a = 1$ and $a' * a = 0$. By uniqueness of complement, a is the complement of a', that is, $a = (a')'$.

(iii) By boundedness law (6a), $0 + 1 = 1$, and by identity axiom (3b), $0 * 1 = 0$. By uniqueness of complement, 1 is the complement of 0, that is, $1 = 0'$. By duality, $0 = 1'$.

12.6. Prove Theorem 12.4 (DeMorgan's laws):

(10a) $(a + b)' = a' * b'$ (10b) $(a * b)' = a' + b'$

(10a) We need to show that $(a + b) + (a' * b') = 1$ and $(a + b) * (a' * b') = 0$; then by uniqueness of complement, $a' * b' = (a + b)'$. We have:

$$(a + b) + (a' * b') = b + a + (a' * b') = b + (a + a') * (a + b')$$
$$= b + 1 * (a + b') = b + a + b' = b + b' + a = 1 + a = 1$$

Also,

$$(a + b) * (a' * b') = ((a + b) * a') * b'$$
$$= ((a * a') + (b * a')) * b' = (0 + (b * a')) * b'$$
$$= (b * a') * b' = (b * b') * a' = 0 * a' = 0$$

Thus $a' * b' = (a + b)'$.

(10b) Principle of duality (Theorem 12.1).

12.7. Prove Theorem 12.5: The following are equivalent in a Boolean algebra:

(1) $a + b = b$, (2) $a * b = a$, (3) $a' + b = 1$, (4) $a * b' = 0$

By Theorem 10.4, (1) and (2) are equivalent. We show that (1) and (3) are equivalent. Suppose (1) holds. Then
$$a' + b = a' + (a + b) = (a' + a) + b = 1 + b = 1$$
Now suppose (3) holds. Then
$$a + b = 1 * (a + b) = (a' + b) * (a + b) = (a' * a) + b = 0 + b = b$$
Thus (1) and (3) are equivalent.

We next show that (3) and (4) are equivalent. Suppose (3) holds. By DeMorgan's law and involution,
$$0 = 1' = (a' + b)' = a'' * b' = a * b'$$
Conversely, if (4) holds then
$$1 = 0' = (a * b')' = a' + b'' = a' + b$$
Thus (3) and (4) are equivalent. Accordingly, all four are equivalent.

12.8. Prove Theorem 12.6: The mapping $f : B \rightarrow P(A)$ is an isomorphism where B is a Boolean algebra, $P(A)$ is the power set of the set A of atoms, and
$$f(x) = \{a_1, a_2, \ldots, a_n\}$$
where $x = a_1 + \cdots + a_n$ is the unique representation of x as a sum of atoms.

Recall (Chapter 10) that if the a's are atoms then $a_i^2 = a_i$ but $a_i a_j = 0$ for $a_i \neq a_j$. Suppose x, y are in B and suppose

$$x = a_1 + \cdots + a_r + b_1 + \cdots + b_s$$
$$y = b_1 + \cdots + b_s + c_1 + \cdots + c_t$$

where

$$A = \{a_1, \ldots, a_r, b_1, \ldots, b_s, c_1, \ldots, c_t, d_1, \ldots, d_k\}$$

is the set of atoms of B.   Then

$$x + y = a_1 + \cdots + a_r + b_1 + \cdots + b_s + c_1 + \cdots + c_t$$
$$xy = b_1 + \cdots + b_s$$

Hence

$$\begin{aligned} f(x + y) &= \{a_1, \ldots, a_r, b_1, \ldots, b_s, c_1, \ldots, c_t\} \\ &= \{a_1, \ldots, a_r, b_1, \ldots, b_s\} \cup \{b_1, \ldots, b_s, c_1, \ldots, c_t\} \\ &= f(x) \cup f(y) \\ f(xy) &= \{b_1, \ldots, b_s\} \\ &= \{a_1, \ldots, a_r, b_1, \ldots, b_s\} \cap \{b_1, \ldots, b_s, c_1, \ldots, c_t\} \\ &= f(x) \cap f(y) \end{aligned}$$

Let $y = c_1 + \cdots + c_t + d_1 + \cdots + d_k$. Then $x + y = 1$ and $xy = 0$, and so $y = x'$.   Thus

$$f(x') = \{c_1, \ldots, c_t, d_1, \ldots, d_k\} = \{a_1, \ldots, a_r, b_1, \ldots, b_s\}^c = (f(x))^c$$

Since the representation is unique, f is one-to-one and onto. Hence f is a Boolean algebra isomorphism.

BOOLEAN EXPRESSION, DNF'S, SWITCHING CIRCUITS

12.9.   Express the following Boolean expressions $E(x, y, z)$ in dnf and in full dnf.
    (a) $E_1 = x(y'z)'$,   (b) $E_2 = z(x' + y) + y'$,   (c) $E_3 = (x' + y)' + x'y$

    (a) $E_1 = x(y'z)' = x(y + z') = xy + xz'$.   Also,

$$\begin{aligned} E_1 &= xy + xz' = xy(z + z') + x(y + y')z' = xyz + xyz' + xyz' + xy'z' \\ &= xyz + xyz' + xy'z' \end{aligned}$$

    is in full dnf.

    (b) $E_2 = z(x' + y) + y' = x'z + yz + y'$.   Also

$$\begin{aligned} E_2 &= x'z + yz + y' = x'z(y + y') + yz(x + x') + y'(x + x')(z + z') \\ &= x'yz + x'y'z + xyz + x'yz + xy'z + xy'z' + x'y'z + x'y'z' \\ &= xyz + xy'z + xy'z' + x'yz + x'y'z + x'y'z' \end{aligned}$$

    (c) $E_3 = (x' + y)' + x'y = xy' + x'y$. This woud be the the full dnf if E_3 were considered a Boolean expression in x and y. However, we are explicitly told that our expressions are in three variables, x, y, z.   Hence,

$$E_3 = xy' + x'y = xy'(z + z') + x'y(z + z') = xy'z + xy'z' + x'yz + x'yz'$$

    is in full dnf.

12.10.   Let $E = xy' + xyz' + x'yz'$. Prove that   (a) $xz' + E = E$, (b) $x + E \neq E$, (c) $z' + E \neq E$. (Thus xz' is a prime implicant of E.)

    First find the full dnf of E:

$$E = xy'(z + z') + xyz' + x'yz' = xy'z + xy'z' + xyz' + x'yz'$$

(a) Find the full dnf of xz':

$$xz' = xz'(y + y') = xyz' + xy'z'$$

Since the summands of xz' are contained in E, we have $xz' + E = E$.

(b) Find the full dnf of x:

$$x = x(y + y')(z + z') = xyz + xyz' + xy'z + xy'z'$$

The summand xyz of x is not a summand of E; hence $x + E \neq E$.

(c) Find the full dnf of z':

$$z' = z'(x + x')(y + y') = xyz' + xy'z' + x'yz' + x'y'z'$$

The summand $x'y'z'$ of z' is not a summand of E; hence $z' + E \neq E$.

12.11. Determine the Boolean expression for each circuit shown in Fig. 12-13.

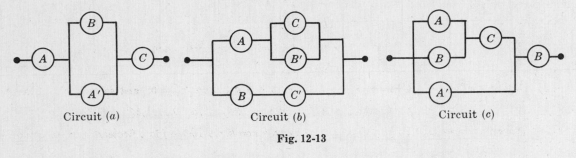

Circuit (a)   Circuit (b)   Circuit (c)

Fig. 12-13

(a) $A \wedge (B \vee A') \wedge C$ (b) $[A \wedge (C \vee B')] \vee (B \wedge C')$ (c) $\{[(A \vee B) \wedge C] \vee A'\} \wedge B$

12.12. Construct a circuit for each of the following Boolean expressions:

(a) $(A \wedge B) \vee [A' \wedge (B' \vee A \vee B)]$,   (b) $(A \vee B) \wedge C \wedge (A' \vee B' \vee C')$

The diagrams in Fig. 12-14 show the required circuits.

(a) Note that the series circuit $A \wedge B$ is in parallel with $A' \wedge (B' \vee A \vee B)$ which is A' in series with the parallel combination $B' \vee A \vee B$.

(b) Note that the parallel circuit $A \vee B$ is in series with C and in series with the parallel circuit $A' \vee B' \vee C'$.

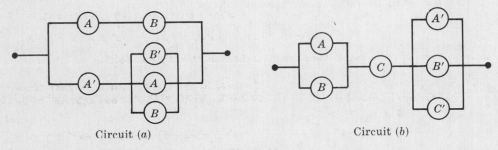

Circuit (a)   Circuit (b)

Fig. 12-14

12.13. Consider the circuit in Fig. 12-15(a). (a) Construct a simpler equivalent circuit. (b) Verify that the circuits are equivalent by finding their "truth" tables.

(a) First write the Boolean expression which represents the circuit:

$$E = (A \wedge B) \vee (A \wedge B') \vee (A' \wedge B')$$

Simplify E by using Karnaugh maps or laws of Boolean algebra, for example,

$$E = [A \wedge (B \vee B')] \vee (A' \wedge B')$$
$$= (A \wedge 1) \vee (A' \wedge B') = A \vee (A' \wedge B')$$
$$= (A \vee A') \wedge (A \vee B') = 1 \wedge (A \vee B')$$
$$= A \vee B'$$

Thus Fig. 12-15(b) is an equivalent circuit.

(b) The required equivalent truth tables follow:

A	B	$(A$	$\wedge$	$B)$	$\vee$	$(A$	$\wedge$	$B')$	$\vee$	$(A'$	$\wedge$	$B')$		A	B	B'	$A \vee B'$
1	1	1	1	1	1	1	0	0	1	0	0	0		1	1	0	1
1	0	1	0	0	1	1	1	1	1	0	0	1		1	0	1	1
0	1	0	0	1	0	0	0	0	0	1	0	0		0	1	0	0
0	0	0	0	0	0	0	0	1	1	1	1	1		0	0	1	1
Step		1	2	1	3	1	2	1	4	1	2	1					

Fig. 12-15

12.14. Prove Theorem 12.8: Every nonzero Boolean expression $E(x_1, \ldots, x_n)$ can be put into full disjunctive normal form and such a representation is unique.

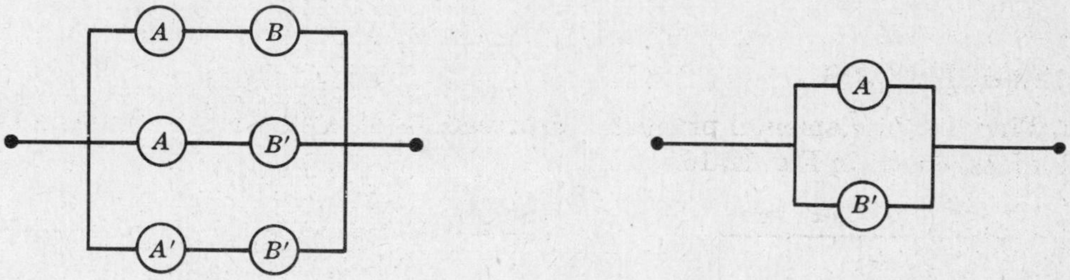

The existence of such a representation follows from the algorithm described on pages 225–226. It remains to show its uniqueness. Note first that if P and Q are distinct fundamental products in the x's, then $PQ = 0$ since one of the variables must be complemented in one of the factors and uncomplemented in the other. Suppose now that

$$E = P_1 + P_2 + \cdots + P_r = Q_1 + Q_2 + \cdots + Q_s$$

are distinct representations of E in full dnf. Then one of the P's must be different from the Q's or one of the Q's must be different from the P's. Say, Q_1 is different from the P's. Then $Q_1 P_i = 0$ for each i. By the absorption law, $Q_1 E = Q_1$. Hence

$$Q_1 = Q_1 E = Q_1(P_1 + \cdots + P_r) = Q_1 P_1 + \cdots + Q_1 P_r = 0 + \cdots + 0 = 0$$

which contradicts the notion that Q_1 is a fundamental product. Hence the representation is unique.

12.15. Use the consensus method to find the prime implicants and a minimal dnf for E where:
(a) $E = xy' + xyz' + x'yz'$,   (b) $E = xy + y't + x'yz' + xy'zt'$.

(a) $\quad E = xy' + xyz' + x'yz' + xz' \qquad$ (Consensus of xy' and xyz')

$\qquad = xy' + x'yz' + xz' \qquad\qquad (xyz'$ includes $xz')$

$\qquad = xy' + x'yz' + xz' + yz' \qquad$ (Consensus of $x'yz'$ and $xz')$

$\qquad = xy' + xz' + yz' \qquad\qquad (x'yz'$ includes $yz')$

Since neither step in the consensus method will change E, we have that xy', xz' and yz' are the prime implicants of E. Writing these prime implicants in full dnf we obtain:

$$xy' = xy'(z + z') = xy'z + xy'z'$$
$$xz' = xz'(y + y') = xyz' + xy'z'$$
$$yz' = yz'(x + x') = xyz' + x'yz'$$

Only the summands xyz' and $xy'z'$ of xz' appear among the other summands and hence xz' can be deleted, i.e. xz' is superfluous. Thus $E = xy' + yz'$ is a minimal dnf for E.

$(b)\quad E = xy + y't + x'yz' + xy'zt' + xzt'$　　　　(Consensus of xy and $xy'zt'$)
$\qquad = xy + y't + x'yz' + xzt'$　　　　　　　　(${xy'zt'}$ includes xzt')
$\qquad \doteq xy + y't + x'yz' + xzt' + yz'$　　　　(Consensus of xy and $x'yz'$)
$\qquad = xy + y't + xzt' + yz'$　　　　　　　　($x'yz'$ includes yz')
$\qquad = xy + y't + xzt' + yz' + xt$　　　　　　(Consensus of xy and $y't$)
$\qquad = xy + y't + xzt' + yz' + xt + xz$　　　(Consensus of xzt' and xt)
$\qquad = xy + y't + yz' + xt + xz$　　　　　　(xzt' includes xz)
$\qquad = xy + y't + yz' + xt + xz + z't$　　　(Consensus of $y't$ and yz')

Since neither step in the consensus method will now change E, the prime implicants of E are xy, $y't$, yz', xt, xz and $z't$. Writing these prime implicants in full dnf and deleting one by one those whose summands appear among the remaining ones, we obtain that $E = y't + xz + yz'$ is a minimal dnf for E.

KARNAUGH MAPS

12.16. Find the fundamental product P represented by each basic rectangle in the Karnaugh maps shown in Fig. 12-16.

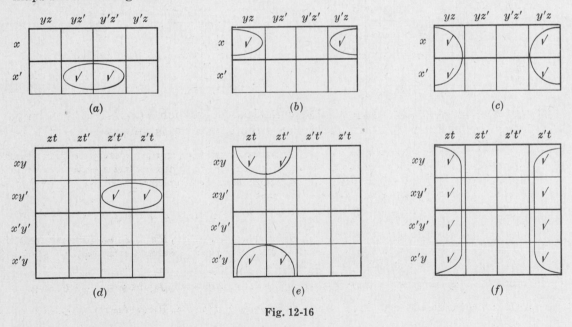

Fig. 12-16

In each case, find those literals which appear in all the squares of the basic rectangle, then P is the product of such literals.

(a) x' and z' appear in both squares; hence $P = x'z'$.

(b) x and z appear in both squares; hence $P = xz$.

(c) Only z appears in all four squares; hence $P = z$.

(d) x, y' and z' appear in both squares; hence $P = xy'z'$.

(e) Only y and z appear in all four squares; hence $P = yz$.

(f) Only t appears in all eight squares; hence $P = t$.

12.17. Find the minimal dnf for each expression E given by the Karnaugh maps shown in Fig. 12-17.

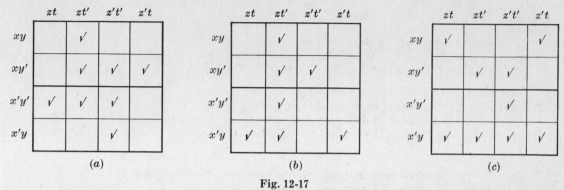

Fig. 12-17

(a) There are five prime implicants designated by the four loops and the dotted circle in Fig. 12-18(a). However, the dotted circle is not needed to cover all the squares, whereas the four loops are required. Thus the four loops give the minimal dnf for E; that is,

$$E = xzt' + xy'z' + x'y'z + x'z't'$$

(b) A minimal cover of E is given by the three loops in Fig. 12-18(b). Hence $E = zt' + xy't' + x'yt$ is a minimal dnf for E.

(c) There are two ways to cover the $x'y'z't'$ square as indicated in Fig. 12-18(c). Hence

$$E = x'y + yt + xy't' + y'z't' = x'y + yt + xy't' + x'z't'$$

are two minimal dnf's for E.

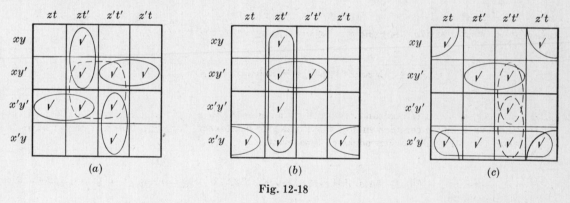

Fig. 12-18

12.18. Find a minimal dnf for $E = xy' + xyz + x'y'z' + x'yzt'$.

We do not need to change E into full dnf in order to represent it by a Karnaugh map. We simply check all four squares which contain xy', the two squares which contain xyz, the two squares which contain $x'y'z'$, and the square $x'yzt'$, as in Fig. 12-19. A minimal cover consists of the three designated basic rectangles. Hence $E = xz + y'z' + yzt'$ is the minimal dnf for E.

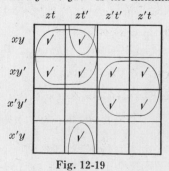

Fig. 12-19

Supplementary Problems

BOOLEAN ALGEBRAS

12.19. Write the dual of each Boolean equation:

(a) $a(a' + b) = ab$,   (b) $(a + 1)(a + 0) = a$,   (c) $(a + b)(b + c) = ac + b$

12.20. Consider the lattice D_m of divisors of m (where $m > 1$). (a) Show that if m is not square free, i.e. m is not the product of distinct primes, then D_m is not a Boolean algebra. (Compare with Problem 12.2.) (b) Which of the following are Boolean algebras: $D_{20}, D_{55}, D_{99}, D_{130}$?

12.21. Let B be a sixteen-element Boolean algebra and let S be an eight-element subalgebra of B. Show that two of the atoms of S must be atoms of B.

12.22. Consider the Boolean algebra D_{110}. (a) List its elements and draw its diagram. (b) Find all its subalgebras. (c) Find the number of all sublattices with four elements. (d) Find the set A of atoms of D_{110}. (e) Give the isomorphic mapping $f : D_{110} \to P(A)$ as defined in Theorem 12.6.

12.23. List the axiom or theorem that is used in each step of the proof of:

(a) Theorem 12.2 (Problem 12.4)   (c) Theorem 12.4 (Problem 12.6)

(b) Theorem 12.3 (Problem 12.5)   (d) Theorem 12.5 (Problem 12.7)

12.24. Prove (5a), (6a), (7a), (8a) and (10b) directly, i.e., without using the principle of duality.

12.25. Prove that $0 \leqq x \leqq 1$ for any element x in a Boolean algebra.

12.26. An element M in a Boolean algebra B is called a *maxterm* if the identity 1 is its only successor. (a) Show that the complements of the atoms are maxterms. (b) Show that any element x in B can be expressed uniquely as a product of maxterms.

12.27. Let $B = (B, +, *, ', 0, 1)$ be a Boolean algebra. Define an operation $\triangle$ on B (called the *symmetric difference*) by

$$x \triangle y = (x * y') + (x' * y)$$

Prove that $R = (B, \triangle, *)$ is a commutative Boolean ring. (See Problems 1.50 and 9.64).

12.28. Let $R = (R, \oplus, \cdot)$ be a Boolean ring with identity $1 \neq 0$. Define:

$$x' = 1 \oplus x, \quad x + y = x \oplus y \oplus x \cdot y, \quad x * y = x \cdot y$$

Prove that $B - (R, +, *, ', 0, 1)$ is a Boolean algebra.

BOOLEAN EXPRESSIONS, SWITCHING CIRCUITS

12.29. Express the following Boolean expressions $E(x, y, z)$ in dnf and in full dnf.

(a) $E_1 = x(xy' + x'y + y'z)$ (d) $E_4 = (x'y)'(x' + xyz')$

(b) $E_2 = (x + y'z)(y + z')$ (e) $E_5 = (x + y)'(xy')'$

(c) $E_3 = (x' + y)' + y'z$ (f) $E_6 = y(x + yz)'$

12.30. Use the consensus method to find the prime implicants of:

(a) $E_1 = xy'z' + x'y + x'y'z' + x'yz$

(b) $E_2 = xy' + x'z't + xyzt' + x'y'zt'$

(c) $E_3 = xyzt + xyz't' + xz't' + x'y'z' + x'yz't$

12.31. Determine the Boolean expression for each of the circuits shown in Fig. 12-20.

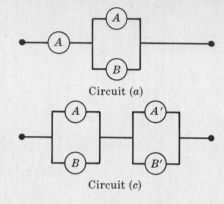

Circuit (a)

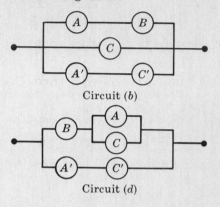

Circuit (b)

Circuit (c)

Circuit (d)

Fig. 12-20

12.32. Construct a switching circuit for each Boolean expression:

(a) $A \vee (B \wedge C)$ (c) $(A \vee B) \wedge (C \vee D)$ (e) $(A \vee B) \wedge [A' \vee (C \wedge B')]$

(b) $A \wedge (B \vee C)$ (d) $(A \wedge B) \vee (C \wedge D)$ (f) $[(A \wedge B) \vee C] \wedge [D \vee (A' \wedge B)]$

MINIMAL DNF'S, KARNAUGH MAPS

12.33. Find all possible minimal dnf's for each Boolean expression E given by the Karnaugh maps in Fig. 12-21.

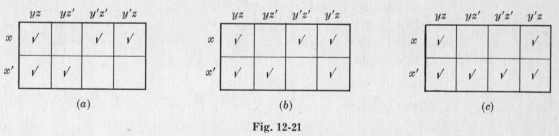

      (a)   (b)   (c)

Fig. 12-21

12.34. Find all possible minimal dnf's for each Boolean expression E given by the Karnaugh maps in Fig. 12-22.

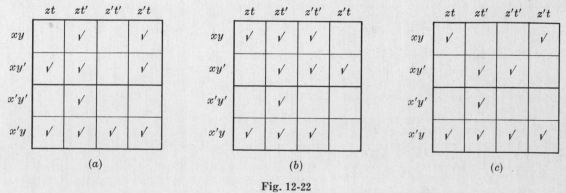

      (a)   (b)   (c)

Fig. 12-22

12.35. Find a minimal dnf for each Boolean expression:

    (a) $E_1 = xy + x'y + x'y'$ (c) $E_3 = y'z + y'z't' + z't$

    (b) $E_2 = x + x'yz + xy'z'$ (d) $E_4 = y'zt + xzt' + xy'z'$

COMPUTER PROGRAMMING PROBLEMS

    Suppose a Boolean expression $E(X, Y, Z)$ in disjunctive normal form is punched on a data card where we let $A = X'$, $B = Y'$ and $C = Z'$. (Note + is simply another alphameric character.)

12.36. Write a program which prints E in full dnf.

12.37.. Write a program which prints the prime implicants of E.

12.38. Write a program which prints E in minimal dnf.

    Test the above programs with the following data:

    (i) $X+AYZ+XBC$ (ii) $BZ+ABC+XC$ (iii) $XBC+AY+ABC+AYZ$

Answers to Supplementary Problems

12.19. (a) $a + a'b = a + b$,   (b) $a \cdot 0 + a \cdot 1 = a$,   (c) $ab + bc = (a + c)b$

12.20. (b) D_{55} and D_{130}

12.21. *Hint*:   Let B be a sixteen-element algebra of sets, i.e. $B = P(A)$ where A has four elements. Show that the atoms of any subalgebra of B must be disjoint sets.

12.22. (a) There are eight elements, 1, 2, 5, 10, 11, 22, 55, 110.

    (b) There are five subalgebras, $\{1, 110\}$, $\{1, 2, 55, 110\}$, $\{1, 5, 22, 110\}$, $\{1, 10, 11, 110\}$, and D_{110}.

    (c) There are fifteen sublattices with four elements, which includes the above three subalgebras.

    (d) $A = \{2, 5, 11\}$

    (e) D_{110}:   1   2   5   11   10   22   55   110
                ↓   ↓   ↓   ↓   ↓   ↓   ↓   ↓
     $P(A)$: $\emptyset$, $\{2\}$, $\{5\}$, $\{11\}$, $\{2, 5\}$, $\{2, 11\}$, $\{5, 11\}$, D_{110}

12.29. (a) $E_1 = xy' + xy'z = xy'z' + xy'z$ (d) $E_4 = xyz' + x'y' = xyz' + x'y'z + x'y'z'$

    (b) $E_2 = xy + xz' = xyz + xyz' + xy'z'$ (e) $E_5 = x'y' = x'y'z + x'y'z'$

    (c) $E_3 = xy' + y'z = xy'z + xy'z' + x'y'z$ (f) $E_6 = x'yz'$

12.30. (a) $x'y, x'z, y'z'$

    (b) $xy', xzt', y'zt', x'z't, y'z't$

    (c) $xyzt, xz't', y'z't', x'y'z', x'z't$

12.31. (b) $(A \wedge B) \vee C \vee (A' \wedge C')$,   (d) $[B \wedge (A \vee C)] \vee (A' \wedge C')$

12.32.   See Fig. 12-23.

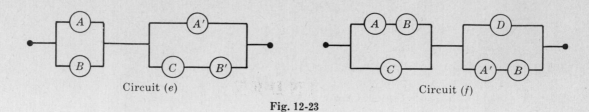

Circuit (e)   Circuit (f)

Fig. 12-23

12.33.   (a) $E = xy' + x'y + yz = xy' + x'y + xz'$

(b) $E = xy' + x'y + z$

(c) $E = x' + z$

12.34.   (a) $E = x'y + zt' + xz't + xy'z = x'y + zt' + xz't + xy't$

(b) $E = yz + yt' + zt' + xy'z'$

(c) $E = x'y + yt + xy't' + x'zt = x'y + yt + xy't' + y'zt$

12.35.   (a) $E_1 = x' + y$ (c) $E_3 = y' + z't$

(b) $E_2 = xz' + yz$ (d) $E_4 = xy' + zt' + y'zt$

INDEX

Catalog

If you are interested in a list of SCHAUM'S
OUTLINE SERIES send your name
and address, requesting your free catalog, to:

SCHAUM'S OUTLINE SERIES, Dept. C
McGRAW-HILL BOOK COMPANY
1221 Avenue of Americas
New York, N.Y. 10020